A Mathematical Approach to Special Relativity

A Mathematical Approach to Special Relativity

Ahmad Shariati
Department of Physics
Faculty of Physics and Chemistry
Alzahra University
Tehran, Iran

Academic Press is an imprint of Elsevier
125 London Wall, London EC2Y 5AS, United Kingdom
525 B Street, Suite 1650, San Diego, CA 92101, United States
50 Hampshire Street, 5th Floor, Cambridge, MA 02139, United States
The Boulevard, Langford Lane, Kidlington, Oxford OX5 1GB, United Kingdom

Notices

Knowledge and best practice in this field are constantly changing. As new research and experience broaden our understanding, changes in research methods, professional practices, or medical treatment may become necessary.

Practitioners and researchers must always rely on their own experience and knowledge in evaluating and using any information, methods, compounds, or experiments described herein. In using such information or methods they should be mindful of their own safety and the safety of others, including parties for whom they have a professional responsibility.

To the fullest extent of the law, neither the Publisher nor the authors, contributors, or editors, assume any liability for any injury and/or damage to persons or property as a matter of products liability, negligence or otherwise, or from any use or operation of any methods, products, instructions, or ideas contained in the material herein.

ISBN: 978-0-323-99708-9

For information on all Academic Press publications
visit our website at https://www.elsevier.com/books-and-journals

Publisher: Katey Birtcher
Editorial Project Manager: Aleksandra Packowska
Publishing Services Manager: Shereen Jameel
Project Manager: Gayathri S
Cover Designer: Victoria Pearson Esser

Typeset by VTeX

To Fereshteh and Settareh

Contents

PART 2 Mathematics

List of figures

Biography

Ahmad Shariati was born in Sirjan, Iran, in 1961. In 1980 he was drafted into Iran's compulsory military service. Upon finishing service, he entered the BS program in Applied Physics at Sharif University of Technology, Tehran, Iran, in 1984. In 1988 he entered the MS program of Physics, also at Sharif University of Technology, and in 1990 was accepted into the Ph.D. program at the same university. Writing a thesis on the deformations of the Poincaré group, he obtained his Ph.D. in physics in 1995 under the supervision of Prof Shahin Rouhani. During his career, he was a researcher at the Institute for Research in Fundamental Sciences (IPM) (1995–2000) and a faculty member at the Institute for Advanced Studies in Basic Sciences, Gava Zang, Zanjan (1995–2003); and since 2003 he has been an associate professor of physics at Alzahra University, Tehran, Iran. He was an editor of the *Iranian Journal of Physics* (1999–2003) and *Gamma* (2004–2010), both of which are Persian journals of physics. During 2003–2005, he was a member of the executive committee of the Physics Society of Iran. He was also a member of the Physics Committee of the Young Scholar's Club, which is the organizer of the National Physics Olympiad in Iran. He is also a member of the American Physical Society and was a member of the American Association for the Advancement of Science.

Preface

This book is about special relativity. I have in mind a student of physics who already knows basic physics and calculus and is learning more physics and mathematics, parallel to his/her study of relativity. Therefore the earlier chapters use more elementary mathematics and physics, in contrast to later chapters.

A great part of learning any subject is to master the vocabulary used by the experts in the field, which has two functions: facilitating thought and communication. I have done my best to build what I think is the necessary vocabulary of relativity.

The text is divided into two parts. In Part 1 the physical basis of special relativity is presented. There is also a chapter on gravitation, which is a bridge between special relativity and general relativity. In Part 2 the mathematical language suitable for relativity—some group theory—is presented. Part 1 is the necessary material for understanding special relativity. Part 2 is the material needed for understanding related topics in quantum mechanics, electrodynamics, and quantum field theories.

I think a good understanding of Galilean relativity is necessary to understand special relativity; therefore I begin Part 1 with Galilean relativity.

Conventions for the metric. Anyone who is learning relativity knows that there are two different conventions for the Minkowski metric. In the *time-like* conventions $\mathrm{d}s^2 = c^2\,\mathrm{d}t^2 - \mathrm{d}x^2 - \mathrm{d}y^2 - \mathrm{d}z^2$, while in the *space-like* conventions $\mathrm{d}s^2 = -c^2\,\mathrm{d}t^2 + \mathrm{d}x^2 + \mathrm{d}y^2 + \mathrm{d}z^2$. To use the literature, one must be able to switch between these two conventions. In this book I introduce the symbol ς (a variant of the Greek letter sigma) and define the metric as follows:

$$\mathrm{d}s^2 = \varsigma\left(-c^2\,\mathrm{d}t^2 + \mathrm{d}x^2 + \mathrm{d}y^2 + \mathrm{d}z^2\right), \tag{1}$$

so that:

- $\varsigma = +1$ for the space-like convention,
- $\varsigma = -1$ for the time-like convention.

A. Shariati

Instructor Resources

For qualified professors, additional, instructor-only teaching materials can be requested here: https://educate.elsevier.com/9780323997089

Acknowledgments

I would like to say my gratitudes to Mohammad Khorrami, Reza Mansouri, Houshang Ardavan, and Mohammad-Reza Khajehpour, whom I learned relativity from. Any misunderstanding or error in concepts offered in this book is completely mine.

PART 1

Physics

CHAPTER

Galilean relativity

1

1.1 The homogeneity of space

We are all familiar with the acceleration of gravity, usually denoted by g. A gravimeter is an apparatus that measures the acceleration of gravity with high enough precision, say $10^{-8}\,\mathrm{m\,s^{-2}}$. Suppose we are given a gravimeter, and we measure the acceleration of gravity in different places. What would be the result? Experiments show that g in different places is not the same. Geologists use gravimeters to map the Earth's inhomogeneities.

Consider the experiment of measuring the magnetic field, B. Obviously, it also depends on location. The same is true for measuring the electric field, E, or temperature. We conclude that generally if we perform an experiment somewhere and then translate that equipment to some other place, the result of the experiment would not be the same. However, in all such cases, we can associate this spatial dependence to some external *source*. In the case of g, this external source is the distribution of mass; in the case of B, it is the distribution of electric currents including magnetized materials. And so on, and so forth.

What if there are no external sources? This question cannot be answered experimentally because we cannot eliminate all the external sources and repeat the experiments. However, we can *postulate* that *if no external sources are present, then the outcome of any experiment should be independent of the position in space.* We call this statement a postulate because we cannot test it directly experimentally.

Postulate. *Space is homogeneous, that is, if the effects of external sources are accounted for properly, the outcome of no experiment depends on the position in space.*

In other words, different regions of space have the same properties, or the laws of physics do not depend on the position in space. This postulate is actually a guideline in our investigation of the physical laws. Whenever we see that the outcome of an experiment is not the same at two different points in space, because of this postulate, we conclude that there should be some external source responsible for that. Up to now, this guideline has worked perfectly well, though it is possible that in the future we might be forced to relax it.

Guide to Literature. One of the best texts available on the subject of symmetry in physical laws is a famous lecture by Richard Feynman, the text and a recording of which are available (Feynman, 1997, pp. 23–47). The article by Philip Morrison on

A Mathematical Approach to Special Relativity. https://doi.org/10.1016/B978-0-32-399708-9.00010-5

the approximate nature of physical symmetries (Morrison, 1958) is both readable and informative.

1.2 The isotropy of space

Consider an experiment. What happens if we rotate the apparatus and repeat the experiment? Of course the result depends on the experiment. For example, consider placing a magnetic compass on the table. The hand of the compass, pointing to the magnetic north pole of the Earth, will make an angle α with the edge of the table. If we rotate the table, this angle will change. But again, we know why this happens: the hand of the compass is affected by the magnetic field of the Earth. Again, we postulate that if no external sources exist, then the hand of a compass is not forced to align to a specific direction. In other words, if we rotate an apparatus, the outcome of the experiment is the same.

Postulate. *Space is isotropic, that is, if the effects of external sources are accounted for properly, the outcome of no experiment depends on the orientation in space. In other words, different directions of space have the same properties. In yet other words, the laws of physics do not depend on the orientation in space.*

Actually, this postulate is meaningless unless we clarify what a rotation is. This will be done in a separate chapter.

1.3 Parity violation

By the statement that space is isotropic, we mean that the laws of physics are not affected by rotating the laboratory—of course by this we mean a fixed rotation, not a "rotating" frame. Put it another way, if we do some experiment with two different apparatuses, A and B if B is the same as A except for a fixed rotation, then the result of the experiment is the same, *provided that there are no external sources*. Now a question follows: What if apparatus B is the mirror image of apparatus A?

Before answering this question, we must clarify the meaning of the question. What do we mean by saying that apparatus B is the mirror image of apparatus A? If the apparatus is in the $z > 0$ half-space and a mirror is placed at the $z = 0$ plane, the image of the apparatus in the mirror would be the result of the transformation $(x, y, z) \mapsto (x, y, -z)$.

Now suppose the experiment is to produce a magnetic field by a right-handed coil. The mirror image of this right-handed coil is a left-handed coil. So if we want to check symmetry under mirror image, we must construct a left-handed coil. By our everyday experience, we are convinced that we can always make the mirror image of any apparatus. But this is not true in the subatomic world. The point is that if we want to make a genuine mirror image of an experiment, we have to use the mirror image type of elementary particles. Consider, for instance, an electron, which classically can be thought of as a charged rotating sphere. If the electron is rotating positive (counterclockwise or right-handed), then its image in the mirror is rotating negative (clockwise or left-handed). So the mirror or parity image of a right-handed electron is a left-handed one. Nature has provided us with both right-handed and left-handed electrons. However, nature has not provided us with the mirror image of all elementary particles. For example, there is a family of particles called neutrinos, which are always left-handed. That is, there are no right-handed neutrinos in nature. Physicists call this "parity violation."

It is important to remember that one cannot decide about rotational symmetry, or any other symmetry, by pure thought. We must do experiments, and it is the result of numerous experiments that we now think that rotational symmetry has been realized in nature, but parity is violated.

Remark. One final remark before leaving this subject: What if in Lab 1, using apparatus A, we do the change of coordinate $(x', y', z') = (x, y, -z)$? Can we do that? Is it legitimate? Yes, it is, and of course we can change the coordinates any way we wish. But we must realize that *a change of coordinate is not a transformation*, that is, a change of coordinate does not affect the behavior of the apparatus. If we use a right-handed Cartesian coordinate system (x, y, z), then we will find that only left-handed neutrinos exist in nature. If we use a left-handed coordinate system (x', y', z'), then we will see that only right-handed neutrinos exist.[1]

Guide to Literature. An excellent readable text is Feynman (1997, pp. 30–35). To see a list of articles about discrete space-time symmetries, see Stuewer (1993). The first experiment indicating parity violation, done in 1928, was done by Cox et al. (1928). The experiment establishing the subject for physicists was done by Chien-Shiung Wu together with a group of her colleagues in 1957 (Wu et al., 1957). For a brief explanation of the experiment, see Weinberg (1995, p. 127).

1.4 Homogeneity and the origin of the coordinate system

Homogeneity of space could be formulated in this way: The laws of nature should not depend on the origin of the coordinate system. To get a feeling of what this means, consider the following example.

Consider two objects of masses m_1 and m_2, connected to each other by a massless spring with constant k, laying on a horizontal table. We call this system S_1. For simplicity, let the motion be only in one dimension, along the x-axis. The governing

[1] The following analogy might be helpful: The human heart is skewed to the left. Changing the definition of right and left has no effect on our physiology. After such a change of words we say that our heart is skewed to the right of our chest.

Newton equations read

$$m_1 \ddot{x}_1 = k(x_2 - x_1) - b_1 \dot{x}_1,$$
$$m_2 \ddot{x}_2 = k(x_1 - x_2) - b_2 \dot{x}_2. \tag{1.1}$$

The terms $b_1 \dot{x}_1$ and $b_2 \dot{x}_2$ are friction forces caused by the interaction of the bodies and the table. In general, b_1 and b_2 depend on the position of the objects on the table—different points of the table may have different roughnesses—which means that, in general, the system does not have translational symmetry. In other words, the behavior of the system depends on where it is placed on the table.

Now if b_1 and b_2 do not depend on the position, that is, different points of the table are equivalent, or the surface of the table is homogeneous, then the system has translational symmetry.

These equations are written with a specific choice for the coordinate origin. Let us translate system S_1 to some other location on the table and name the translated system S_2. If we use the original coordinate system, which we have used to describe S_1, then the governing equations for S_2 are obtained by changing $x_1 \to x_1 + a$ and $x_2 \to x_2 + a$:

$$m_1 \ddot{x}_1 = k(x_2 + a - (x_1 + a)) - b_1 \dot{x}_1 = k(x_2 - x_1) - b_1 \dot{x}_1, \tag{1.2}$$
$$m_2 \ddot{x}_2 = k(x_1 + a - (x_2 + a)) - b_2 \dot{x}_2 = k(x_1 - x_2) - b_2 \dot{x}_2. \tag{1.3}$$

Note that a does not appear in this equation, provided, of course, that b_1 and b_2 do not depend on position. So this system has translational symmetry: The form of the differential equations governing S_2 is the same as that of those governing S_1. (In §1.14 we will see that this system does not have Galilean symmetry.)

1.5 Uniform rectilinear motion

If we are in a car or train or ship or plane, moving with constant velocity (relative to the ground) in a horizontal direction, then we do not sense the motion. In other words, if we do not look outside, we cannot feel we are moving. If we perform any experiment in that moving train, the result would be the same as a result obtained if the experiment was performed in a stationary train. Although this is a common experience, it was not noticed before Galileo[2] in the early 17th century. How Galileo was led to this discovery has a long and interesting history. Briefly, the sequence of events was this: First, Nicolaus Copernicus[3] invented a heliocentric model for the celestial objects, the Sun, and the planets, which we now call the Solar System. Galileo accepted this idea. But this idea had powerful opponents. One argument of

[2] Galileo Galilei (1564–1642) was known, both during and after his life, by his given name Galileo.

[3] Nicolaus Copernicus (1473–1543).

the opponents was this: If the Earth is both rotating around its own axis and moving around the Sun, why do we not feel this motion? Thinking about this problem led Galileo to formulate what we now call the *Galilean principle of relativity*.

Guide to Literature. For a comprehensive text on the history of the concepts of inertia and the Galilean principle of relativity, see Cohen (1960).

1.6 Galilean principle of relativity

Consider a device S_1 stationary in a lab L_1. Perform an experiment with S_1 and observe the result. Then construct a duplicate of the lab and the device, which we name L_2 and S_2, respectively. Put L_2 and consequently S_2 in *uniform* motion relative to L_1. This could be done by putting S_2 on board a train, say. Now perform the same experiment. The Galilean principle of relativity states that the result of this experiment would be exactly the same as the result obtained in S_1. At this point it is better to fix the nomenclature and think about the meaning of uniform motion.

In any lab L, we fix a point O, called the origin of coordinates, and consider three mutually orthogonal Cartesian axes, x_1, x_2, and x_3. Now for any particle we consider the vector connecting O to that particle. The components of this vector $\boldsymbol{r}$ are (x_1, x_2, x_3). By the velocity of the particle relative to lab L we mean the vector $\boldsymbol{v} = \dot{\boldsymbol{r}} = (\dot{x}_1, \dot{x}_2, \dot{x}_3)$. The particle is fixed in L if $\boldsymbol{v} = 0$, that is if the coordinates of its position are fixed numbers.

Definition. Lab L′ is in uniform motion relative to lab L if all points fixed in L′ have the same vector velocity relative to L.

If L′ is in uniform motion relative to L, then L is also in uniform motion relative to L′.

Postulate. ***Galilean principle of relativity*** *If labs L_1 and L_2 are related by a boost, which means one is moving with constant velocity relative to the other, then the result of any experiment performed in L_2 is the same as the result of the same experiment performed in L_1. Therefore, it is impossible to distinguish the two labs by any experiment. Therefore, the laws of nature should have the same mathematical form in both L_1 and L_2.*

Of course, it is possible to measure the velocity of L_2 relative to L_1 by observing a particle attached to L_2—fixed in L_2—using the devices in L_1. What the Galilean principle of relativity states is that it is impossible by an experiment *inside* L_2 to say that L_2 is in uniform motion relative to L_1.

Guide to Literature. An accessible article about Galilean covariance is by Rosen (1972).

1.7 Inertial vs. noninertial frames

A frame of reference K is said to be inertial if Newton's first law is fulfilled, that is if a free particle in K moves on a straight line with constant velocity. What we saw in the previous section means that if K' is a system moving with constant velocity v with respect to K, then K' is also inertial, for if Newton's first law is not fulfilled in K', then one can distinguish the two by experiments inside them.

Now let $\tilde{K}$ be a rotating frame. A very good example is the Earth itself, which is rotating around its polar axis. In $\tilde{K}$ Newton's first law is not valid—a free particle in $\tilde{K}$ moves not along straight lines with constant velocity. Such frames are called noninertial. It is important to note that an observer in a noninertial frame $\tilde{K}$ could find that $\tilde{K}$ is noninertial by mechanical or optical experiments *inside* $\tilde{K}$. An example of a mechanical experiment of this kind is the famous experiment of Foucault, demonstrating that the Earth is rotating.

So we have to realize that there is an important difference between systems in relative uniform rectilinear motion and systems in relative uniform rotation: Systems with relative rectilinear motion are not distinguishable by experiments inside them, while the system in relative uniform rotations are. Absolute rectilinear motion in space has no meaning, but absolute rotation in space has!

We will consider noninertial frames later in this book, using the mathematics of curvilinear coordinates in special relativity.

Guide to Literature. Mechanics in rotating frames is taught in textbooks on mechanics. See, for example, Symon (1971, pp. 271–285), Kleppner and Kolenkow (2014, pp. 356–359), and Goldstein (1980, pp. 174–182).

1.8 Time and clock

Let us think about time. What is time? Time is a real number we use to order the events. The world around us is continuously changing, and we want to organize the events. To do this, we make clocks. A clock is a device that has a moving part—the arm of the clock—which we can use to show us the time. For example, we can make a simple pendulum of a fixed length and make it oscillating. We can make a mechanism to count the number of oscillations and thus make a clock. We can also use the Earth's rotation as a clock. Or, we can use the motion of the Earth around the Sun as our clock. Or, we can use the pulsations of our heart as a clock. Any such system could be used as a clock.

Now suppose we make several clocks with different internal mechanisms. We can compare them, and usually, they do not agree with each other. For example, if we use the spinning Earth as our clock and define each rotational period to be exactly 86164.1006 s, and we also make an accurate atomic clock, we will see that the ticks of the spinning Earth are not of the same duration as those of the atomic clock. Now the question is, which one of these two clocks, the spinning Earth or the atomic clock, should we use to define time? The answer is this: Use the clock which makes the equations of physics more simple. To understand this statement, let us think about what would happen if we used the rate of our heartbeats to define time. If we do this, the equation governing a projectile would depend on the state of our body because when we run, the rate of our heartbeat increases, which, if we define the unit of time by our heartbeat rate, means that when we run, all physical phenomena around us would slow! So we see that using our heart as a clock makes physics more complicated. This example is perhaps very artificial. Let us think about another more realistic case. Up to the early 20th century, the second was defined as 1/86164.1006 of the rotational period of the Earth. When technology advanced, we were able to make atomic clocks with better accuracy. We decided to change the definition of a second by using atomic clocks because it is more simple to explain nonuniformity in the rotation of the Earth than to theorize that the rate of the atomic clocks depends on time. It is now meaningful to say that after the Sumatran earthquake of 26 Dec 2004, the period of Earth's rotation shortened by 3 μs (see Hopkin, 2004).

Guide to Literature. An excellent text about time as we understand it today is Audoin and Guinot (2001).

1.9 Newton's absolute time

The Galilean principle of relativity alone does not lead to the form of transformation relating L_2 to L_1. We must invoke some other fact about space and time. At this point we introduce Newton's absolute time, which is consistent with our everyday experience. Later, we will show that it leads to unacceptable results and therefore should be abandoned. Let us begin by thinking about the notion of time.

Consider a system composed of more than one point particle; for example, the Solar System. It has several dynamical variables. The goal of physics is to find these variables (or observables) as functions of time. As a simple example, consider a system of only two point particles moving along the x-axis. The position of particle 1 could be specified by $x_1(t)$ and that of particle 2 by $x_2(t)$. What is the meaning of saying $x_1(t_0) = a$? It means that at time t_0, particle 1 is at the point with coordinate a. The meaning of this last sentence is that "when clock C shows t_0, particle 1 is at $x = a$." That is, this statement is about the *simultaneity* of two *events*: Event 1 (particle 1 is at $x = a$) and Event 2 (the position of the clock's hand shows t_0). We owe this explanation to Einstein (1905).

Now remember that clock C is not at point $x = a$ but is usually located at another point. The notion of absolute time emerges from postulating that one can find simultaneity of these two events, even though there is a finite spatial distance between the two. That is, we postulate that one can use one single clock for every event in the world, for any observer. This is the postulate, or axiom, or principle of absolute time. In the words of Newton:

> *Absolute, true, and mathematical time, in and of itself and of its own nature, without reference to anything external, flows uniformly and by another name is called*

duration. Relative, apparent, and common time is any sensible and external measure (precise or imprecise) of duration by means of motion: such a measure—for example, an hour, a day, a month, a year—is commonly used instead of true time.

Principia, Definitions, Scholium (Cohen et al., 1999, p. 408)

So we state the following postulate, which is valid in Galilean relativity and is violated in Einstein's special relativity (as we will see in §2.3).

Postulate. *There is an absolute time, by which we can order events. This absolute time is independent of observer and position—that is, one single time for ordering anything in the universe, for any observer.*

1.10 Causal structure in Galilean relativity

Let E be an event, that is, something happening at a single point of space, at a single instant of time, or in mathematical terms, a point in space-time. Space-time is the totality of all events. We use Cartesian coordinates of space and read the absolute time on an ideal clock so that event E is characterized by $E = (t, x, y, z)$. Let us denote the time of event E by $t(E)$. In Galilean relativity, (x, y, z) is not absolute—it depends on the frame of reference and the coordinate system used by the observer. However, t is absolute! That is, it does not depend on the frame.

Consider two events, E_1 and E_2. We say that E_1 precedes E_2 if $t(E_1) < t(E_2)$. We denote this by writing $E_1 \prec E_2$. If $t(E_1) \leq t(E(2)$, we write $E_1 \preceq E_2$. If $t(E_1) = t(E_2)$, we write $E_1 \doteq E_2$ and we say that E_1 and E_2 are simultaneous. It is obvious that for any two events E_1 and E_2, we have $E_1 \prec E_2$, $E_1 \doteq E_2$, or $E_2 \prec E_1$.

Now consider two events and assume that E_1 is somehow the *cause* of E_2 (the *effect*). For example, E_1 may be the shooting of a bullet from a gun and E_2 may be the breaking of a vase. The bullet goes with velocity v. If the vase is a distance L from the gun, it takes $t = L/v$ for the bullet to reach the vase. Obviously, if v is finite, $t(E_1) < t(E_2)$, or in our notation $E_1 \prec E_2$. For $v \to \infty$ we get $E_1 \doteq E_2$, that is, *cause* and *effect* would be simultaneous. For $v < \infty$, *cause* always precedes the *effect*. In summary, E_1 could only be the cause of E_2 if $E_1 \preceq E_2$.

1.11 Galilean boosts

In lab L_1 we use orthonormal Cartesian coordinates (x_1, x_2, x_3) which we denote briefly by K. Lab L_2 is moving with constant velocity relative to L_1. Let $\boldsymbol{v} = (v_1, v_2, v_3)$ be this velocity. In lab L_2 we also use orthonormal Cartesian coordinates (x'_1, x'_2, x'_3), which we briefly denote by K'. In general these axes are not parallel to the coordinate axes of L_1, but let us choose the coordinates in L_2 to be parallel to those of L_1.

To describe the motion of any particle, in both labs L_1 and L_2 we use the Newtonian absolute time t, which is read on any clock synchronized with our standard clock C (for example, the Earth). For reasons that will become clear later, it is better to write t for the absolute time as read by an observer in L_1 and t' for the same absolute time as read by an observer in L_2. Note that we have

$$t' = t. \tag{1.4}$$

Now we want to find the transformation relating (t, x_1, x_2, x_3) to (t', x_1', x_2', x_3'). Suppose that at time $t = t' = 0$ the origin of the K' coordinates is at point (a_1, a_2, a_3). Note that these three real numbers must be considered as defining a point in the K coordinate system. So at $t = t' = 0$ we have

$$\begin{cases} x_1' = x_1 - a_1, \\ x_2' = x_2 - a_2, \\ x_3' = x_3 - a_3, \end{cases} \tag{1.5}$$

which is a translation of coordinates.

As the absolute time passes, the origin of K' moves so that at time t the origin of K' is at point $(a_1 + v_1 t, a_2 + v_2 t, a_3 + v_3 t)$. Therefore, at time t we would have

$$\begin{cases} x_1' = x_1 - a_1 - v_1 t, \\ x_2' = x_2 - a_2 - v_2 t, \\ x_3' = x_3 - a_3 - v_3 t. \end{cases} \tag{1.6}$$

Adding these to the relation $t' = t$ we get the so-called Galilean boost transformations:

$$\begin{cases} t' = t, \\ x_1' = x_1 - a_1 - v_1 t, \\ x_2' = x_2 - a_2 - v_2 t, \\ x_3' = x_3 - a_3 - v_3 t. \end{cases} \tag{1.7}$$

To get a graphical picture of these relations let us consider the case $a_1 = a_2 = a_3 = v_2 = v_3 = 0$ and consider $(t' = t, x' = x - v\,t)$ as a change of coordinates in the (t, x)-plane. It is customary to draw the x-axis as abscissa (horizontal) and the t-axis as the ordinate (vertical). We first note that since $t' = t$, the x'-axis coincides with the x-axis. This is because the x'-axis is the set $t' = 0$, which is the same set $t = 0$. The t'-axis however is not the same as the t-axis because the t'-axis is the set $x' = 0$, which is the same as the set $x = v\,t$, which is a line passing through the origin.

Note that although the t'-axis is not the same as the t-axis, the unit of time on t' is such that the line $t =$ const coincides with the line $t' =$ const. Thus, for example, in Fig. 1.1 the segment AB on the t'-axis has the same duration as the segment AC on the t-axis. The length of segment AB on the t'-axis is the time duration measured by the moving observer K', and the length of segment AC is the time duration measured by the stationary observer K. These two time durations are the same, and both are equal to the absolute time duration.

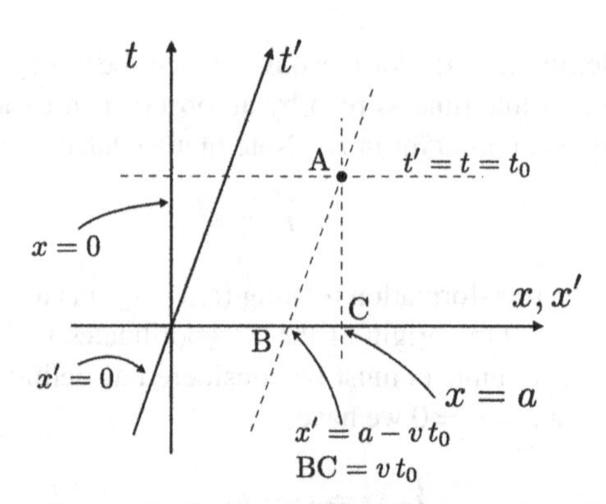

FIGURE 1.1 Space-time diagram for the simple 2D Galilean transformation.

Galilean boost. The stationary observer uses the coordinates (t, x), while the moving observer uses the coordinates (t', x'). Note that the x'-axis coincides with the x-axis; but the t'-axis is not the same as the t-axis. Also note that the unit of time on the t'-axis is such that $t = t'$ for any event.

That the x'-axis is the same as the x-axis is nothing but the absoluteness of time, since now if two events E_1 and E_2 are simultaneous in the K-frame, they are simultaneous in the K'-frame.

1.12 Galilean addition of velocities

Let K' be moving with velocity $\boldsymbol{v}$ relative to the stationary frame K. Consider a particle moving in space. The observer in K describes the motion by the set of three functions $\boldsymbol{x}(t) = (x_1(t), x_2(t), x_3(t))$. The velocity of the particle as defined and measured in K is $\boldsymbol{u} = \mathrm{d}\boldsymbol{x}/\mathrm{d}t$. According to an observer in frame K', the same particle is described by $\boldsymbol{x}'(t') = (x_1'(t'), x_2'(t'), x_3'(t'))$, with velocity vector $\boldsymbol{u}' = \mathrm{d}\boldsymbol{x}'/\mathrm{d}t'$. From our Galilean boost formulas we could obtain

$$\mathrm{d}t = \mathrm{d}t', \qquad \mathrm{d}\boldsymbol{x} = \mathrm{d}\boldsymbol{x}' + \boldsymbol{v}\,\mathrm{d}t' \tag{1.8}$$

and consequently

$$\boldsymbol{u} = \frac{\mathrm{d}\boldsymbol{x}}{\mathrm{d}t} = \frac{\mathrm{d}\boldsymbol{x}' + \boldsymbol{v}\,\mathrm{d}t'}{\mathrm{d}t'} = \boldsymbol{u}' + \boldsymbol{v}. \tag{1.9}$$

This is called the classical law of addition of velocities. It is derived by simple direct differentiation from the Galilean boosts, so it is sometimes called the addition of velocities theorem.

Motion of an airplane. As an example of the application of the Galilean addition of velocities, let us consider the motion of an airplane. The airplane moves in the air. Usually the air is not stationary with respect to the ground. If the wind is blowing

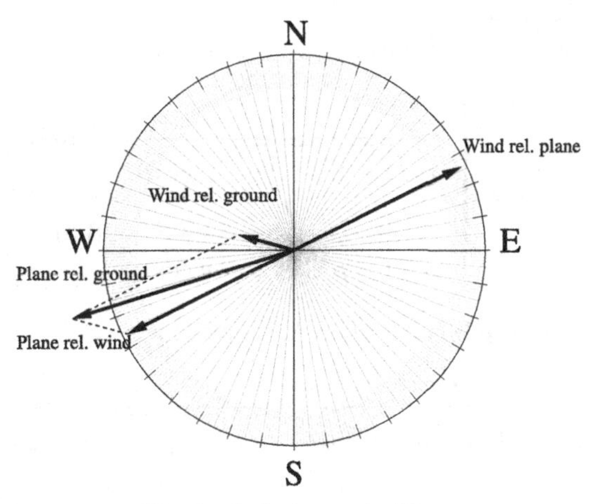

FIGURE 1.2 Typical geometry of Earth–wind–plane velocities.

with velocity $\boldsymbol{v}_{\text{wind}}$ relative to the ground and the plane moves with velocity $\boldsymbol{u}'_{\text{plane}}$ relative to the wind, then the velocity of the plane relative to the ground is

$$\boldsymbol{u}_{\text{plane}} = \boldsymbol{u}'_{\text{plane}} + \boldsymbol{v}_{\text{wind}}. \tag{1.10}$$

A typical situation is shown in Fig. 1.2.

Now we investigate the round trip travel time of a plane between two points when the wind is blowing. We will use this calculation later.

Worked Example 1. Consider an airplane moving along the x-axis (say eastward) from A to B and then returning back to A. Suppose that the velocity of the plane is always the constant value c with respect to the air and that the wind is blowing with velocity v in the direction $(\cos\theta, \sin\theta, 0)$. What is the time of flight (neglecting the time needed to reverse the direction of motion)?

Solution. To find this time, we use the theorem of addition of velocities. If $\boldsymbol{u}$ is the velocity of the plane with respect to the Earth and $\boldsymbol{u}'$ is its velocity relative to the air, then $\boldsymbol{u} = \boldsymbol{u}' + \boldsymbol{v}$. Therefore, when going from A to B (see Fig. 1.3),

$$\begin{cases} u_x = u'_x + v_x = c\,\cos\alpha + v\,\cos\theta, \\ u_y = u'_y + v_y = c\,\sin\alpha + v\,\sin\theta = 0. \end{cases} \tag{1.11}$$

We have $u_y = 0$, because the plane is moving along the x-axis. From this it follows that $\sin\alpha = -\beta\sin\theta$, where $\beta := v/c$. Now from the first equation we have $u_x = c\sqrt{1-\beta^2\sin^2\theta} + v\,\sin\theta$. We note also that in the return trip,

$|u_x| = c\sqrt{1-\beta^2 \sin^2\theta} - v \sin\theta$ (see Fig. 1.3), so we have

$$t = \frac{L}{|u_x|} = \begin{cases} \dfrac{L}{c\sqrt{1-\beta^2 \sin^2\theta} + v \cos\theta} & \text{A to B,} \\[2ex] \dfrac{L}{c\sqrt{1-\beta^2 \sin^2\theta} - v \cos\theta} & \text{B to A.} \end{cases} \tag{1.12}$$

The return time, from A to B and back to A, is given by

$$\begin{aligned} t &= \frac{L}{c\sqrt{1-\beta^2 \sin^2\theta} + v \cos\theta} + \frac{L}{c\sqrt{1-\beta^2 \sin^2\theta} - v \cos\theta} \\ &= \frac{L}{2c} f(\beta, \theta), \end{aligned} \tag{1.13}$$

where

$$f(\beta, \theta) = \frac{\sqrt{1-\beta^2 \sin^2\theta}}{1-\beta^2}. \tag{1.14}$$

For further reference,

$$f(\beta, 0) = \frac{1}{1-\beta^2}, \qquad f(\beta, \pi/2) = \frac{1}{\sqrt{1-\beta^2}}. \tag{1.15}$$

Problem 2. The wind is blowing eastward with velocity $20\,\mathrm{m\,s^{-1}}$. An airplane can move with velocity $50\,\mathrm{m\,s^{-1}}$ relative to the air. The plane is supposed to travel a distance of 100 km to the north. What is the velocity of the plane with respect to the Earth? And how much time does it take (to travel the given distance)?

Remark. In aether theories of the 19th century, it was assumed that light always moves with velocity c in aether.[4] Now, if a "wind of aether" is blowing with velocity $\boldsymbol{v}$ and a pulse of light moves along the x-axis from A to B and then back from B to A, the return travel time is given by the same formula $t = (L/2c) f(\beta, \theta)$, where $\beta = v/c$ and c is the velocity of light in aether. This is the basis of the nonrelativistic interpretation of the famous Michelson–Morley experiment (see §2.2).

Worked Example 3. Let K and K' be two inertial frames with parallel coordinate axes, K' moving with velocity $(0, 0, v)$ relative to K. Consider a light ray moving with velocity vector $c\,(\sin\theta \cos\varphi, \sin\theta \sin\varphi, \cos\theta)$ in K, where θ and φ are the usual angles of the spherical polar coordinates. Describe the light ray as seen in K', that is, determine both the speed and direction of the ray in K'.

---

[4] We use c here for the speed of light; previously we used the same symbol for the speed of sound. The reader is asked to pay attention to the context.

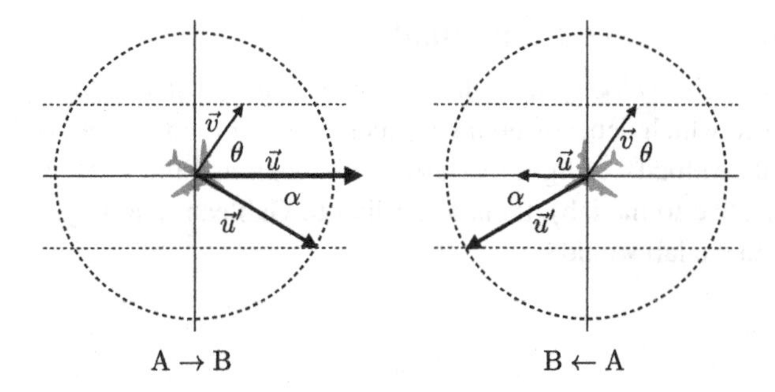

FIGURE 1.3

A plane is flying from A to B (left) and then back from B to A (right). $\boldsymbol{v}$ is the velocity of the wind relative to the Earth, $\boldsymbol{u}'$ is the velocity of the plane relative to the air. It is assumed that $|\boldsymbol{u}'|$ is always c, indicated by a circle of radius c. The vector $\boldsymbol{u}$ is the velocity of the plane relative to the Earth. θ is the angle of the wind with respect to the AB line, α is the angle of $\boldsymbol{u}'$ with the AB line. Two horizontal dashed lines are drawn with equal distance $v\cos\theta$ from the v_x-axis.

Remark. Here we are using Galilean relativity to study the aberration of light. The correct treatment will be presented later, using special relativity.

Solution. Let us denote the direction of the ray in K' by the angles θ' and φ' and the magnitude of the velocity vector of the light ray in K' by c'. Writing the additions of velocities in the form $\boldsymbol{u}' = \boldsymbol{u} - \boldsymbol{v}$, we have

$$c' \sin\theta' \cos\varphi' = c \sin\theta \cos\varphi, \tag{1.16}$$

$$c' \sin\theta' \sin\varphi' = c \sin\theta \sin\varphi, \tag{1.17}$$

$$c' \cos\theta' = c \cos\theta - v. \tag{1.18}$$

Dividing the first two equations we get $\varphi' = \varphi$. Squaring and adding all three equations we get

$$c' = c\sqrt{1 - 2\beta \cos\theta + \beta^2}, \tag{1.19}$$

where $\beta = v/c$. Now, using this result in the third equation we get

$$\cos\theta' = \frac{\cos\theta - \beta}{\sqrt{1 - 2\beta\cos\theta + \beta^2}}, \tag{1.20}$$

$$\sin\theta' = \frac{\sin\theta}{\sqrt{1 - 2\beta\cos\theta + \beta^2}}, \tag{1.21}$$

$$\tan\theta' = \frac{\sin\theta}{\cos\theta - \beta}. \tag{1.22}$$

Problem 4. Investigate the $\beta \to 0$ limit (that is, the $v \ll c$ limit) of the above equations.

1.12.1 Failure: Fizeau experiment

It is known that the velocity of light in matter is not c; it is c/n, where n is the refractive index, which depends on the frequency of light. For water and visible light, $n \simeq 1.33$. If the velocity of light in water is $u' = c/n$ and the water is moving with velocity v (relative to the lab), then, according to Galilean relativity, the velocity of light relative to the lab would be

$$u = \frac{c}{n} + v. \tag{1.23}$$

In 1851, Fizeau[5] performed an experiment and found that this is not correct. He found

$$u = \frac{c}{n} + v\left(1 - \frac{1}{n^2}\right). \tag{1.24}$$

In §3.9 we will see that special relativity has a very simple explanation for this—see also the remark note on page 61.

1.13 Transformation of kinetic energy and momentum

Let us look at the dependence of the kinetic energy and the momentum on the frame of reference. Consider a particle of mass m. We postulate that the mass of the particle does not depend on its velocity; in other words, we suppose that the mass of a particle is the same in frames K and K'.

The kinetic energy of this particle in K' is $E' = \frac{1}{2} m\, u'^2$, and its momentum is $\boldsymbol{P}' = m\,\boldsymbol{u}'$. From the addition of velocity theorem we find

$$E = \frac{1}{2}mu^2 = \frac{1}{2}m\left(u'^2 + 2\boldsymbol{u}'\cdot\boldsymbol{v} + v^2\right) = E' + \boldsymbol{v}\cdot\boldsymbol{P}' + \frac{1}{2}m\,v^2, \tag{1.25}$$

$$\boldsymbol{P} = m\,\boldsymbol{u} = m\left(\boldsymbol{u}' + \boldsymbol{v}\right) = \boldsymbol{P}' + m\,\boldsymbol{v}. \tag{1.26}$$

The inverse of these equations are obtained by changing the primed and un-primed variables and changing $\boldsymbol{v}$ to $-\boldsymbol{v}$, that is,

$$E' = E - \boldsymbol{v}\cdot\boldsymbol{P} + \frac{1}{2}m\,v^2, \tag{1.27}$$

$$\boldsymbol{P}' = \boldsymbol{P} - m\,\boldsymbol{v}. \tag{1.28}$$

[5] Armand Hippolyte Louis Fizeau (1819–1896).

1.14 Preferred frame of reference

Consider the example presented in §1.4: Two particles on a horizontal table, attached to each other by a spring, and there is a friction between the objects and the table:

$$\begin{aligned} m_1 \ddot{x}_1 &= k\,(x_2 - x_1) - b_1\,\dot{x}_1, \\ m_2 \ddot{x}_2 &= k\,(x_1 - x_2) - b_2\,\dot{x}_2. \end{aligned} \tag{1.29}$$

The equations of motion do not depend on the origin of coordinates, that is, the system has translational symmetry. However, consider the same system in a moving frame K' where $x' = x - v\,t$. It is easy to see that $\ddot{x}_1 = \ddot{x}'_1$, $\ddot{x}_2 = \ddot{x}'_2$, $x'_2 - x'_1 = x_2 - x_1$, $\dot{x}_1 = \dot{x}'_1 + v$, and $\dot{x}_2 = \dot{x}'_2 + v$. Therefore, in frame K' the system is governed by the following system of differential equations:

$$\begin{aligned} m_1 \ddot{x}'_1 &= k\left(x'_2 - x'_1\right) - b_1\left(\dot{x}'_1 + v\right), \\ m_2 \ddot{x}'_2 &= k\left(x'_1 - x'_2\right) - b_2\left(\dot{x}'_2 + v\right). \end{aligned} \tag{1.30}$$

Obviously this system of differential equations is not the same as (1.29). Frame K, in which the table is at rest, is the *preferred* frame—the friction is proportional to the velocity of the object relative to the table ($\dot{x}_1$ and $\dot{x}_2$).

When we state that the empty space has Galilean symmetry, we mean that there is no such preferred frame.

Remark. This example shows that Galilean symmetry is not derived from homogeneity of space!

1.15 Classical Doppler effect from addition of velocities

If there is no wind, the speed of sound in air, c, is the same in all directions, say $c = 340\,\mathrm{m\,s^{-1}}$. If a stationary source emits sound with frequency ν, a stationary detector placed a distance from the source detects the same frequency. However, if either the source or the detector moves, the detected frequency is not the same as ν. We will discuss the general case later; here we investigate the simple 1D case, that is, when the source and the detector move along the same line toward each other. Let us discuss three cases: (1) the source moves toward a stationary detector; (2) the detector moves toward a stationary source; and (3) both move toward each other with known velocities. In the general case 3, if v_s is the velocity of source (positive if toward the detector) and v_d is the velocity of the detector (positive if toward the source), both relative to the air (or ground), then the observed frequency ν' is given by

$$\nu' = \frac{c + v_\mathrm{d}}{c - v_\mathrm{s}}\,\nu. \tag{1.31}$$

Here, we show the special cases 1 and 2, where $v_d = 0$ or $v_s = 0$, respectively, and the general proof is left for the reader.

The space-time diagram of case 1 is shown in Fig. 1.4 left. The source is moving with velocity v_s, emitting pulses at times 0 and τ, where $\nu = \frac{1}{\tau}$ is the frequency. The equations of the rays (dashed lines) are

$$\text{ray 1} \qquad x = c\,t, \tag{1.32}$$

$$\text{ray 2} \qquad x = c\,(t - \tau) + v_s\,\tau. \tag{1.33}$$

To justify the second of these equations, note that at time τ the source is at $x = v_s\,\tau$, and the slope of the line is c. Now, suppose the detector is at $x = D$. The first sound pulse reaches the detector at time t_1, which solves $D = c\,t_1$. Similarly, the second pulse reaches the detector at time t_2, which solves $D = c\,(t_2 - \tau) + v_s\,\tau$. Equating these two we get

$$\nu' = \frac{1}{t_2 - t_1} = \frac{c}{c - v_s} \cdot \frac{1}{\tau} = \frac{c}{c - v_s}\,\nu. \tag{1.34}$$

For case 2 (Fig. 1.4 right), when the detector is moving ($x = D - v_d\,t$), the equations of the rays read

$$\text{ray 1} \qquad x = c\,t, \tag{1.35}$$

$$\text{ray 2} \qquad x = c\,(t - \tau). \tag{1.36}$$

Justification of the second equation is that its slope is c, and at $t = \tau$ it is at $x = 0$. Using these together with the equation of motion of the detector ($x = D - v_d\,t$) we get

$$c\,t_1 = D - v_d\,t_1, \tag{1.37}$$

$$c\,(t_2 - \tau) = D - v_d\,t_2, \tag{1.38}$$

which lead to

$$\nu' = \frac{1}{t_2 - t_1} = \frac{c + v_d}{c\,\tau} = \frac{c + v_d}{c}\,\nu. \tag{1.39}$$

Problem 5. A formula 1 car is moving in a straight line with velocity $v = 320\,\mathrm{km\,h^{-1}}$. The engine is making a sound of frequency $\nu_0 = 200\,\mathrm{Hz}$. Determine the frequency observed by someone near the road, both for when the car is approaching or receding. Take $c = 340\,\mathrm{m\,s^{-1}}$.

The classical, that is, nonrelativistic, Doppler effect of sound has very important applications. One is Doppler ultrasonography, where sound with a frequency of $\sim 40\,\mathrm{MHz}$ is produced and sent into the body of an animal and the echo of this sound is received and analyzed. The echo is due to the scattering of the sound from the moving red blood cells. Since the velocity of sound in the blood is $\sim 1.5 \times 10^3\,\mathrm{m\,s^{-1}}$, the wavelength of the sound with a frequency of $\sim 400\,\mathrm{MHz}$ is $\sim 4\,\mu\mathrm{m}$, almost half the size of blood cells.

Guide to Literature. For the physical basis and the engineering of the Doppler ultrasound, see Routh (1996).

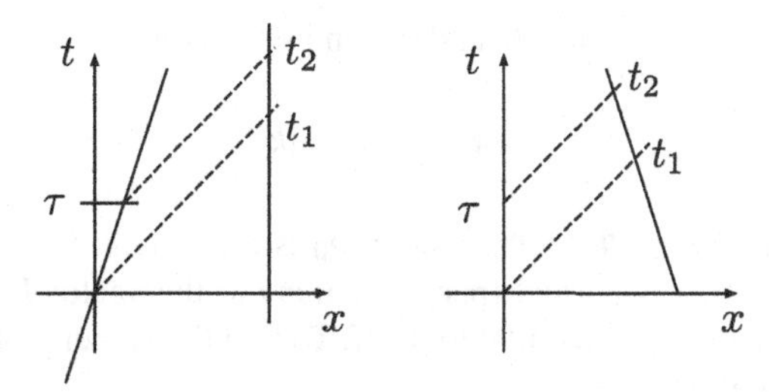

FIGURE 1.4 Space-time diagram for the classical Doppler effect.

Dashed lines show sound pulses moving with velocity c relative to the absolute space, which is the stationary air. Two pulses are shown. In the left diagram, the source is moving with velocity v_s relative to the air; in the right diagram, the detector is moving with velocity v_d relative to the air. The time interval between the emission of the pulses is $\tau = \frac{1}{\nu}$, where ν is the frequency of the source. The time interval between the detection of the pulses by the detector is $\tau' = t_2 - t_1 = \frac{1}{\nu'}$, where ν' is the frequency observed by the detector.

1.16 Plane sound waves

Sound is acoustic waves propagating in air. Let K be the lab system in which air is stationary (no winds). It is known that a sufficiently small disturbance in the air pressure p satisfies the following equation, which is called the wave equation:

$$\frac{1}{c^2} \cdot \frac{\partial^2 p}{\partial t^2} - \nabla^2 p = 0, \tag{1.40}$$

$$c := \sqrt{\left(\frac{\partial p}{\partial \rho}\right)_S}. \tag{1.41}$$

Here ρ is the density, p is the pressure, and subscript S means adiabatic change of state. For air at room temperature, $c \simeq 340\,\mathrm{m\,s^{-1}}$.

Guide to Literature. For a rigorous derivation of the wave equation of sound and its velocity, see Landau and Lifshitz (1987, pp. 251–252). For a more elementary derivation, see Faber (1995, pp. 78–81).

Remark. In the aether theories of the 19th century, light was considered as a wave of the aether. We can use the formulas of this section for aether theories of light by substituting the sound velocity with the light velocity.

A plane wave is described by a disturbance in pressure or density of the air with the functional form $p = A \cos \Phi$, where A is a constant called the amplitude of the wave and $\Phi = \omega t - \boldsymbol{k} \cdot \boldsymbol{x}$ is called the phase of the wave. Moreover, ω is the angular frequency, $\nu = \omega/(2\pi)$ is the frequency, and $\boldsymbol{k}$ is the wave vector. Inserting $A \cos(\omega t - \boldsymbol{k} \cdot \boldsymbol{x})$ in the wave equation, we see that there has to be a relation between the magnitude of the wave vector, k, and the angular frequency:

$$\frac{\omega}{k} = c, \qquad k = |\boldsymbol{k}|. \tag{1.42}$$

At any instant t, the set $\Phi = \Phi_0$, where Φ_0 is a constant, is a plane—the plane $\boldsymbol{k} \cdot \boldsymbol{x} = \omega t + \Phi_0$, which is normal (perpendicular) to the vector $\boldsymbol{k}$. As time passes, this plane will move in space, parallel to itself. To find the velocity of this motion, we have to differentiate the phase

$$\mathrm{d}\Phi = \omega \,\mathrm{d}t - \boldsymbol{k} \cdot \mathrm{d}\boldsymbol{x}. \tag{1.43}$$

Since we are considering a constant phase plane, we have $\mathrm{d}\Phi = 0$, and the phase velocity is defined by $\mathrm{d}\boldsymbol{x} = \boldsymbol{v}_p \,\mathrm{d}t$. Hence we get

$$\boldsymbol{k} \cdot \frac{\mathrm{d}\boldsymbol{x}}{\mathrm{d}t} = \omega. \tag{1.44}$$

We define the *phase* velocity $\boldsymbol{v}_p$, being a vector parallel to $\boldsymbol{k}$, with magnitude

$$v_p = \frac{\omega}{k} = c. \tag{1.45}$$

Note that a component normal to $\boldsymbol{k}$ will lead to motion along the $\Phi = \Phi_0$ plane, which is not measurable.

1.16.1 Doppler effect and the aberration from the invariance of phase

Let the frame K' be moving with velocity $\boldsymbol{v}$ relative to the stationary frame K. What is the description of the same plane sound wave from the point of view of a moving observer K'? To answer this question let us recall the Galilean transformation $t = t'$ and $\boldsymbol{x} = \boldsymbol{x}' + \boldsymbol{v}\,t'$, from which it follows that

$$\Phi = \omega t - \boldsymbol{k} \cdot \boldsymbol{x} \tag{1.46}$$

$$= \omega t' - \boldsymbol{k} \cdot \left(\boldsymbol{x}' + \boldsymbol{v}\,t'\right) \tag{1.47}$$

$$= (\omega - \boldsymbol{k} \cdot \boldsymbol{v})\, t' - \boldsymbol{k} \cdot \boldsymbol{x}' \tag{1.48}$$

$$= \omega \left(1 - \boldsymbol{\beta} \cdot \hat{n}\right) t' - \boldsymbol{k} \cdot \boldsymbol{x}', \tag{1.49}$$

where $\hat{n}$ is the direction of $\boldsymbol{k}$ and $\boldsymbol{\beta} = \boldsymbol{v}/c$ is the dimensionless vector velocity. We can therefore define

$$\omega' := \omega \left(1 - \boldsymbol{\beta} \cdot \hat{n}\right), \quad \boldsymbol{k}' = \boldsymbol{k}, \tag{1.50}$$

which leads to

$$\Phi = \omega' t' - \boldsymbol{k}' \cdot \boldsymbol{r}' = \Phi'. \tag{1.51}$$

Definition (1.50) has a physical basis—the invariance of the phase of a plane wave. The phase of a plane wave could be observed by finding the maxima of the pressure p—at the maxima of p, the phase is $2\pi\, n$ for integer n. Both observers (in K or K') would agree on whether or not a point is a maximum; they may disagree on its value but not on whether it is a maximum. They may also assign different integers n and n', writing $\Phi = 2\pi\, n$ and $\Phi' = 2\pi\, n'$. But this is not important. We therefore conclude that *because of the invariance of the phase of a plane wave, the transformation of angular frequency and wave vector is as (1.50).*

Therefore, the frequency of the wave, as measured by the moving observer K', differs from the frequency measured by the stationary observer. Note that to measure the frequency of a plane wave, the observer must use a *stationary* device, that is, a device with constant spatial coordinates. For observer K' this means constant $\boldsymbol{x}'$, but for the observer K this means constant $\boldsymbol{x}$.

From $\Phi = \omega' t' - \boldsymbol{k} \cdot \boldsymbol{x}'$ we see that the surfaces of constant phase are again planes normal to the vector $\boldsymbol{k}$, but these planes are moving with another phase velocity:

$$\mathrm{d}\Phi = 0 \quad \wedge \quad \mathrm{d}\boldsymbol{x}' = \boldsymbol{v}_{\mathrm{p}}' \mathrm{d}t' \quad \Rightarrow \quad \boldsymbol{k} \cdot \boldsymbol{v}_{\mathrm{p}}' = \omega - \boldsymbol{k} \cdot \boldsymbol{v}. \tag{1.52}$$

We rewrite this last equation by writing $\boldsymbol{k} = \dfrac{\omega}{c}\,\hat{\boldsymbol{n}}$; the result is

$$\hat{n} \cdot \left(\boldsymbol{v}_{\mathrm{p}}' + \boldsymbol{v}\right) = c = v_{\mathrm{p}} = v_{\mathrm{p}}\,\hat{\boldsymbol{n}} \cdot \hat{\boldsymbol{n}}, \tag{1.53}$$

from which it follows that $\boldsymbol{v}'_p = \boldsymbol{v}_p - \boldsymbol{v} + \boldsymbol{w}$, where $\boldsymbol{w}$ could be any vector normal to $\boldsymbol{n}$. The choice $\boldsymbol{w} = 0$ is the natural choice. Note that $\boldsymbol{w}$ is not observable, because it is a velocity of the $\Phi = \Phi_0$ plane sliding on itself. Therefore we have

$$\boldsymbol{v}_{\mathrm{p}}' = \boldsymbol{v}_{\mathrm{p}} - \boldsymbol{v}. \tag{1.54}$$

This equation has a natural interpretation: Consider a point of the $\Phi = \Phi_0$ plane. The velocity of this point relative to K is $\boldsymbol{v}_{\mathrm{p}}$; the velocity of frame K' relative to K is $\boldsymbol{v}$, so that, by the law of addition of velocities, the velocity of this point relative to K' is $\boldsymbol{v}_{\mathrm{p}} - \boldsymbol{v}$, which is nothing but $\boldsymbol{v}_{\mathrm{p}}'$. Therefore, if, as in quantum mechanics, we associate a particle to the plane wave, moving with velocity $\boldsymbol{v}_{\mathrm{p}} = c\,\hat{n}$, this particle, observed in frame K', has velocity $\boldsymbol{v}_{\mathrm{p}}'$. From this vector we can find the formula for the so-called Galilean *aberration*. Note that in general $\boldsymbol{v}_{\mathrm{p}}'$ is not parallel to $\hat{n}$, which means it is not normal to the $\Phi = \Phi_0$ plane! (See Fig. 1.5.)

As an illustrative example, consider the following case. Let frame K' be moving along the x-axis with velocity v, that is, $\boldsymbol{v} = (v, 0, 0)$, and let the plane wave vector be $\boldsymbol{k} = k\,\hat{n}$, where $\hat{n} = (\cos\theta, \sin\theta, 0)$. We have

$$\boldsymbol{v}_{\mathrm{p}}' = \boldsymbol{v}_{\mathrm{p}} - \boldsymbol{v} = c\,\hat{n} - \boldsymbol{v} = (c\,\cos\theta - v, c\,\sin\theta, 0)\,, \tag{1.55}$$

$$v_{\mathrm{p}}' = \sqrt{c^2 - 2\,c\,v\,\cos\theta + v^2}, \tag{1.56}$$

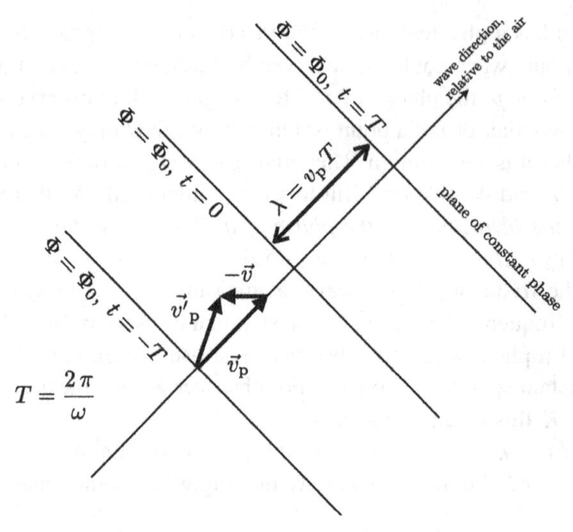

FIGURE 1.5 Aberration of a plane wave.

The direction of $\boldsymbol{v}'_p$ is not the same as the direction of $\boldsymbol{v}_p$. This is the aberration.

$$\begin{cases} \theta = 0 & v'_{\mathrm{p}} = c - v & < c, \\ \theta = \frac{\pi}{2} & v'_{\mathrm{p}} = \sqrt{c^2 + v^2} & > c, \\ \theta = \pi & v'_{\mathrm{p}} = c + v & > c. \end{cases} \tag{1.57}$$

Note that in frame K, the wave normal has angle θ with the x_1-axis (or that the corresponding particle, photon say, moves in that direction). In frame K', the vector $\boldsymbol{v}'_{\mathrm{p}}$ has an angle θ' with the x'_1-axis (or the corresponding particle moves in that direction). Here θ' could be obtained from θ:

$$\tan\theta' = \frac{\sin\theta}{\cos\theta - \beta}, \qquad \beta := \frac{v}{c}. \tag{1.58}$$

Guide to Literature. A very good text for learning the Doppler effect, both non-relativistic and relativistic, is Kleppner and Kolenkow (2014, pp. 466–470). For a somehow different derivation and discussion of the classical Doppler effect, see Jackson (1999, pp. 518–524). For a discussion of the propagation of sound waves in moving media see Landau and Lifshitz (1987, pp. 263–266).

1.17 Partial derivatives

Consider the atmosphere of the Earth. Fix a Cartesian coordinate system, stationary with respect to the ground. At time t the temperature of the atmosphere at point

(x_1, x_2, x_3) is $f(t, x_1, x_2, x_3)$. To this function we associate four partial derivatives

$$\partial_0 f := \frac{\partial f}{\partial t}, \quad \partial_1 f := \frac{\partial f}{\partial x_1}, \quad \partial_2 f := \frac{\partial f}{\partial x_2}, \quad \partial_3 f := \frac{\partial f}{\partial x_3}. \tag{1.59}$$

Recall the meaning of these partial derivatives: $\partial_0 f$ is the rate of change of f at the fixed point (x_1, x_2, x_3) in space; $\partial_1 f$ is the rate of change of f at fixed moment of time t, in the direction of the x_1-axis; etc.

Now consider an observer $\mathcal{O}'$ moving with velocity $\boldsymbol{v} = (v_1, v_2, v_3)$ relative to K, who uses Cartesian coordinates (x'_1, x'_2, x'_3) and time $t' = t$, which we briefly denote by K'. From the previous section we know that

$$\begin{cases} t' = t, \\ x'_1 = x_1 - a_1 - v_1 t, \\ x'_2 = x_2 - a_2 - v_2 t, \\ x'_3 = x_3 - a_3 - v_3 t. \end{cases} \tag{1.60}$$

The temperature field as described in K' is the function

$$f'(t', x'_1, x'_2, x'_3) = f(t, x_1 - a_1 - v_1 t, x_2 - a_2 - v_2 t, x_3 - a_3 - v_3 t). \tag{1.61}$$

To this function we associate the following partial derivatives:

$$\partial_{0'} f' = \frac{\partial f'}{\partial t'}, \quad \partial_{1'} f' = \frac{\partial f'}{\partial x'_1}, \quad \partial_{2'} f' = \frac{\partial f'}{\partial x'_2}, \quad \partial_{3'} f' = \frac{\partial f'}{\partial x'_3}. \tag{1.62}$$

Recall the meaning of these: $\partial_{0'} f'$ is the rate of change of the temperature at a fixed point in K', that is, at a fixed point relative to $\mathcal{O}'$ (of course, a point fixed relative to $\mathcal{O}'$ is not fixed relative to the ground); $\partial_{1'} f'$ is the rate of change of the temperature at a fixed moment of time, in direction x'_1. Remember that the time is considered to be absolute, so that fixed t' is the same as fixed t. Now we are ready to apply mathematics:

$$\partial_0 f = \frac{\partial f}{\partial t} = \frac{\partial t'}{\partial t} \cdot \frac{\partial f'}{\partial t'} + \frac{\partial x'_1}{\partial t} \cdot \frac{\partial f'}{\partial x'_1} + \frac{\partial x'_2}{\partial t} \cdot \frac{\partial f'}{\partial x'_2} + \frac{\partial x'_3}{\partial t} \cdot \frac{\partial f'}{\partial x'_3} \tag{1.63}$$

$$= \partial_{0'} f' - v_1 \partial_{1'} f' - v_2 \partial_{2'} f' - v_3 \partial_{3'} f' \tag{1.64}$$

$$= \partial_{0'} f' - \boldsymbol{v} \cdot \nabla' f'. \tag{1.65}$$

This identity is customarily written as

$$\partial_0 = \partial_{0'} - \boldsymbol{v} \cdot \nabla'. \tag{1.66}$$

In this form it should be remembered that the left-hand side, ∂_0, acts on $f(t, x_1, x_2, x_3)$, while the right-hand side acts on $f'(t', x'_1, x'_2, x'_3)$. I leave it to the reader to show that $\nabla f = \nabla' f'$, or equivalently

$$\nabla = \nabla'. \tag{1.67}$$

The inverse relations are

$$\partial_{0'} = \partial_0 + \boldsymbol{v} \cdot \nabla, \tag{1.68}$$

$$\nabla' = \nabla. \tag{1.69}$$

The equation $\partial_{0'} = \partial_0 + \boldsymbol{v} \cdot \nabla$ is familiar in fluid mechanics. If $\boldsymbol{v}$ is the velocity of an element of fluid, then $\partial_{0'} f$ is the temporal rate of change of the quantity f in the element's rest frame, customarily named the material or convective derivative, sometimes written as

$$\frac{Df}{Dt} = \frac{\partial f}{\partial t} + \boldsymbol{v} \cdot \nabla f. \tag{1.70}$$

Guide to Literature. The mathematical theory of functions of several real variables is discussed in all textbooks on advanced calculus. See, for example, Duistermaat and Kolk (2004); and Apostol (1957). A concise exposition of partial differentiation could be found in Levine (1997, pp. 3–16).

1.18 Galilean transformations of some partial differential equations

1.18.1 The continuity equation

Consider a fluid with density ρ and velocity field $\boldsymbol{u}$. The vector field $\boldsymbol{J} = \rho \boldsymbol{u}$ is called the current density. If $\mathrm{d}a$ is an infinitesimal surface with normal $\hat{\boldsymbol{n}}$, during time interval $\mathrm{d}t$ the amount $\boldsymbol{J} \cdot \hat{\boldsymbol{n}}\, da\, \mathrm{d}t$ of mass flows across this infinitesimal surface—see Fig. 1.6. Using this, the total amount of mass flowing out of the closed surface S is $\left(\oint_S \boldsymbol{J} \cdot \hat{\boldsymbol{n}}\, da\right) \mathrm{d}t$. The law of conservation of mass says that this must be equal to $-\mathrm{d}M = -\int_V (\partial_t \rho) \mathrm{d}V\, \mathrm{d}t$, *i.e.*, the amount of mass moving out of the closed surface S. Therefore, the law of conservation of mass is equivalent to saying that the following identity must hold for any volume V with boundary S:

$$\oint_S \boldsymbol{J} \cdot \hat{\boldsymbol{n}}\, \mathrm{d}a = -\int_V \frac{\partial \rho}{\partial t} \mathrm{d}V. \tag{1.71}$$

Now if we use Gauss' theorem,

$$\oint_S \boldsymbol{J} \cdot \hat{\boldsymbol{n}}\, \mathrm{d}a = \int_V \nabla \cdot \boldsymbol{J}\, \mathrm{d}V, \tag{1.72}$$

we get

$$\int_V \left\{ \frac{\partial \rho}{\partial t} + \nabla \cdot \boldsymbol{J} \right\} \mathrm{d}V = 0. \tag{1.73}$$

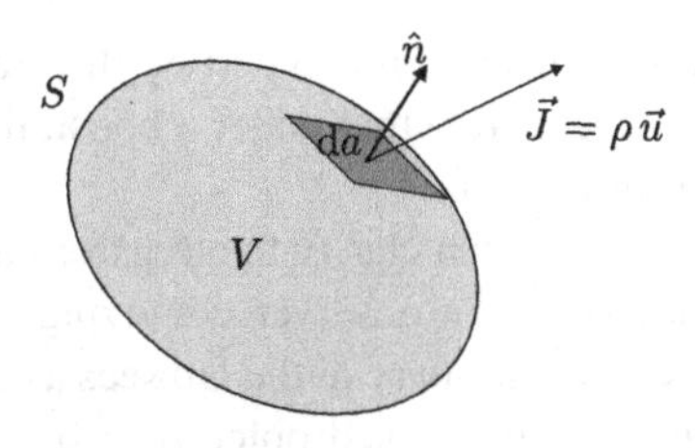

FIGURE 1.6 Conservation of mass.

Conservation of mass states that the amount of mass in volume V changes only if mass is going out of the surface S or coming into it.

Noting that volume V is arbitrary, that is, V could be any volume anywhere in space, we conclude that

$$\frac{\partial \rho}{\partial t} + \nabla \cdot \boldsymbol{J} = 0. \tag{1.74}$$

Applying the Galilean principle. According to the Galilean principle of relativity, the law of conservation of mass, being a law of nature, must have the same form in any two frames related by a Galilean boost. In other words, if K' is related to K by a Galilean boost, then the law of conservation of mass must be true in K'; otherwise one can check the law of conservation of mass in K and in K' and distinguish these two frames. So, the Galilean principle of relativity and the experimental fact that mass is neither created nor destroyed lead us to the following logical statement:

$$\frac{\partial \rho'}{\partial t'} + \nabla' \cdot \boldsymbol{J}' = 0 \iff \frac{\partial \rho}{\partial t} + \nabla \cdot \boldsymbol{J} = 0. \tag{1.75}$$

Let us write the left-hand equation, in the K' frame, and describe it in terms of (t, x_1, x_2, x_3):

$$0 = \frac{\partial \rho'}{\partial t'} + \nabla' \cdot \boldsymbol{J}' \tag{1.76}$$

$$= \frac{\partial \rho'}{\partial t} + \boldsymbol{v} \cdot \nabla \rho' + \nabla \cdot \boldsymbol{J}' \tag{1.77}$$

$$= \frac{\partial \rho'}{\partial t} + \nabla \cdot \left(\boldsymbol{v}\, \rho' + \boldsymbol{J}'\right). \tag{1.78}$$

Now the above proposition reads

$$\frac{\partial \rho'}{\partial t} + \nabla \cdot \left(\boldsymbol{v}\, \rho' + \boldsymbol{J}'\right) = 0 \iff \frac{\partial \rho}{\partial t} + \nabla \cdot \boldsymbol{J} = 0. \tag{1.79}$$

This statement is always true if we have

$$\begin{cases} \rho = \rho', \\ \boldsymbol{J} = \boldsymbol{J}' + \boldsymbol{v}\, \rho'. \end{cases} \tag{1.80}$$

This is the transformation law of $(\rho, \boldsymbol{J})$, dictated by the Galilean principle of relativity, or the Galilean covariance. Note that under a boost, the four quantities $(\rho, \boldsymbol{J})$ transform as the four quantities $(t, \boldsymbol{x})$.

Perhaps the following example can clarify the transformation. Consider a droplet of water moving with velocity $\boldsymbol{v}$. An observer comoving with the droplet sees no current, $\boldsymbol{J}' = 0$; but an observer stationary in the lab sees a current of water, $\boldsymbol{J} = \rho\, \boldsymbol{v}$. Both observers agree on the density of the droplet, $\rho' = \rho$.

1.18.2 The diffusion equation

Consider a drop of ink being inserted in a tank of water. Let f be the density of the ink molecules; f is a function of t and $x = (x_1, x_2, x_3)$. It can be shown that f satisfies the following partial differential equation:

$$\frac{\partial f}{\partial t} = \kappa\, \nabla^2 f, \tag{1.81}$$

where κ is called the diffusion coefficient and has the dimension of $L^2 T^{-1}$. Now consider a moving tank of water. In the moving frame K', the differential equation of diffusion reads

$$\frac{\partial f'}{\partial t'} = \kappa\, \nabla^2 f'. \tag{1.82}$$

This differential equation, written in terms of the stationary coordinates K, reads

$$\frac{\partial f'}{\partial t} + \boldsymbol{v} \cdot \nabla f' = \kappa\, \nabla^2 f'. \tag{1.83}$$

The density of ink is a scalar, that is, $f'(t', x') = f(t, x - v t)$. It is customary in physics to drop the prime from f' and write the above equation as

$$\frac{\partial f}{\partial t} + \boldsymbol{v} \cdot \nabla f = \kappa\, \nabla^2 f. \tag{1.84}$$

Note that this equation is not of the same form as (1.81). We say that the diffusion equation is *not* Galilean-covariant. It is not strange because now the tank of water determines a preferred frame. The diffusion equation has the simple standard form (1.82) only in the rest frame of the tank. If the tank is moving with velocity $\boldsymbol{v}$ in the lab, then in the lab frame K the diffusion equation is (1.84).

1.18.3 The Schrödinger equation

Note. The material in this section requires a knowledge of basic quantum mechanics.

Let K' be a moving frame moving with velocity $\boldsymbol{v}$ relative to the stationary frame K. In K', the wave function of a free particle with mass m is $\psi'(t', \boldsymbol{x}')$, which satisfies the Schrödinger equation

$$-\frac{\hbar^2}{2m}\nabla'^2\psi = i\,\hbar\frac{\partial\psi'}{\partial t'}. \tag{1.85}$$

Before discussing the Galilean covariance of this equation, recall that if the energy of the particle is E' and its momentum is $\boldsymbol{P}'$, then the wave function is

$$\psi'(t',\boldsymbol{x}') = A\,e^{-\frac{i}{\hbar}\Phi'}, \qquad \Phi' = E'\,t' - \boldsymbol{P}'\cdot\boldsymbol{x}'. \tag{1.86}$$

Using the transformation laws of the kinetic energy and the linear momentum $E' = E - \boldsymbol{v}\cdot\boldsymbol{P} + \frac{1}{2}m\,v^2$ and $\boldsymbol{P}' = \boldsymbol{P} - m\,\boldsymbol{v}$, we get

$$\Phi' = \left(E - \boldsymbol{v}\cdot\boldsymbol{P} + \frac{1}{2}m\,v^2\right)t - (\boldsymbol{P} - m\,\boldsymbol{v})\cdot(\boldsymbol{x} - \boldsymbol{v}\,t) \tag{1.87}$$

$$= E\,t - \boldsymbol{P}\cdot\boldsymbol{x} - \boldsymbol{v}\cdot\boldsymbol{P}\,t + \frac{1}{2}m\,v^2\,t + \boldsymbol{P}\cdot\boldsymbol{v}\,t + m\,\boldsymbol{v}\cdot\boldsymbol{x} - m\,v^2\,t \tag{1.88}$$

$$= E\,t - \boldsymbol{P}\cdot\boldsymbol{x} + m\,\boldsymbol{v}\cdot\boldsymbol{x} - \frac{1}{2}m\,v^2\,t. \tag{1.89}$$

Now let us transform the Schrödinger equation using $\partial_{0'} = \partial_0 + \boldsymbol{v}\cdot\nabla$ and $\nabla' = \nabla$. The result is

$$-\frac{\hbar^2}{2m}\nabla^2\psi'(t,\boldsymbol{x} - \boldsymbol{v}\,t) = i\,\hbar\left(\frac{\partial}{\partial t} + \boldsymbol{v}\cdot\nabla\right)\psi'(t,\boldsymbol{x} - \boldsymbol{v}\,t). \tag{1.90}$$

It is customary to drop the prime from ψ'. It must be understood that

$$\psi(t,\boldsymbol{x}) = \psi'(t' = t, \boldsymbol{x}' = \boldsymbol{x} - \boldsymbol{v}\,t). \tag{1.91}$$

With this convention we would have

$$-\frac{\hbar^2}{2m}\nabla^2\psi = i\,\hbar\frac{\partial\psi}{\partial t} + i\,\hbar\boldsymbol{v}\cdot\boldsymbol{\nabla}\psi. \tag{1.92}$$

This equation is not of the same form as (1.85). However, if we define

$$\tilde{\psi}(t,\boldsymbol{x}) := e^{-\frac{i}{\hbar}\left(\frac{1}{2}m\,v^2\,t - m\,\boldsymbol{v}\cdot\boldsymbol{x}\right)}\psi(t,\boldsymbol{x}), \tag{1.93}$$

then we would get

$$-\frac{\hbar^2}{2m}\nabla^2\tilde{\psi} = i\,\hbar\frac{\partial\tilde{\psi}}{\partial t}. \tag{1.94}$$

In words, we could say that the Schrödinger equations have Galilean symmetry, provided that we change the wave function as well. Since the wave function itself is not observable—only $|\psi|^2 \, \mathrm{d}^3x$ is observable—this change of the wave function is legitimate and we say that the Schrödinger equation respects Galilean symmetry. Note that

$$|\psi(t,x,y,z)|^2 \, \mathrm{d}x\,\mathrm{d}y\,\mathrm{d}z = \left|\tilde{\psi}(t,x,y,z)\right|^2 \mathrm{d}x\,\mathrm{d}y\,\mathrm{d}z. \tag{1.95}$$

Problem 6. Investigate the Galilean transformation of the wave equation

$$-\frac{1}{c^2}\frac{\partial^2 f}{\partial t^2} + \nabla^2 f = 0, \tag{1.96}$$

where c is the velocity of sound.

References

Apostol, T.M., 1957. Mathematical Analysis. A Modern Approach to Advanced Calculus. Addison-Wesley, Reading, MA.

Audoin, C., Guinot, B., 2001. The Measurement of Time. Time, Frequency and the Atomic Clock. Cambridge University Press, Cambridge.

Cohen, I.B., 1960. The Birth of a New Physics. Anchor Books, New York.

Cohen, I.B., Whitman, A., Budenz, J., 1999. The Principia: Mathematical Principles of Natural Philosophy, 1st edition. University of California Press, Cambridge.

Cox, R.T., Mcllwraith, C.G., Kurrelmeyer, B., 1928. Apparent evidence of polarization in a beam of β-rays. Proceedings of the National Academy of Sciences, USA 14, 544–549.

Duistermaat, J.J., Kolk, J.A.C., 2004. Multidimensional Real Analysis I. Differentiation. Cambridge University Press, Cambridge.

Einstein, A., 1905. Zur elektrodynamik bewegter körper. Annalen der Physics 17, 891–921.

Faber, T.E., 1995. Fluid Dynamics for Physicists. Cambridge University Press, Cambridge.

Feynman, R.P., 1997. Six Not-so-Easy Pieces: Einstein's Relativity, Symmetry, and Space-Time. Addison-Wesley, Reading, MA.

Goldstein, H., 1980. Classical Mechanics, 2nd edition. Addison-Wesley, Reading, MA.

Hopkin, M., 2004. Sumatran quake sped up Earth's rotation. Nature. https://doi.org/10.1038/news041229-6.

Jackson, J.D., 1999. Classical Electrodynamics, 3rd edition. John Wiley & Sons, New York.

Kleppner, D., Kolenkow, R., 2014. An Introduction to Mechanics, 2nd edition. Cambridge University Press, Cambridge.

Landau, L.D., Lifshitz, E.M., 1987. Fluid Mechanics, 2nd edition. Landau and Lifshitz Course of Theoretical Physics, vol. 6. Pergamon Press, Oxford.

Levine, H., 1997. Partial Differential Equations. Studies in Advanced Mathematics, vol. 6. American Mathematical Society, International Press, Providence, Rhode Island, USA.

Morrison, P., 1958. Approximate nature of physical symmetries. American Journal of Physics 26 (6), 358–368.

Rosen, G., 1972. Galilen invariance and the general covariance of nonrelativistic laws. American Journal of Physics 40 (5), 683–687.

Routh, H., 1996. Doppler ultrasound. IEEE Engineering in Medicine and Biology Magazine 15 (6), 31–40.

Stuewer, R.H., 1993. Resource letter ETDSTS-1: experimental tests of the discrete space-time symmetries. American Journal of Physics 61 (9), 778–788.
Symon, K.R., 1971. Mechanics, 3rd edition. Addison-Wesley, Reading, MA.
Weinberg, S., 1995. The Quantum Theory of Fields, vol. I, Foundations. Cambridge University Press, Cambridge.
Wu, C.S., Ambler, E., Hayward, R.W., Hoppes, D.D., Hudson, R.P., 1957. Experimental test of parity conservation in beta decay. Physical Review 105, 1413–1415.

CHAPTER

Failure of Newton's absolute time

2

2.1 Relativity and light

At the end of the 19th century, physics was divided into three disciplines:

1. Classical mechanics, the subject of which was the motion of material bodies, including fluids and celestial bodies.
2. Classical electrodynamics, whose subject was electric and magnetic phenomena. It was known that light is a wave of electromagnetic fields.
3. Thermodynamics, whose subject was heat and related phenomena.

Of these three disciplines, classical mechanics fulfilled completely the requirements of Galilean principle of relativity *and* the Newtonian absolute time. Electrodynamics did not. Thermodynamics was almost silent about this principle.

To see why Maxwell's equations are not consistent with Galilean transformations, we must briefly, but rigorously review their content.

Electrodynamics. Some bodies in nature show a property that we now attribute to their having *electric charge*. From the experiment it is known that electric charge could be described by a real number, positive, zero, or negative. Two particles a distance r apart exert a force on each other, which is due to their being charged. The force is inversely proportional to the square of distance, r^2, and linearly proportional to the charge of each particle. That is, one can write

$$F = K_1 \frac{q_1 q_2}{r^2}. \tag{2.1}$$

Fixing a value for K_1 is equivalent to choosing or fixing the unit of charge. For example, in the old cgs system of units, K_1 is fixed to be 1, which means that the unit of charge in cgs, 1 esu, is that amount of charge for which if $q_1 = q_2 = 1$ esu and $r = 1$ cm, then the force exerted on each of the particles would be 1 dyn.

All particles of nature are charged (a neutral particle is a particle with zero charge). One can define a density, ρ_e, for the charge, in exact analogy with the definition of mass density—the charge inside an infinitesimal volume $\mathrm{d}x\,\mathrm{d}y\,\mathrm{d}z$ around the point (x, y, z) at time t is $\mathrm{d}q$, where we have

$$\mathrm{d}q = \rho_e(t, x, y, z)\,\mathrm{d}x\,\mathrm{d}y\,\mathrm{d}z. \tag{2.2}$$

It is an experimental fact that current-carrying wires also exert force on each other. By current we mean charge passing through the cross-section of a wire in unit time.

A Mathematical Approach to Special Relativity. https://doi.org/10.1016/B978-0-32-399708-9.00011-7

It is known that if the wires are straight and parallel, carrying currents I_1 and I_2, a distance r apart, the force per unit length exerted on each wire is

$$f = K_2 \frac{I_1 I_2}{r}. \tag{2.3}$$

Here K_2 is another constant and f is force per unit length, that is, total force divided by the total length. If we have fixed K_1, then K_2 would be measurable by experiment. The logic of the modern conventions is that we fix the value of K_2, not that of K_1. In the SI system of units, by definition $K_2 = 2 \times 10^{-7}\,\mathrm{N\,A^{-2}}$.

Let us denote the dimension of length by L, that of time by T, that of charge by Q, that of current by $I = Q\,T^{-1}$, and that of force by F. Since f is force divided by length we have $K_2\,I^2 = F$. We also have $K_1\,Q^2\,L^{-2} = F$. From these two equations it follows that $K_1\,K_2^{-1}$ has the dimension of $L^2\,T^{-2}$, that is, the square of velocity.

In SI units two constants, ϵ_0, the permittivity of free space, and μ_0, the permeability of free space are defined as follows:

$$K_1 = \frac{1}{4\pi\,\epsilon_0}, \qquad K_2 = \frac{\mu_0}{2\pi}. \tag{2.4}$$

From the above analysis, we know that in SI units the quantity $\sqrt{\mu_0\,\epsilon_0}$ has a dimension of inverse velocity. We define

$$c := \frac{1}{\sqrt{\mu_0\,\epsilon_0}}. \tag{2.5}$$

If a particle with charge q is at point (x, y, z) of the laboratory, a force will be exerted on it. If the particle is at rest, relative to the laboratory, the force is $\boldsymbol{F} = q\,\boldsymbol{E}$. Here $\boldsymbol{E}$ is called the electric field. If the same particle is moving with velocity $\boldsymbol{u}$, relative to the laboratory, the force exerted on it is $\boldsymbol{F} = q\,\boldsymbol{E} + q\,\boldsymbol{u} \times \boldsymbol{B}$, where $\boldsymbol{B}$ is the magnetic field. The two fields $(\boldsymbol{E}, \boldsymbol{B})$ are called electromagnetic fields.

The electromagnetic fields are themselves produced by material content of the world, through the charge density ρ_e and current density $\boldsymbol{J}_e$.

Electric charge is a conserved quantity; like the mass in nonrelativistic mechanics, it obeys the continuity equation

$$\nabla \cdot \boldsymbol{J}_e + \frac{\partial \rho_e}{\partial t} = 0. \tag{2.6}$$

Maxwell's equations, describing the electromagnetic fields in vacuum, are a set of eight first-order partial differential equations which, by modern notation, are written as follows:

$$\nabla \cdot \boldsymbol{E} = \frac{\rho_e}{\epsilon_0}, \tag{2.7}$$

$$\nabla \cdot \boldsymbol{B} = 0, \tag{2.8}$$

$$\nabla\times \boldsymbol{E}+\frac{\partial \boldsymbol{B}}{\partial t}=0, \tag{2.9}$$

$$\nabla\times \boldsymbol{B}-\mu_0\,\epsilon_0\,\frac{\partial \boldsymbol{E}}{\partial t}=\mu_0\,\boldsymbol{J}_e. \tag{2.10}$$

Taking the divergence of the last equation, we get (2.6), so conservation of charge is a consequence of the Maxwell equations.

Taking the curl of the last two equations of the above set and using

$$\nabla\times(\nabla\times \boldsymbol{A})=\nabla(\nabla\cdot \boldsymbol{A})-\nabla^2\boldsymbol{A}, \tag{2.11}$$

it is easy to show that

$$\frac{1}{c^2}\frac{\partial^2 \boldsymbol{E}}{\partial t^2}-\nabla^2\boldsymbol{E}=\mu_0\left(\frac{\partial \boldsymbol{J}_e}{\partial t}+c^2\,\nabla\rho_e\right), \tag{2.12}$$

$$\frac{1}{c^2}\frac{\partial^2 \boldsymbol{B}}{\partial t^2}-\nabla^2\boldsymbol{B}=-\mu_0\,\nabla\times \boldsymbol{J}_e. \tag{2.13}$$

These six equations say that all Cartesian components of the electric and magnetic fields satisfy the inhomogeneous wave equation, with wave velocity c. The source, that is, the right-hand side of the above equations, is determined by the charge and current densities. In vacuum, where there is no matter, we still have waves propagating with velocity c.

When Maxwell realized the wave solution of his equations, he claimed that light is a form of electromagnetic waves. Later, Heinrich Hertz[1] experimentally produced electromagnetic waves, and ever since, we believe that light is nothing but electromagnetic waves.

In the 19th century, all physicists believed that any wave needs a substrate, like the surface waves of water or sound waves. So Maxwell proposed that these electromagnetic waves are waves of a material medium called the aether. In other words, according to Maxwell and other physicists, Maxwell's equations in the above form are valid in the rest frame of the aether.

Dilemma. We know that the wave equation derived from Maxwell's equations is not invariant under Galilean transformations. Therefore we have two choices:

1. Maxwell's equations are valid only in some specific frame, the so-called aether frame. This leads to the prediction that one can find the velocity of a frame relative to the aether frame by studying the velocity of electromagnetic waves in different directions. This was the way out of the dilemma accepted by almost all physicists before Einstein.
2. Maxwell's equations are valid in any inertial frame, and the Galilean principle of relativity is respected by the laws of electrodynamics. However, the Galilean boosts are not the correct transformations relating two inertial frames moving

[1] Heinrich Rudolf Hertz (1857–1894).

relative to each other. This choice has the strange consequence that the velocity of light is the same in all inertial frames, which is in contradiction to the ordinary Galilean theorem of addition of velocities. This is the way out of the dilemma proposed by Einstein in 1905 and is universally accepted by physicists ever since.

These two ways out of the dilemma have different observational predictions, so one can decide between them by experiment. For example, the first way, the so-called aether theories, is ruled out by the famous experiment of Michelson and Morley; the second way out, the so-called Einstein special relativity, has predictions such as time dilation and many many others, which so far are all in agreement with experiments.

Let us think about the speed of light from another perspective which can clarify this dilemma. In lab L_1 (our lab) the observer, which is at rest in the lab, fixes the values of kilogram (unit of mass), second (unit of time interval), and meter (unit of length), by some standard conventions. He then fixes the value of μ_0, which means fixing the unit of current (Ampere). Then he can measure, in an intrinsically static experiment, the value of ϵ_0. He therefore gets the value of $c = 1/\sqrt{\mu_0 \epsilon_0}$. Now think of another lab, L_2, which is moving relative to L_1 with velocity v. The observer in L_2 can perform the same experiments. If she finds some other value for ϵ_0 and hence for c, then she has found by an *internal* experiment that her frame of reference is in absolute motion relative to the aether, and we have a method to find absolute motion. So, if the Galilean principle of relativity is respected by the laws of electrodynamics, then she must find the same value for c. But her value for c is simply the speed of light in L_2!

Now suppose the observer in L_1 sends an electromagnetic wave signal to lab L_2. This wave, being an electromagnetic wave in L_1, is moving with velocity c relative to L_1. The same wave, being a wave of electromagnetic fields seen from L_2, must be moving with the same speed c relative to L_2! This is a counterintuitive result. If a material particle (for example, an electron) is being ejected from L_1 with a velocity u, the same particle, when observed by devices in L_2, will have a different velocity $u' \neq u$. Later we will see the mathematics behind this *seemingly odd* prediction.

2.2 The Michelson–Morley experiment

In the last decades of the 19th century, the majority of physicists were in favor of the so-called *aether* theories, which state that the Maxwell equations in the usual form are valid only in the rest frame of the aether, and therefore, the speed of light is the constant value c only in the aether rest frame. Since we know that the Earth is rotating around its axis, we know that relative to the center of the Earth a point on Earth at latitude λ is moving with a velocity

$$v_1 = R\omega \cos\lambda = 6.4 \times 10^6\,\mathrm{m}\left(\frac{2\pi}{86164\,\mathrm{s}}\right)\cos\lambda = 4.7 \times 10^2 \cos\lambda\,\mathrm{m\,s^{-1}}. \quad (2.14)$$

Since we know that the Earth is orbiting the Sun, we know that the center of the Earth is moving with velocity

$$v_2 = 1.5 \times 10^{11}\,\mathrm{m}\left(\frac{2\pi}{365.25\,\mathrm{d}}\right) = 3.0 \times 10^4\,\mathrm{m\,s^{-1}} \tag{2.15}$$

relative to the Sun.

The Earth could not be in rest relative to the aether at all moments. It is reasonable to think that (at least at some times) a laboratory on Earth is moving with velocity $\sim 3.0 \times 10^4\,\mathrm{m\,s^{-1}}$ through the aether. Equivalently, there should be a wind of aether on Earth with velocity $\sim v_2$ or $\beta = v/c \simeq 10^{-4}$. This could be observed by a Michelson interferometer. Michelson[2] and Morley[3] did the experiment.

The following analogy could help us understand the situation. Consider several ships on a calm sea. Each ship is a frame—K, K', K'', etc. If someone disturbs the seawater near a vessel, a surface wave of water will be produced and is propagated. This wave moves with velocity c_w with respect to the stationary water, which is a preferred frame. Maxwell's aether is analogous to this sea. A sailor in a ship can measure the velocity of the surface wave of water in different directions. If they are all the same value c_w, the ship is stationary with respect to the sea; otherwise the ship is moving relative to the sea and its velocity could be inferred from the values of the speed of wave in different directions.

Michelson interferometer. The Michelson interferometer has two perpendicular arms (see Fig. 2.1). A monochromatic beam of light with wavelength λ enters the device. The beam splitter splits the beam in two, passing through the two arms. At the end of each arm, there is a mirror causing the beam to return. The two beams would interfere after returning to the splitter. If both arms are of the same length L and the device is stationary in the aether rest frame, the time of flight of the light in both arms is $t_0 = 2L/c$. However, as we saw in Worked Example 1 on page 13, if there is a wind of aether with velocity v with an angle θ to the first arm, the time of flight in the first arm would be $t_1 = t_0\, f(\beta, \theta)$, where $\beta = v/c$ and

$$f(\beta, \theta) = \frac{\sqrt{1 - \beta^2 \sin^2\theta}}{1 - \beta^2}. \tag{2.16}$$

The time of flight of the light in the second arm would be $t_2 = t_0\, f(\beta, \pi/2 - \theta)$. The phase difference of the two beams is proportional to

$$\frac{c\,\Delta t}{\lambda} = \frac{c\,(t_2 - t_1)}{\lambda} = \frac{c\,t_0}{\lambda} g(\beta, \theta), \tag{2.17}$$

[2] Albert A. Michelson (1852–1931), 1907 Nobel Prize in Physics.

[3] Edward Morley (1838–1923).

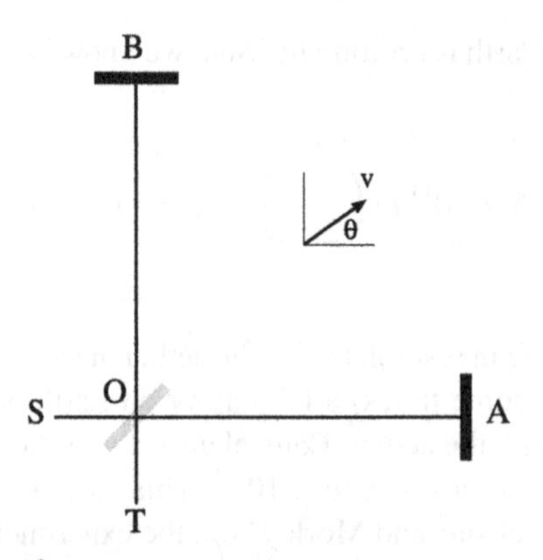

FIGURE 2.1 Michelson interferometer.

The light source is at S. The light moves from S to the beam splitter at O, and it then splits into two beams. Beam 1 goes from O to mirror A, returns to O, and then (by reflection) goes to T. Beam 2 goes from O to mirror B, returns to O, and then goes to T. $\boldsymbol{v}$ is the velocity of the "aether wind."

where

$$g(\beta,\theta) = \frac{\sqrt{1-\beta^2\sin^2\theta} - \sqrt{1-\beta^2\cos^2\theta}}{1-\beta^2}. \tag{2.18}$$

The maximum change in g is

$$g(\beta,0) - g(\beta,\pi/2) = 2\left(\frac{1}{1-\beta^2} - \frac{1}{\sqrt{1-\beta^2}}\right) \simeq \beta^2. \tag{2.19}$$

Therefore, if one rotates the device 90 degrees, there should be a fringe shift of the order $n = \frac{c\,t_0}{\lambda}\beta^2$. We have $t_0 = 2L/c$ so that

$$n = \frac{2L}{\lambda}\beta^2. \tag{2.20}$$

For $L = 10\,\text{m}$, $\lambda = 500\,\text{nm}$, and $\beta \simeq 10^{-4}$, we have $n \simeq 0.4$. Michelson and Morley did the experiment and found that there is no appreciable fringe shift, showing that the velocity of the lab in the aether is much less than $10^{-4}\,c$.

The accuracy of the experiment depends on the length L. Michelson and Morley used a set of mirrors to extend the length by a factor of 8. Among the difficulties of the experiment is rotating the interferometer such that the different parts (mirrors and splitter) do not move relative to each other to the accuracy of a fraction of the wavelength. To achieve this, Michelson and Morley set up the interferometer on a stone table and floated the table on a pool of mercury so that they can rotate the heavy table. So the experiment is actually setting up the device (with a monochromatic coherent source of light), rotating the table, and observing the fringe shift.

In the last few decades, some large-scale Michelson interferometers were built, not to test Einstein's special theory of relativity but rather to detect tiny distortions of space-time as predicted by Einstein's general theory of relativity. To get a feeling of the scale, the arms of one such interferometer, which is located in Washington,[4] USA, are 4 km long.

2.3 Synchronization of clocks

The objective of physics is to understand mathematically what happens in nature (lab). To describe the motion of a system of particles, we must be able to specify the location and the time. To specify the location, we need a rigid coordinate system. To assign time to events, we need clocks. In Galilean relativity, we use Newton's absolute time, which means that we have assumed that there is a single time for organizing all events of nature. Einstein suspected this. In his 1905 paper (Einstein, 1905), he wrote:

> *If we wish to describe the* motion *of a material point, we give the values of its co-ordinates as functions of the time. Now we must bear carefully in mind that a mathematical description of this kind has no physical meaning unless we are quite clear as to what we understand by "time." We have to take into account that all our judgments in which time plays a part are always judgments of* simultaneous events. *If for instance, I say, "that train arrives here at 7 o'clock," I mean something like this: "The pointing of the small hand of my watch to 7 and the arrival of the train are simultaneous events."*
>
> **Einstein et al. (1952, pp. 38–39)**

Now let us have a closer look at this. Suppose two events happen in two different locations, for example, let event E_1 happen at point A and event E_2 at point B. How can we say that these two events are simultaneous or one precedes the other? Obviously, we have to have two clocks, one close enough to A, the other close enough to B. We record two times t_1 and t_2, which are the times E_1 and E_2 happen at. But these two numbers, t_1 and t_2, are measured by two different clocks, and using them for ordering the events is meaningless unless we can say that:

1. the two clocks are identical and use the same standard of time,
2. the two clocks are *synchronized.*

For example, if A is a city in Europe, B is a city in Asia, and t_1 and t_2 are reported by two local police officers to be 10:00 a.m., we cannot say that the two events were simultaneous—we have to consider that the time zones are not the same. So not only

[4] http://www.ligo.caltech.edu.

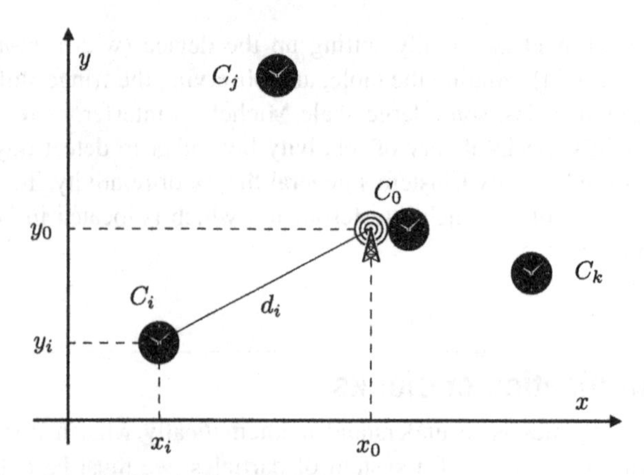

FIGURE 2.2 Einstein's synchronization of clocks.

do we have to have many clocks—one for each spatial position—the clocks must be reasonably synchronized. And now we see that we are facing a serious problem: How can we set many different clocks in different locations to be *reasonably synchronized*?

Einstein's solution is to use the constancy of the speed of light. Remember that if we accept that Maxwell's equations are valid in all inertial frames, then the speed of light must be the same in all inertial frames. But at this stage, we do not need the whole content of the Maxwell equations; all we need is to assume that the velocity of light is the same in all inertial frames. Using this, as we will see in a moment, we can synchronize all clocks in any inertial frame. This synchronization would result in a change in the Galilean transformations—resulting in the Lorentz transformations, which themselves lead to various observable predictions, all of which are verified in experiments.

Let K be an inertial frame, for example, the Earth.[5] We have several atomic clocks, $C_1, C_2, \ldots, C_N$, stationary with respect to the Earth, at various positions. By means of devices such as theodolites and measuring rods, we can find the coordinates of the clocks. Let (x_i, y_i, z_i) be the Cartesian coordinate of the clock C_i. Let a radio station equipped with the atomic clock C_0 be located at point (x_0, y_0, z_0). Clock C_i is a distance d_i from the radio station—see Fig. 2.2. We have

$$d_i = \sqrt{(x_i - x_0)^2 + (y_i - y_0)^2 + (z_i - z_0)^2}. \tag{2.21}$$

If the velocity of light is c, it takes

$$\Delta t_i = \frac{d_i}{c} \tag{2.22}$$

[5] Of course, the rotating Earth is not an inertial frame, but for the moment, let us forget about this rotation.

for the wave emitted at the radio station to reach the clock C_i. Einstein's synchronization procedure is as follows. When clock C_0 (at the radio station) shows time t_0, announce the time t_0 by sending an electromagnetic wave by the radio station, and when the clock C_i receives the signal, set C_i to show the time $t_i = t_0 + \Delta t_i$. Obviously we now have

$$c\,\Delta t_i = c\,(t_i - t_0) = \sqrt{(x_i - x_0)^2 + (y_i - y_0)^2 + (z_i - z_0)^2}, \tag{2.23}$$

or

$$c^2\,(t_i - t_0)^2 - (x_i - x_0)^2 + (y_i - y_0)^2 + (z_i - z_0)^2 = 0. \tag{2.24}$$

This equation is nothing but the statement that the velocity of light in any direction in frame K is c so that it takes d/c for the light signal to go from any point to any point a distance d from it.

Remark. To implement this synchronization, we have *defined* the velocity of light to be c. The reason is that to find the velocity of light going from point A to point B, we have to have synchronized clocks, but we have synchronized these clocks by the velocity of light being c! In the SI system of units, the definition is

$$c = 299{,}792{,}458\,\mathrm{m\,s^{-1}}. \tag{2.25}$$

In the SI system of units we also define the second by some atomic phenomena, and then the meter is defined to be the length light travels in $1/299{,}792{,}458\,\mathrm{s}$.

Now that we have synchronized clocks at various positions, we can measure the velocity of a particle moving from A to B. To do this, we measure the distance d between A and B, record the departure time t_1 by a clock at A, record the arrival time t_2 by a synchronized clock at B, and calculate

$$v = \frac{d}{t_2 - t_1}. \tag{2.26}$$

Thus far, all observations show that nothing moves faster than light. Note that in principle it is possible to observe some particles moving faster than light, but, as we will see later, the existence of such particles, called tachyons, has strange consequences for physics. No tachyon has been detected yet.[6]

[6] In early 2011, a group of CERN scientists reported that they have measured the time of flight of τ-neutrinos in a distance of about 730 km and found this time was almost 60 ns shorter than d/c, which means that $(v - c)/c \simeq 2.5 \times 10^{-5}$. Later on, after carefully reanalyzing the data, they announced that they found (Adam et al., 2012) $-3.7 \times 10^{-6} \le \frac{v-c}{c} \le 9.2 \times 10^{-6}$, which is not as strong as their earlier claim and the majority of physicists do not consider this range as indicating that τ-neutrinos are tachyons.

2.4 Maximum velocity of propagation of interactions

It is an experimental fact that we cannot influence distant objects instantaneously. For example, a commander can shout "fire" at time t_0, but the subordinate soldier hears the shout at time $t_0 + d/c_{\text{sound}}$, where d is the distance between the commander and the soldier and c_{sound} is the velocity of sound in air. The commander can use light signals to command, but light also moves with a finite velocity.[7] Logically, there are only two possibilities:

1. The maximum velocity of propagation of interactions is ∞.
2. The maximum velocity of propagation of interactions is finite.

Whichever of these two possibilities is true, it is a law of nature. Combining with the Galilean principle of relativity, if it is valid in frame K, it must be valid in frame K' which is moving with velocity v relative to K (otherwise one can distinguish being in K or in K' with no reference to the other frame, and that is a violation of the Galilean principle of relativity).

If the maximum velocity of propagation of interactions is ∞, then we can use interactions propagating with ∞ velocity to synchronize clocks and get the Galilean relativity. If the maximum velocity of propagation of interactions is finite, $c_{\max}$ say, then we can use this kind of interaction to synchronize clocks and that leads to transformations which we are going to study in this book. It can be proved that:

1. The velocity $c_{\max}$ is invariant, that is, the same in any frame.
2. No other velocity could be invariant, that is, any velocity either smaller than $c_{\max}$ or greater than that could not be the same in frames that are moving relative to each other.

For the proof, see the theorem at the end of §3.8.

The maximum velocity of propagation of interactions, $c_{\max}$, is the velocity of light in vacuum, denoted by c.

Remark. Even in matter, $c_{\max} = c$. This might seem confusing. For example, in water with refractive index $n \simeq 1.3$, the speed of light is $c/n \simeq 0.75\,c$, but still we have $c_{\max} = c$! When we say that the velocity of light in water is c/n, we should note that this is the velocity of a wave in a set of *macroscopic, average* quantities—the electric displacement and magnetization, which are manifestations of the electric charge distribution and electric current in the molecules and groups of molecules of water. In water (or in any matter), we could have particles moving with velocity greater than c/n but smaller than c. For example, we can inject an electron with

---

[7] The speed of light is so high compared with anything else around us that we feel it is infinite. Measuring the speed of light has been a challenge to experimentalists. The first acceptable value for the speed of light was the result of an observation by Ole Christensen Rømer. According to Rømer's findings, the velocity of light is almost $\frac{2\,\text{AU}}{22\,\text{min}}$, which, for the currently accepted value of the astronomical unit, means almost 2.3×10^8 m/s.

velocity $0.99\,c$ into water. Of course, this particle will experience deceleration, and eventually, its velocity will be less than c/n, but for a period of time, it will have a velocity larger than $c/n = 0.75\,c$. So we can have particles moving faster than c/n, but still slower than $c_{\max} = c$.

Guide to Literature. If a *charged* particle moves in matter faster than c/n, it emits light. This phenomenon—first observed in 1934 by P. Cherenkov[8] and explained in 1937 by I. Tamm[9] and I. Frank[10]—is so important that it brought them the 1958 Nobel Prize in Physics. Understanding the physics of Cherenkov radiation requires some advanced electrodynamics. The interested reader is referred to Jackson (1999, pp. 637–640) and Landau and Lifshitz (1984, pp. 406–408).

2.5 No perfect rigid body

As we saw in §2.4, no interaction could propagate faster than light in vacuum. We now argue that this means that there are no idealized rigid bodies.

Suppose we have the following system (see Fig. 2.3): two mass points attached to the ends of a massless spring of rest length L. By rest length, we mean that when the length of the spring is L there is no tension in it. Suppose the masses are at positions $x_0 = 0$ and $x_2 = L$. At time $t_1 = 0$, we apply an impulse to particle 1 so that it begins to move to the right with velocity v. There would be a mechanical wave in the spring, propagating to the right toward particle 2. This mechanical wave propagates with a velocity $c' \leq c$, where c is the velocity of light in vacuum. It takes L/c' for the wave to reach point 2 at $x_2 = L$. During this period, particle 1 has moved a distance $(L/c')\,v = L\,v/c'$. The maximum possible velocity of propagation of a physical wave is c—the velocity of light in vacuum—so that the minimum amount the spring length changes is $L\,\beta$, where $\beta = v/c$.

A metallic rod is a collection of atoms attached to each other by the electromagnetic force of its nuclei and electrons. When we exert a force on one end of a rod, we actually apply force (or impulse) to those atoms. The distant atoms (for example, those at the other end of the rod) do not experience the force until *the disturbance wave* reaches them. This takes time, a time at least equal to L/c. So if we apply a force such that one end of the rod moves with velocity v, there would be a change of length of order $L\,\beta$ of the rod.

In classical nonrelativistic mechanics, a rigid body is a body whose points are fixed relative to each other, that is, for any two points 1 and 2 of the body, the distance is always the same. As we saw in the preceding paragraph, this is impossible in special relativity. When we apply a force to the point of an almost rigid body, only if the velocity of that point is much less than c can we ignore the change in the distance

[8] Pavel Alekseyevich Cherenkov (1904–1990).

[9] Igor Yevgenyevich Tamm (1895–1971).

[10] Ilya Mikhailovich Frank (1908–1990).

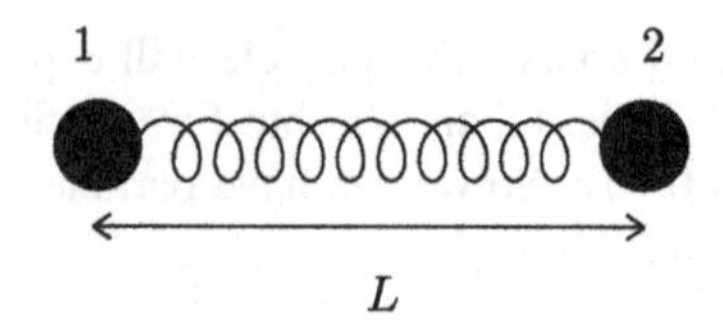

FIGURE 2.3

Two particles are attached by a spring of rest length L. If we push particle 1 to the right, a wave will propagate along the spring with velocity $c' \leq c$. Particle 2 experiences the force a time $T = L/c' \geq L/c$ later. In words, the interaction between them propagates with a velocity $\leq c$.

between that point and other points of the body. For example, suppose a rod of length 30 cm is hit by a heavy object (hammer) moving with velocity $v = 100\,\mathrm{m\,s^{-1}}$. It takes 1 ns for the interaction to reach the other end of the rod. During this time the hitting object moves

$$100\,\frac{\mathrm{m}}{\mathrm{s}} \times 1 \times 10^{-9}\,\mathrm{s} = 0.1\,\mu\mathrm{m}. \tag{2.27}$$

That is, before the other end experiences anything, the rod has shrunken by $0.1\,\mu\mathrm{m} \sim 3 \times 10^{-7} L$. Now suppose the rod is hit by a hammer moving with velocity $0.6c$. Again the other end of the rod experiences a force after 1 ns; but during this time the hammer has moved 18 cm! That is before the other end moves, the rod has shrunken $18\,\mathrm{cm} = 0.6L$!

2.5.1 Superluminal "velocities"

Whenever we divide a length by a time, we get a quantity with the dimensions of velocity. But such a quantity is not necessarily the velocity of a particle, and it could easily be greater than c. As an example, if two particles A and B start moving from the origin with velocities $0.6c$, A moving along $\hat{x}$ and B moving along $-\hat{x}$, at time t particle A is at $x_A = 0.6ct$ and particle B is at $x_B = -0.6ct$. The distance between A and B is $D = 1.2ct$ and dividing by t we get $1.2c$. This is greater than c, but it is not the velocity of any particle. In §3.8, we will see that in this situation, the velocity of B relative to A is $-0.88c$.

Problem 7. A star is at distance D from us. Around this star, there is a spherical shell of dust of radius $R \ll D$. The star explodes at time t_0, illuminating the dust shell at time $t_0 + R/c$. The light of this illuminated dust shell reaches the Earth. An astronomer observes an expanding ring. Find the "velocity" the astronomer assigns to this expansion.

Problem 8. An object is a distance D from us. At time t_0 a part of the object is separated and ejected toward us, such that the velocity of the ejected particle has an angle θ with respect to the radial vector ($\theta = 0$ if the particle is ejected directly toward us). An astronomer on Earth could measure the angular distance of the object, which increases as time passes, and, knowing the distance D, defines the "velocity" of the ejection by

$$v = \frac{D\alpha}{\Delta t}, \tag{2.28}$$

where Δt is the time increment between two successive observations. Show that

$$\Delta t = \delta t \left(1 - \frac{u}{c} \cos\theta\right), \tag{2.29}$$

where δt is the increment of time between the ejection and the emission of light that leads to our second observation. Therefore,

$$v = \frac{D\alpha}{\Delta t} = \frac{D\alpha}{\delta t} \cdot \frac{1}{1 - \frac{u}{c}\cos\theta} = \frac{u \sin\theta}{1 - \frac{u}{c}\cos\theta} \underset{u \to c^-}{\longrightarrow} c \cot\frac{\theta}{2}. \tag{2.30}$$

For $\theta = 12$ degrees we have

$$v \simeq 9.5\, c. \tag{2.31}$$

An example of such a superluminal velocity is the quasar 3C273 (see Pearson et al., 1981).

2.6 Relativity of simultaneity

Now consider two frames moving relative to each other. For example, as shown in Fig. 2.4, let K be the ground frame, defined by the objects fixed on the ground, and let K' be the moving frame, defined by the solid body of the train. In each of these frames, we can synchronize the clocks stationary in that frame by Einstein's procedure. But now something unintuitive happens: If two events E_1 and E_2 are simultaneous by clocks stationary in K, they would not be simultaneous by clocks stationary in K'.

To see this, consider a train moving with velocity v along the x-axis. This we call frame K'. Let the length of the train (as measured by the observer stationary in the train) be L_0. Someone standing in the middle of the train turns on a flash so that two beams of light are sent to the two ends of the train. These two light signals will reach simultaneously, in K', the two ends of the train, at time

$$t_1' = t_2' = \frac{L_0}{2c}. \tag{2.32}$$

Let us think about the time these two signals reach the two ends of the train *as observed by the observer stationary in the station*, that is, frame K. Let us denote the train's length by L. This is the length of the train as measured in K. Later we will see that L is not the same as L_0. For the moment, we just denote this length by L. At time t_0 (by synchronized stationary clocks in K) the flash is being fired. If t_1 is the

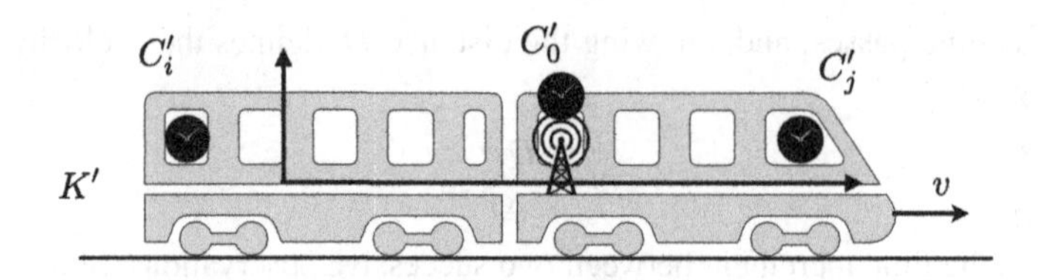

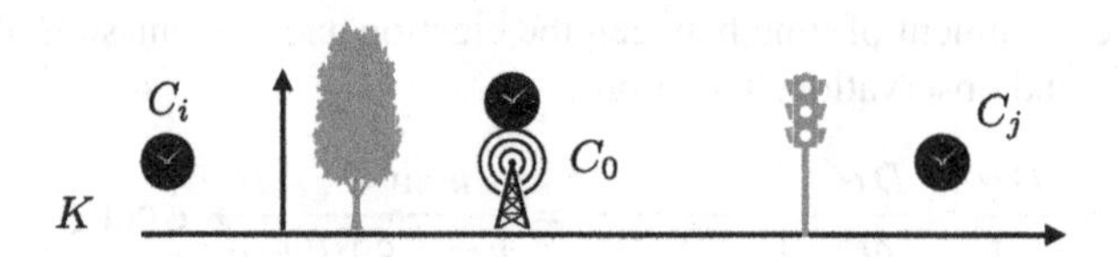

FIGURE 2.4

In frame K', the train is moving with velocity v relative to frame K, the ground. In both frames we use Einstein's synchronization procedure.

time the light signal reaches the head of the train and t_2 is the time the light signal reaches the tail of the train, we have the times of flight defined by

$$t_1 = t_0 + \Delta t_1, \tag{2.33}$$

$$t_2 = t_0 + \Delta t_2. \tag{2.34}$$

Now, from simple geometry we have

$$c\,\Delta t_1 = \frac{L}{2} + v\,\Delta t_1, \tag{2.35}$$

$$c\,\Delta t_2 = \frac{L}{2} - v\,\Delta t_2. \tag{2.36}$$

From these two relations it follows that

$$\Delta t_1 = \frac{L}{2(c-v)}, \qquad \Delta t_2 = \frac{L}{2(c+v)}, \tag{2.37}$$

from which it follows that these two events are not simultaneous in K. Some comments:

1. Although these two time intervals could be derived by the Galilean addition of velocities, note that our derivation does not rely on that. What we are using is simply addition or subtraction of two lengths. For the light signal moving toward the head of the train, the length the light signal travels is equal to $L/2$ (half the length of the train) plus $v\,\Delta t_1$ (the distance the head of the train has moved in time Δt_1). For the light signal moving toward the tail of the train, the length the

light signal travels is equal to $L/2$ minus $v\,\Delta t_2$ (the length the tail of the train has moved in time Δt_2).

2. This reasoning does not rely on what L is. The only requirement is that we know that the flashing device is fired at the center of the train.

2.7 Synchronization is a convention

Consider a rigid frame K with a set of clocks all over the space. To describe the events—motions of particles—we use a coordinate system to specify the positions and use a set of stationary clocks being spread all over the space. Usually, we use orthonormal Cartesian coordinates, but this is not a requisite—one can use curvilinear coordinates. Using orthonormal Cartesian coordinates will make the description of distance between points very simple. The same is true for the synchronization of the clocks. Einstein's synchronization is not a requisite, but it leads to simplification of our physical theories because:

1. It can be implemented in any inertial frame, with the clock C_0 being set anywhere.
2. No other informations besides the spatial locations of the clocks are needed.
3. If the clocks are synchronized by Einstein's scheme, the speed of light going from any point A to any other point B is always the same c.
4. Maxwell's equations governing the electrodynamics would be valid in their standard form.

To see what happens if we use another synchronization scheme, let us change the settings of the clocks which are already synchronized by Einstein's scheme. For simplicity, let us shift the clock, which is at position (x, y, z) by $f(x, y, z)$, which could be a given arbitrary continuous function. This means that we define the "local" time $\tilde{t}$ by

$$\tilde{t} = t + f(x, y, z). \tag{2.38}$$

Surely the physics is not affected by introducing this local time—only our description of the events changes. To get a feeling, consider only two clocks at points $A = (0, 0, 0)$ and $B = (L, 0, 0)$, and let us use $f(0, 0, 0) = a$ and $f(L, 0, 0) = b$, which means defining

$$\tilde{t}_0 = t + a, \qquad \tilde{t}_1 = t + b. \tag{2.39}$$

Now if at Einstein time t we send a light signal to B, the signal will arrive at B at Einstein time $t + L/c$, which in terms of the "local" time $\tilde{t}$ means

$$\tilde{t}_1 = t + \frac{L}{c} + b. \tag{2.40}$$

The local time it takes for the light to go from A to B is then

$$\Delta\tilde{t}_{A\to B} = t + \frac{L}{c} + b - (t + a) = \frac{L}{c} + b - a. \tag{2.41}$$

It is easy to see that the local time for the light to go from B to A is similarly

$$\Delta\tilde{t}_{B\to A} = \frac{L}{c} + a - b. \tag{2.42}$$

The coordinate velocity of light, defined as $L/\Delta\tilde{t}$, depends on the direction! Note that the two-way speed of light, defined by the local time it takes for light to go from A to B and return back to A, is now independent of offsets a and b:

$$\Delta\tilde{t}_{A\to B\to A} = \Delta\tilde{t}_{A\to B} + \Delta\tilde{t}_{B\to A} = \frac{2L}{c}. \tag{2.43}$$

This is due to the fact that we have not changed the rate of clocks, *i.e.*, the definition of a second. We have only changed the origin of the time on each clock. All that we have done is to introduce a new definition of synchronization, which could be called $\tilde{t}$-synchronization. But is this definition useful?

Guide to Literature. For a discussion of the conventionality of synchronization, the interested reader is referred to Sexl and Urbantke (1992, pp. 43–46) and Mansouri and Sexl (1977).

References

Adam, T., et al., 2012. Measurement of the neutrino velocity with the OPERA detector in the CNGS beam. The Journal of High Energy Physics 2012 (10), 93.

Einstein, A., 1905. Zur elektrodynamik bewegter körper. Annalen der Physics 17, 891–921.

Einstein, A., Lorentz, H.A., Weyl, H., Minkowski, H., 1952. The Principle of Relativity. Dover, New York.

Jackson, J.D., 1999. Classical Electrodynamics, 3rd edition. John Wiley & Sons, New York.

Landau, L.D., Lifshitz, E.M., 1984. Electrodynamics of Continuous Media. Landau and Lifshitz Course of Theoretical Physics, vol. 8. Pergamon Press, Oxford.

Mansouri, R., Sexl, R.U., 1977. A test theory of special relativity: I. Simultaneity and clock synchronization. General Relativity and Gravitation 8 (7), 497–513.

Pearson, T.J., Unwin, S.C., Cohen, M.H., Linfield, R.P., Readhead, A.C.S., Seielstad, G.A., Simon, R.S., Walker, R.C., 1981. Superluminal expansion of quasar 3C273. Nature 290 (5805), 365–368.

Sexl, R.U., Urbantke, H.K., 1992. Relativity, Groups, Particles. Special Relativity and Relativistic Symmetry in Field and Particle Physics. Springer-Verlag, Wien.

CHAPTER

Lorentz boosts

3

3.1 Pure boost along the x-axis

Laboratory K' is moving with velocity v relative to laboratory K. In both K and K', the observers use the same system of units, say SI units, and both use orthonormal Cartesian coordinates. The axes are chosen such that K sees K' moving in the x-direction; and besides, the similar axes are parallel, that is, the x-axis of K is parallel to the x'-axis of K', its y-axis is parallel to the y'-axis of K', and its z-axis is parallel to the z'-axis of K'.

From our study of the Galilean transformations, we know that if we assume the absolute time of Newton, then there is no way out of the Galilean addition of velocities, and the velocity of light could not be the same in both K and K'. Therefore, we have to abandon Newton's absolute time. Abandoning Newton's absolute time was a very important and difficult task. Some great physicists such as Henry Poincaré and Hendrik Lorentz had serious problems in grasping the idea, which shows the importance of what Albert Einstein did. The core of Einstein's special relativity is a set of transformations, replacing the Galilean boosts; transformations which are named after Hendrik Antoon Lorentz. Let us postpone the derivation of Lorentz transformations, but instead let us see the consequences of them.

Lorentz transformation connecting K and K'

It is better to first introduce the function γ:

$$\gamma(v) = \frac{1}{\sqrt{1 - \frac{v^2}{c^2}}}. \tag{3.1}$$

The Lorentz transformation connecting K and K' is defined by

$$t' = \gamma(v)\left(t - \frac{v}{c^2}x\right), \tag{3.2}$$

$$x' = \gamma(v)\,(x - v\,t), \tag{3.3}$$

$$y' = y, \tag{3.4}$$

$$z' = z. \tag{3.5}$$

A Mathematical Approach to Special Relativity. https://doi.org/10.1016/B978-0-32-399708-9.00012-9

This transformation has some properties which we now list.

Observation 1. Only for $|v| < c$ the value of $\gamma(v)$ is real. In words, if v is not strictly less than the speed of light this transformation is meaningless. For $v = c$, dividing by zero is meaningless, and for $v > c$ the coordinates and the time in the transformed system would be imaginary, which is also meaningless. Therefore, these transformations are saying, from the very outset, that the relative velocity of two frames can only be a value strictly less than the speed of light.

Observation 2. A point fixed in K' has fixed coordinates (x', y', z'). From the above equations we deduce that this point, described in K, has the equations of motion

$$x = \frac{x'}{\gamma(v)} + v\,t, \quad y = y', \quad z = z', \tag{3.6}$$

which means that any point fixed in K' is moving with velocity v relative to K.

Observation 3. The inverse transformation could be found by solving the system of linear equations. The result is

$$t = \gamma(v)\left(t' + \frac{v}{c^2}x'\right), \tag{3.7}$$

$$x = \gamma(v)\left(x' + v\,t'\right), \tag{3.8}$$

$$y = y', \tag{3.9}$$

$$z = z'. \tag{3.10}$$

This means that to get the inverse transformation we just change the primed and unprimed coordinates and change v to $-v$.

Observation 4. A point fixed in K has fixed coordinates (x, y, z), which, described in K', is

$$x' = \frac{x}{\gamma(v)} - v\,t', \quad y' = y. \quad z' = z. \tag{3.11}$$

This means that any point fixed in K is moving with velocity $-v$ relative to K'.

Observation 5. If $v \ll c$ the transformation is simply the Galilean transformation. This means that for velocities small compared to the speed of light, the predictions of these transformations are very close to the predictions of the Galilean transformations.

Observation 6. From $t' = \gamma(v)\left(t - v\,x/c^2\right)$, we see that if E_1 and E_2 are two *simultaneous* events in K, that is, two events with equal t, but with different x, the same two events are not *simultaneous* in K', that is, their t' would be different. This is in sharp contrast to Newton's absolute time notion. Therefore accepting transformation (3.2), we have to reject the notions of absolute simultaneity and absolute time, and there emerges a new causal structure (see §2.3).

Observation 7. Since the transformation is linear in the coordinates, the differential of the transformation relating $(\mathrm{d}t, \mathrm{d}x, \mathrm{d}y, \mathrm{d}z)$ to $(\mathrm{d}t', \mathrm{d}x', \mathrm{d}y', \mathrm{d}z')$ has the same form:

$$\mathrm{d}t' = \gamma(v)\left(\mathrm{d}t - \frac{v}{c^2}\,\mathrm{d}x\right), \tag{3.12}$$

$$\mathrm{d}x' = \gamma(v)\,(\mathrm{d}x - v\,\mathrm{d}t), \tag{3.13}$$

$$\mathrm{d}y' = \mathrm{d}y, \tag{3.14}$$

$$\mathrm{d}z' = \mathrm{d}z. \tag{3.15}$$

Observation 8. It can be easily shown that the following quadratic relation holds:

$$-c^2\,\mathrm{d}t^2 + \mathrm{d}x^2 + \mathrm{d}y^2 + \mathrm{d}z^2 = -c^2\,\mathrm{d}t'^2 + \mathrm{d}x'^2 + \mathrm{d}y'^2 + \mathrm{d}z'^2, \tag{3.16}$$

that is, the quantity $-c^2\,\mathrm{d}t^2 + \mathrm{d}x^2 + \mathrm{d}y^2 + \mathrm{d}z^2$ is invariant, which means that both observers, in K and in K', agree on its value (provided they use the same system of units). We will return to this very important identity later.

Observation 9. As we saw in §A.3, the identity

$$\gamma^2 - \beta^2\gamma^2 = 1, \tag{3.17}$$

which follows from the definitions $\beta = v/c$ and $\gamma(v) = \left(1-\beta^2\right)^{-1/2}$, indicates that we can define a real parameter α such that

$$\beta\gamma = \sinh\alpha, \quad \gamma = \cosh\alpha, \quad \beta = \tanh\alpha. \tag{3.18}$$

The Lorentz boost therefore can be written as follows (in which for brevity, we have not written the transformation for y and z, which is trivial):

$$\begin{bmatrix} ct \\ x \end{bmatrix} = \begin{bmatrix} \cosh\alpha & \sinh\alpha \\ \sinh\alpha & \cosh\alpha \end{bmatrix} \begin{bmatrix} ct' \\ x' \end{bmatrix}. \tag{3.19}$$

Parameter α, which is called the *rapidity*, unambiguously determines the velocity $v = \beta c$, and vice versa. This form of transformation is called *hyperbolic rotation* (see §A.3 on page 285). One benefit of using this form for the transformation is that the combination of two such transformations in this form is very simple:

$$\begin{bmatrix} \cosh\alpha & \sinh\alpha \\ \sinh\alpha & \cosh\alpha \end{bmatrix} \begin{bmatrix} \cosh\alpha' & \sinh\alpha' \\ \sinh\alpha' & \cosh\alpha' \end{bmatrix} = \begin{bmatrix} \cosh(\alpha+\alpha') & \sinh(\alpha+\alpha') \\ \sinh(\alpha+\alpha') & \cosh(\alpha+\alpha') \end{bmatrix}.$$

That is, the rapidity of the combination is simply the sum of the rapidities of the two transformations.

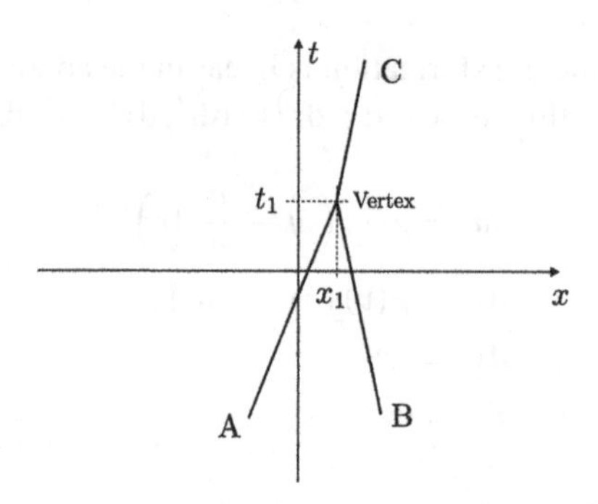

FIGURE 3.1 Collision of two particles.

Particles A and B are moving to the right and left, respectively, along the x-axis, with constant velocities. They collide at time t_1 at point x_1. After the collision, a single particle (C) emerges and is moving with constant velocity to the right. The event (t_1, x_1) is called the *vertex* of this interaction.

3.2 World-lines

Recollect that by an event, we mean the point (t, x, y, z) in space-time. For example, "the collision of two particles happening in (x, y, z) at time t" is a single event (see Fig. 3.1). Now consider a particle moving in space. The particle is at $(x(t), y(t), z(t))$ at time t. This is a single event. The curve

$$\{(t, x(t), y(t), z(t)) \mid t \in \mathbb{R}\} \tag{3.20}$$

is a set consisting of infinitely many events. We call this curve the world-line of the particle. If the particle is stationary at point (x_0, y_0, z_0), the world-line is the line $\{(t, x_0, y_0, z_0) \mid t \in \mathbb{R}\}$, parallel to the t-axis.

The vector $\boldsymbol{u} = d\boldsymbol{r}/dt$ with Cartesian components $(\dot{x}, \dot{y}, \dot{z})$ is tangent to the world-line. Later we will see that for all particles, we have $|u| \leq c$, so that the world-line of particles should fulfill this requirement: $|d\boldsymbol{r}/dt| \leq c$. In other words, if $\boldsymbol{r}(t)$ is a curve in space-time such that $|d\boldsymbol{r}/dt| > c$ on some interval $I = (t_1, t_2)$, then, on that interval, this curve is not the world-line of a real particle.

Problem 9. An excited atom is moving with constant velocity to the right, along the x-axis. At time t_1 and position x_1 it emits a photon to the right. Draw the world-line diagram.

Problem 10. Describe the following world-lines:

1. The center of the Earth (a helix). Consider the Sun to be stationary.
2. A simple harmonic oscillator.
3. A photon moving between two parallel mirrors.

3.3 Causal structure in special relativity

In §1.10 we studied the causal structure of Galilean relativity. The causal structure of special relativity is quite different and is one of the important differences between Galilean relativity and special relativity.

Let $E_1 = (t_1, x_1, y_1, z_1)$ and $E_2 = (t_2, x_2, y_2, z_2)$ be two events. We define $\Delta t = t_2 - t_1$, $\Delta x = x_2 - x_1$, $\Delta y = y_2 - y_1$, and $\Delta z = z_2 - z_1$. Naturally, if $\Delta t > 0$, we say "E_2 happens after E_1"; and if $\Delta t = 0$, we say "the two events are simultaneous."

For any two events E_1 and E_2, we define

$$D(E_1, E_2) = -c^2\,(t_2 - t_1)^2 + (x_2 - x_1)^2 + (y_2 - y_1)^2 + (z_2 - z_1)^2. \qquad (3.21)$$

This quantity, which could be negative, zero, or positive, has the dimensions of $(\text{length})^2$. As mentioned in Observation 8, $D(E_1, E_2)$ is invariant under a Lorentz boost; therefore, all observers agree on its value.

Theorem. *Let E_1 and E_2 be two events, such that in frame K event E_2 occurs later than E_1, that is, $t(E_1) < t(E_2)$, where $t(E)$ is the time component of event E in frame K.*

1. *If $D(E_1, E_2) > 0$, then there exists a frame of reference in which E_1 and E_2 are simultaneous, and there exist frames of reference in which $t'(E_2) < t'(E_1)$, that is, E_2 occurs before E_1!*
2. *If $D(E_1, E_2) < 0$, there exists a frame of reference in which E_1 and E_2 occur at the same spatial position, and for any frame, $t'(E_1) < t'(E_2)$, that is, in any frame E_2 occurs later than E_1.*
3. *If $D(E_1, E_2) = 0$, in any frame,*

$$c\,(t_2 - t_1) = \sqrt{(x_2 - x_1)^2 + (y_2 - y_1)^2 + (z_2 - z_1)^2}. \qquad (3.22)$$

Proof. Without loss of generality, we can assume that $E_1 = (0, 0, 0, 0)$ occurs at the origin of the coordinate system and $E_2 = (T, w\,T, 0, 0)$ occurs at time T at the point $x = w\,T$ on the x-axis. Here w is a quantity of dimensions of velocity. For positive $D(E_1, E_2)$, since $D(E_1, E_2) = (w^2 - c^2)\,T^2 > 0$, we have $w > c$. In this case we define $v = c^2/w$, $\beta = v/c = c/w < 1$. From the Lorentz transformation we have

$$t'(E_2) = \gamma(\beta)\left(t(E_2) - \frac{v}{c^2}\,x(E_2)\right) = \gamma(\beta)\left(T - \frac{1}{w}\,w\,T\right) = 0. \qquad (3.23)$$

It is easy to see that for $c^2/w < v < c$, we get $t'(E_2) < 0$, that is, for such frames $t'(E_2) < t'(E_1)$.

For negative $D(E_1, E_2)$ we have $w < c$, and we define $v = w$. Now from the Lorentz transformation we have

$$x'(E_2) = \gamma(\beta)\,(x(E_2) - w\,t(E_2)) = \gamma(\beta)\,(w\,T - w\,T) = 0. \qquad (3.24)$$

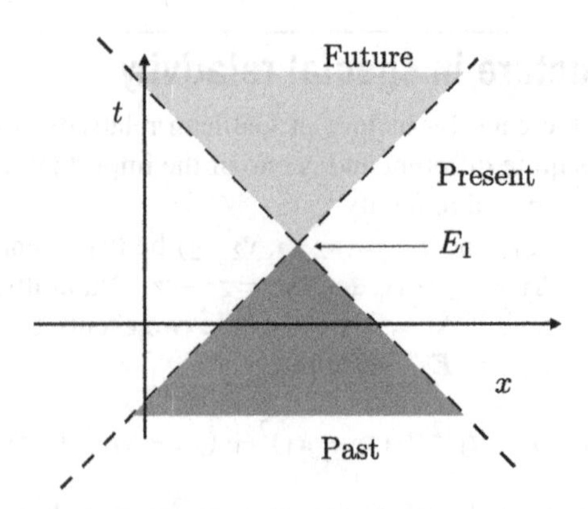

FIGURE 3.2 Light-cone, future and past of an event.

In 2D space-time, the light-cone with vertex $E_1 = (t_1, x_1)$ is the set $x - x_1 = \pm c(t - t_1)$ which for $c = 1$ is $x - x_1 = \pm(t - t_1)$. The light-shaded region is "inside or on the future light-cone of E_1" or "the causal future of E_1" and consists of those events E such that $E_1 \preceq E$, that is events that could be affected by E_1. The dark-shaded region is "inside or on the past light-cone of E_1" or "the causal past of E_1" and consists of those events that $E \preceq E_1$, that is, events that could affect E_1. The region outside the light-cone (not hashed) is the "present" and is sometimes called *elsewhere*, and consists of events that have no causal relation with E_1. It is called "present" because for any event E, there is a frame in which E and E_1 are simultaneous.

The last part of the theorem (item 3) follows immediately from the definition of $D(E_1, E_2)$. □

Definition. If $D(E_1, E_2) \leq 0$ and $t(E_1) \leq t(E_2)$, we say E_2 succeeds E_1 and we write $E_1 \preceq E_2$ or $E_2 \succeq E_1$.

If $E_1 \npreceq E_2$, then, from the previous theorem, we know that there exist frames in which $t'(E_2) < t'(E_1)$ and there exist frames in which $t'(E_1) < t'(E_2)$. Therefore, there could be no *cause-and-effect* relation between the two. Therefore, only if $E_1 \preceq E_2$ there could be a cause-and-effect relation (E_1 being the cause and E_2 being the effect). In Galilean relativity, an event E_1 could be the cause to any event if $t(E_2) \geq t(E_1)$, and because of the existence of absolute time, there is no limit on the spatial distance between the cause and the effect. However, in special relativity, event E_1 could cause event E_2 only if $D(E_1, E_2) \leq 0$ and $t(E_1) \leq t(E_2)$. (See Figs. 3.2 and 3.3.)

Given E_1, the set of events E such that $D(E_1, E) = 0$ is a *cone* in space-time. For example, if $E_1 = (0, 0, 0, 0)$ and $E = (t, x, y, z)$, then this cone is given by

$$x^2 + y^2 + z^2 - c^2 t^2 = 0. \tag{3.25}$$

Definition. For the event E_1:

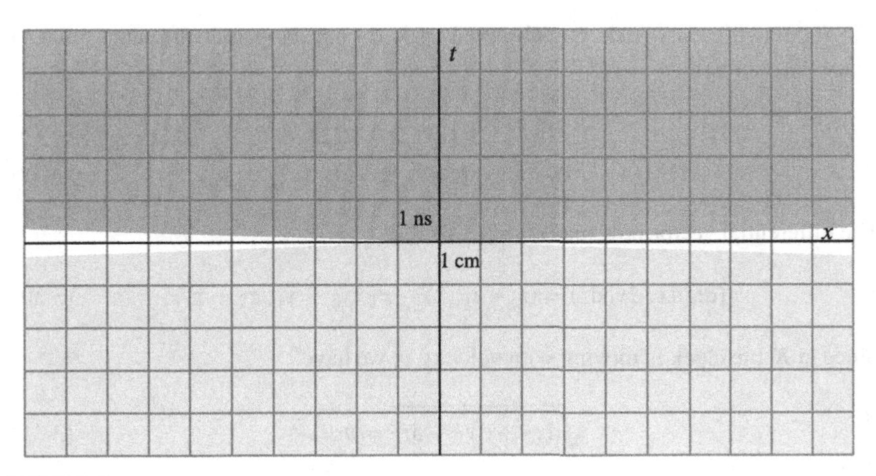

FIGURE 3.3

Causal structure of the 2D space-time, on scale. The x-axis scale is 1 cm and the t-axis scale is 1 ns. Note that the light-cone is very close to the x-axis. The reader is asked to compare this with the causal structure of the 2D Galilean space-time.

- The set of events E such that $E_1 \prec E$ is called "the chronological future of E_1" and is denoted by $I^+(E_1)$.
- The set of events E such that $E_1 \preceq E$ is called "the causal future of E_1" and is denoted by $J^+(E_1)$.
- The set of events E such that $E \prec E_1$ is called "the chronological past of E_1" and is denoted by $I^-(E_1)$.
- The set of events E such that $E \preceq E_1$ is called "the causal past of E_1" and is denoted by $J^-(E_1)$.

It is obvious that

$$I^+ \subset J^+, \qquad I^- \subset J^-. \tag{3.26}$$

Guide to Literature. The terminology and symbols $I^\pm$ and $J^\pm$ are adopted from Hawking and Ellis (1973, p. 183).

3.4 Time dilation

Consider a clock, stationary in K' at a fixed point (x', y', z'). This clock is ticking, that is, it is announcing passing units of time. Consider two successive ticks. The first tick is an event (t'_1, x', y', z'), and the second tick is another event (t'_2, x', y', z'). In K' both ticks happen at the same point in space. We can construct the differential vector

$$\left(\mathrm{d}t' = t'_2 - t'_1,\ \mathrm{d}x' = 0,\ \mathrm{d}y' = 0,\ \mathrm{d}z' = 0\right). \tag{3.27}$$

The same two events (ticks) as observed in K do not happen at the same point in space. We can write

$$\text{tick 1:} \quad (t_1, x_1, y_1, z_1)\,, \tag{3.28}$$

$$\text{tick 2:} \quad (t_2, x_2, y_2, z_2)\,. \tag{3.29}$$

The differential vector now reads

$$(\mathrm{d}t, \mathrm{d}x, \mathrm{d}y, \mathrm{d}z) = (t_2 - t_1,\; x_2 - x_1,\; y_2 - y_1,\; z_2 - z_1)\,. \tag{3.30}$$

Since in K the clock is moving with velocity v, we have

$$\sqrt{\mathrm{d}x^2 + \mathrm{d}y^2 + \mathrm{d}z^2} = v\,\mathrm{d}t\,. \tag{3.31}$$

We can now write the invariant quadratic form $c^2\,\mathrm{d}t^2 - v^2\,\mathrm{d}t^2 = c^2\,\mathrm{d}t'^2$, from which we conclude

$$\mathrm{d}t' = \sqrt{1 - \frac{v^2}{c^2}}\,\mathrm{d}t \qquad \text{or} \qquad \mathrm{d}t = \gamma(v)\,\mathrm{d}t'\,. \tag{3.32}$$

This result is known as "special relativistic time dilation," which means that the moving clock's successive ticks are separated longer with a factor of $\gamma(v)$, *if they are measured by stationary clocks.*

The same relation could be derived directly from the Lorentz transformation. Thus, we know that $\mathrm{d}t = \gamma(v)\left(\mathrm{d}t' - v\,\mathrm{d}x'\right)$. For the two successive ticks we have $\mathrm{d}x' = 0$, which leads to $\mathrm{d}t = \gamma(v)\,\mathrm{d}t'$.

3.5 Proper time

The time registered by the moving clock is called *the proper time*. From the invariant quadratic form we can deduce a mathematical form for the increment of the proper time, which we denote $\mathrm{d}\tau$.

$$\mathrm{d}\tau^2 = \mathrm{d}t^2 - \frac{1}{c^2}\left(\mathrm{d}x^2 + \mathrm{d}y^2 + \mathrm{d}z^2\right). \tag{3.33}$$

To be convinced, consider the frame K' which is comoving with the clock. In this frame $\mathrm{d}t' = \mathrm{d}\tau$ and $\mathrm{d}x' = \mathrm{d}y' = \mathrm{d}z' = 0$ (because in K' the ticks of the clock happen at the same position). For a clock moving with a time-varying velocity $\boldsymbol{v}(t)$, as observed in frame K, divide the time interval $[t_i, t_f]$ into N parts: $t_0 = t_i$, $t_1 = t_0 + \mathrm{d}t$, $\ldots$, $t_N = t_0 + N\,\mathrm{d}t = t_f$. If N is sufficiently large, the interval $\mathrm{d}t$ is small enough so that we can consider the motion of the clock in the interval $[t_n, t_{n+1}]$ as having the constant velocity $\boldsymbol{v}_n := \boldsymbol{v}(t_n)$. By the same reasoning as above, we know that the clock itself will show an interval $dt'_n = dt\sqrt{(1-\beta_n^2)}$, where $\beta_n = |\boldsymbol{v}_n|/c$. As $N \to \infty$, the sum of these intervals approaches the following integral:

$$\Delta t' = \int_{t_i}^{t_f} \sqrt{1 - \frac{v^2(t)}{c^2}}\, dt. \tag{3.34}$$

What we are doing here has an analogy in Euclidean geometry—defining the arc length of a curve. Remember that to define the arc length of a curve, we first approximate it with a polygon. The length of each segment of the polygon, which is a straight line, is derived by the Pythagorean relation $\Delta s_i = \sqrt{(\Delta x_i^2 + \Delta y_i^2)} = \Delta x_i \sqrt{(1 + y'^2(x_i))}$. Going to the $N \to \infty$ limit we get the familiar formula

$$s = \int_{x_0}^{x} \sqrt{1 + y'^2(x')}\, dx'. \tag{3.35}$$

Now we can ask: Why, in Euclidean geometry, is the length of a polygon equal to the sum of the lengths of its segments? If we define the length of a polygon this way, the quantity so defined, the length of a polygon, would be a useful one, that is, we can prove several useful theorems about that. In the case of the clocks, the assumption is not so obvious because this assumption implies that the time elapse read on a moving clock does not depend on its acceleration. To see that this assumption is not obvious or trivial, think of a clock that is at rest in the interval $[t_0, t_1]$ and then starts to move with velocity v, such that in the interval $(t_1, t_2]$, it is moving with velocity v. Note that at time t_1 the acceleration of the clock is infinite. When we say that the time elapse measured by the clock is simply $\Delta\tau = (t_1 - t_0) + \sqrt{1 - c^{-2}v^2}\,(t_2 - t_1)$, we are assuming that this infinite acceleration has no effect on the clock. This surely is not correct for an ordinary mechanical clock. (Think of a watch being hit by a golf club to be thrown 50 m. Do you expect it to work properly after the strike?) However, we assume that in principle it is possible to make strong enough clocks, such that they are not affected by the acceleration, and it is time read on these idealized clocks that we are accepting as the definition of time. We assume the following postulate.

Postulate. *The time elapse measured by an ideal moving clock is* $\int d\tau$.

Guide to Literature. For a more detailed discussion of this *clock hypothesis* see Sexl and Urbantke (1992, pp. 37–38).

Remark. In §8.8 we will see that in some sense acceleration (or gravity) affects clocks, but still a clock shows $\int d\tau$, which must be carefully written.

Round trip of a clock. Consider two clocks at the same position in an inertial frame. Clock 1 is at rest, and clock 2 goes to some other point and returns to the same point. Obviously clock 2 must experience some acceleration. On return, we can compare the two time elapses—the one measured on clock 1 and the one measured on clock 2. Let us denote these two time elapses by ΔT_1 and ΔT_2. We can prove that $\Delta T_1 > \Delta T_2$.

The proof is simple:

$$\Delta T_1 = \int_{t_1}^{t_2} d\tau = \int_{t_1}^{t_2} \sqrt{1-0}\, dt = \int_{t_1}^{t_2} dt = (t_2 - t_1), \tag{3.36}$$

$$\Delta T_2 = \int_{t_1}^{t_2} d\tau = \int_{t_1}^{t_2} \sqrt{1-c^{-2}\, v^2(t)}\, dt \leq \int_{t_1}^{t_2} dt = \Delta T_1. \tag{3.37}$$

The inequality follows from the fact that $|v| < c$. The equality of ΔT_1 and ΔT_2 holds only if the second clock is also stationary.

There is nothing paradoxical here, but after the introduction of special relativity by Einstein in 1905, some opponents of the theory argued as follows:

> *From the point of view of an observer fixed relative to the second clock, it is clock 1 that has made a round trip. So, by* relativity *it must be true that clock 1 registers a shorter time. This is a paradox (could not be true) and therefore special relativity is wrong.*

The resolution of the seeming paradox is this:

> *The situation is not symmetrical. Clock 1 is always in an inertial frame, but clock 2 experiences acceleration. It is the accelerated clock that registers the shorter time.*

Guide to Literature. For a detailed discussion about the twin paradox and its history, see Debs and Redhead (1996). Considering the effect of acceleration (or gravity) on clocks, one can calculate the difference of the time readings of a stationary clock and one on a return trip, which results in the same result as discussed above. The interested reader is referred to Grøn and Hervik (2007, pp. 101–102).

Problem 11. Consider two clocks A and B. A is stationary, and B goes with velocity v along the x-axis to a point a distance L from A and then returns with the same velocity. Let each clock send light signals with frequency 1. Draw the space-time diagram showing both clocks and the signals. Explain the "twin paradox" from this graph. Caution: Pay attention to the unit of time on the t- and t'-axes.

3.6 Length contraction

Suppose two particles are moving in space. What is the distance between them? Of course, the distance between the two could depend on time if, for example, the two are moving with different velocities. The trajectory of the two particles could be described mathematically as follows:

$$\text{particle 1:} \quad (x_1(t),\ y_1(t), z_1(t))\,, \tag{3.38}$$

$$\text{particle 2:} \quad (x_2(t),\ y_2(t), z_2(t))\,. \tag{3.39}$$

We can form the *simultaneous* differential

$$(\Delta x,\ \Delta y,\ \Delta z) = (x_2(t) - x_1(t),\ y_2(t) - y_1(t),\ z_2(t) - z_1(t))\,. \tag{3.40}$$

At time t, the distance between the two particles is simply the length of this *simultaneous* differential,

$$\Delta s = \sqrt{(\Delta x)^2 + (\Delta y)^2 + (\Delta z)^2}. \tag{3.41}$$

Now consider a moving rod. What could be the length of this moving rod? The answer is simple: We consider terminal points, the head and the tail of the rod, as two moving particles. The length of the rod is the *simultaneous* distance between these two points. Let K' be a frame, comoving with the rod, with coordinate axes parallel to the coordinate axes of K. In K' the coordinates of different points of the rod are constant in time. Therefore, the length of the rod in its rest frame K' (its so-called proper length) is

$$L_0 = \sqrt{(\Delta x')^2 + (\Delta y')^2 + (\Delta z')^2}. \tag{3.42}$$

We stress that since the rod is stationary in K', these differentials need not to be *simultaneous*.

Definition. The length of a moving rod is $\sqrt{(\Delta x)_s^2 + (\Delta y)_s^2 + (\Delta z)_s^2}$, where $(\Delta x_i)_s$ is the *simultanous* differential.

By the Lorentz boosts for the differentials we have

$$\Delta x' = \gamma(v)\,(\Delta x - v\,\Delta t)\,, \tag{3.43}$$
$$\Delta y' = \Delta y, \tag{3.44}$$
$$\Delta z' = \Delta z. \tag{3.45}$$

Setting $\Delta t = 0$, we get

$$(\Delta x)_s = \gamma^{-1}\,\Delta x', \tag{3.46}$$
$$(\Delta y)_s = \Delta y', \tag{3.47}$$
$$(\Delta z)_s = \Delta z'), \tag{3.48}$$

from which it follows that

$$\begin{aligned} L^2 &= (\Delta x)_s^2 + (\Delta y)_s^2 + (\Delta z)_s^2 \\ &= \frac{1}{\gamma^2}\,(\Delta x')^2 + (\Delta y')^2 + (\Delta z')^2 \\ &= \left(1 - \frac{v^2}{c^2}\right)(\Delta x')^2 + (\Delta y')^2 + (\Delta z')^2 \\ &= L_0^2 - \frac{v^2}{c^2}\,L_0^2\cos^2\theta', \end{aligned} \tag{3.49}$$

where θ' is the angle (in frame K') between the rod and the x'-axis. We can write this equation in the compact form

$$L = L_0 \sqrt{1 - \beta^2 \cos^2 \theta'}. \tag{3.50}$$

Problem 12. A train of proper length L_0 is moving with velocity $0.6\,c$, approaching a stationary tunnel of proper length $0.8\,L_0$. Let K be the rest frame of the tunnel and K' the rest frame of the train. At $t = t' = 0$, the head of the train reaches the entrance of the tunnel. This we call event E_1. Find the time of the following events, in both K and K', and order them. Draw a space-time diagram showing everything.

1. E_2: The head of the train reaches the exit of the tunnel.
2. E_3: The tail of the train reaches the entrance of the tunnel.
3. E_4: The tail of the train exits the tunnel.

Guide to Literature. The preceding problem is usually called the length contraction paradox or Rindler's paradox. It first appeared in Rindler (1961).

3.7 Boosts in other directions

Consider two frames, K, which is the stationary one, and K', which is moving relative to K. Suppose that the axes of both frames are "parallel" and suppose K' is moving with velocity $\boldsymbol{v} = (v_1, v_2, v_3) = v\,\hat{\boldsymbol{n}}$, where v is the absolute value of the velocity and $\hat{\boldsymbol{n}} = (n_1, n_2, n_3)$ is the corresponding unit vector. We can write the Lorentz boost by a simple argument: The vector $\boldsymbol{r}$ could be written as $\boldsymbol{r} = \boldsymbol{r}_\perp + \boldsymbol{r}_\parallel$ (see Fig. 3.4), where $\boldsymbol{r}_\parallel = \boldsymbol{r} \cdot \hat{\boldsymbol{n}}\,\hat{\boldsymbol{n}}$ and $\boldsymbol{r}_\perp := \boldsymbol{r} - \boldsymbol{r}_\parallel$. Note also that

$$\left|\boldsymbol{r}_\parallel\right| = \boldsymbol{r} \cdot \hat{\boldsymbol{n}}. \tag{3.51}$$

Before proceeding, note that both observers agree on the vector $\boldsymbol{v}$—both its absolute value and its direction are observable for both observers. Therefore, the observer in K' can decompose $\boldsymbol{r}' = \boldsymbol{r}'_\perp + \boldsymbol{r}'_\parallel$, where $\boldsymbol{r}'_\parallel = \boldsymbol{r}' \cdot \hat{\boldsymbol{n}}\,\hat{\boldsymbol{n}}$ and $\boldsymbol{r}'_\perp := \boldsymbol{r}' - \boldsymbol{r}'_\parallel$. Now the generalization of the Lorentz boost is obvious:

$$ct = \gamma(v)\left(ct' + \frac{v}{c}\left|\boldsymbol{r}'_\parallel\right|\right), \tag{3.52}$$

$$\left|\boldsymbol{r}_\parallel\right| = \gamma(v)\left(\left|\boldsymbol{r}'_\parallel\right| + v\,t'\right), \tag{3.53}$$

$$\boldsymbol{r}_\perp = \boldsymbol{r}'_\perp. \tag{3.54}$$

The first equation is simply the usual transformation of the time coordinate, properly written in terms of the time and the coordinate parallel to the direction of motion. The second equation is the usual Lorentz transformation of the spatial coordinate parallel to the direction of motion. The third equation is simply the statement that the coordinate normal to $\vec{v}$ is the same in both frames.

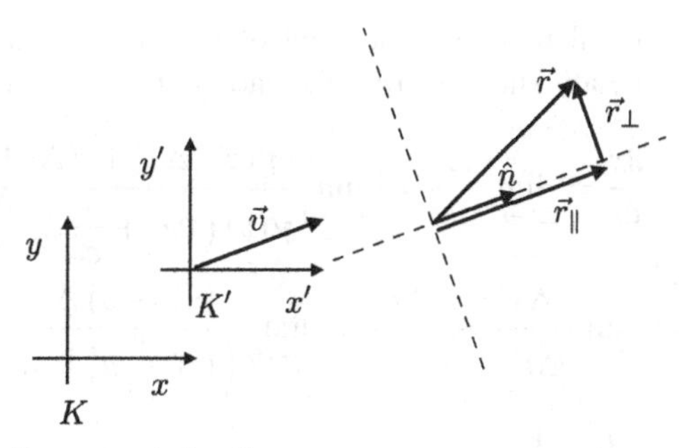

FIGURE 3.4 Boost in the general direction.

The vector $\boldsymbol{r}$ could be decomposed parallel and normal to the direction of $\boldsymbol{v}$, and thus we can find the general boost.

Now, multiplying the second equation by $\hat{\boldsymbol{n}}$ we get $\boldsymbol{r}_\parallel = \gamma(v)\left(\boldsymbol{r}'_\parallel + \boldsymbol{v}\,t'\right)$. Now we have

$$\begin{aligned}\boldsymbol{r} &= \boldsymbol{r}_\perp + \boldsymbol{r}_\parallel \\ &= \boldsymbol{r}'_\perp + \gamma(v)\,\boldsymbol{r}'_\parallel + \gamma(v)\,\boldsymbol{v}\,t' \\ &= \boldsymbol{r}'_\perp + \boldsymbol{r}'_\parallel - \boldsymbol{r}'_\parallel + \gamma(v)\,\boldsymbol{r}'_\parallel + \gamma(v)\,\boldsymbol{v}\,t' \\ &= \boldsymbol{r}' + (\gamma(v) - 1)\,\hat{\boldsymbol{n}}\left(\hat{\boldsymbol{n}}\cdot\boldsymbol{r}'\right) + \gamma(v)\,\boldsymbol{v}\,t'.\end{aligned} \tag{3.55}$$

It is now straightforward to see that

$$\begin{bmatrix} c\,t \\ x \\ y \\ z \end{bmatrix} = \begin{bmatrix} \gamma & \gamma\,v_1 & \gamma\,v_2 & \gamma\,v_3 \\ \gamma\,v_1 & 1+\dfrac{\gamma-1}{v^2}v_1\,v_1 & \dfrac{\gamma-1}{v^2}v_1\,v_2 & \dfrac{\gamma-1}{v^2}v_1\,v_3 \\ \gamma\,v_2 & \dfrac{\gamma-1}{v^2}v_2\,v_1 & 1+\dfrac{\gamma-1}{v^2}v_2\,v_2 & \dfrac{\gamma-1}{v^2}v_2\,v_3 \\ \gamma\,v_3 & \dfrac{\gamma-1}{v^2}v_3\,v_1 & \dfrac{\gamma-1}{v^2}v_3\,v_2 & 1+\dfrac{\gamma-1}{v^2}v_3\,v_3 \end{bmatrix} \begin{bmatrix} c\,t' \\ x' \\ y' \\ z' \end{bmatrix}. \tag{3.56}$$

3.8 Addition of velocities

The definition of velocity in both frames K and K' is as usual:

$$u_x := \frac{\mathrm{d}x}{\mathrm{d}t}, \quad u_y := \frac{\mathrm{d}y}{\mathrm{d}t}, \quad u_z := \frac{\mathrm{d}z}{\mathrm{d}t}, \tag{3.57}$$

$$u'_x := \frac{\mathrm{d}x'}{\mathrm{d}t'}, \quad u'_y := \frac{\mathrm{d}y'}{\mathrm{d}t'}, \quad u'_z := \frac{\mathrm{d}z'}{\mathrm{d}t'}. \tag{3.58}$$

We note that when we calculate a component of velocity, we actually divide two differentials, and we already know the transformation of the differentials:

$$u_x = \frac{dx}{dt} = \lim_{\Delta t \to 0} \frac{\Delta x}{\Delta t} = \lim_{\Delta t \to 0} \frac{\gamma(v)\left(\Delta x' + v\,\Delta t'\right)}{\gamma(v)\left(\Delta t' + \frac{v}{c^2}\Delta x'\right)} \tag{3.59}$$

$$= \lim_{\Delta t \to 0} \frac{\Delta x' + v\,\Delta t'}{\Delta t' + \frac{v}{c^2}\Delta x'} = \lim_{\Delta t \to 0} \frac{\left(u'_x + v\right)\Delta t'}{\left(1 + \frac{v}{c^2}u'_x\right)\Delta t'} \tag{3.60}$$

$$= \frac{u'_x + v}{1 + \frac{v}{c^2}u'_x}, \tag{3.61}$$

$$u_y = \frac{dy}{dt} = \lim_{\Delta t \to 0} \frac{\Delta y}{\Delta t} = \lim_{\Delta t \to 0} \frac{\Delta y'}{\gamma(v)\left(\Delta t' + \frac{v}{c^2}\Delta x'\right)} \tag{3.62}$$

$$= \frac{u'_y}{\gamma(v)\left(1 + \frac{v}{c^2}u'_x\right)} \tag{3.63}$$

$$u'_z = \frac{u'_z}{\gamma(v)\left(1 + \frac{v}{c^2}u'_x\right)}. \tag{3.64}$$

Problem 13. Prove that the following identity holds:

$$1 - \frac{u^2}{c^2} = \frac{\left(1 - \frac{u'^2}{c^2}\right)\left(1 - \frac{v^2}{c^2}\right)}{\left(1 + \frac{v}{c^2}u'_x\right)^2}, \tag{3.65}$$

where $u^2 = u_x^2 + u_y^2 + u_z^2$ and $u'^2 = \left(u'_x\right)^2 + \left(u'_y\right)^2 + \left(u'_z\right)^2$. The same identity, if we invert and take the square root, reads

$$\gamma(u) = \gamma(v)\,\gamma(u')\left(1 + \frac{v}{c^2}u'_x\right). \tag{3.66}$$

We know that $|v| < c$, that is, $1 - \frac{v^2}{c^2}$ is always positive. The denominator of (3.65) is also always positive. We therefore conclude the following important theorem, the proof of which follows easily from the above identity.

Theorem. *We have $|u| < c$ if and only if $|u'| < c$, $|u| = c$ if and only if $|u'| = c$, and $|u| > c$ if and only if $|u| > c$. Besides, $u = u' \neq c$ if and only if $v = 0$.*

In words, if a particle is moving with a velocity less than the speed of light in K, its velocity in K' would also be less than the speed of light. If a particle moves with the speed for light in K, its velocity would be the same c in frame K'. And, if a particle is moving with a velocity greater than c in K, its velocity in K' would also be greater than c. The last part of the theorem states that c is the only invariant velocity.

Problem 14. Write the proof of this theorem.

Up to now we have not observed superluminal particles, but this does not invalidate the above theorem. This theorem shows that Lorentz transformations are consistent with Einstein's postulate that the speed of light is the same in all inertial frames.

3.9 Success: Fizeau experiment

We mentioned in §1.12.1 (page 16) that the Fizeau experiment showed that the Galilean rule of addition of velocities does not explain the velocity of light in a beam of moving water. Fizeau himself thought that his result shows the validity of the aether theories. Now let us see how special relativity explains this. According to the addition of velocities in special relativity, if a particle is moving with velocity $u' = c/n$ relative to the water (in the same direction) and the water is moving with velocity v relative to the lab, then the velocity of the particle relative to the lab is

$$u = \frac{u' + v}{1 + \frac{u' v}{c^2}} = \frac{\frac{c}{n} + v}{1 + \frac{v}{n c}}. \tag{3.67}$$

Expanding this—noting that $v/c \ll 1$—we get

$$u = \left(\frac{c}{n} + v\right)\left(1 - \frac{v}{n c}\right) = v + \frac{c}{n} - \frac{v}{n^2} \tag{3.68}$$

$$= \frac{c}{n} + v\left(1 - \frac{1}{n^2}\right). \tag{3.69}$$

Remark. This is the standard interpretation of the result of the Fizeau experiment. However, it should be noted that there had been works to explain it in the context of Galilean relativity, both between 1851 and 1905 and after 1905. It is possible to invent models for the propagation of light in matter such that the above formula is consistent with Galilean relativity; see for example Clément (1980).

3.10 Aberration of light

Consider a light ray. The observer in K' observes that it is moving in the $x'y'$-plane, making an angle θ' with the x'-axis. This means that the velocity vector of the light particle in K' reads

$$\boldsymbol{u}' = \left(u'_x, u'_y, u'_z\right) = \left(c \cos\theta', c \sin\theta', 0\right). \tag{3.70}$$

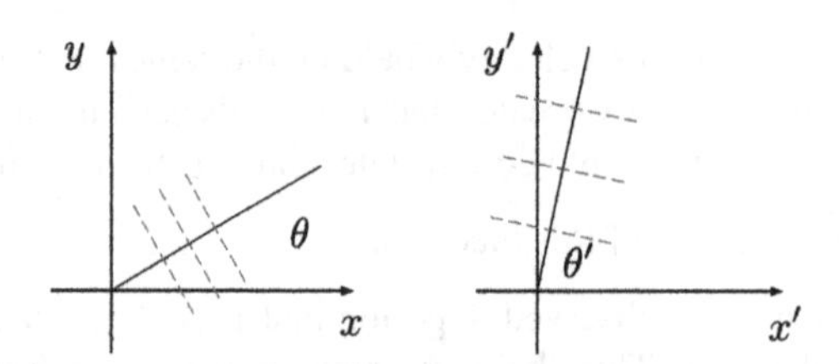

FIGURE 3.5 Aberration of a plane wave.

Aberration of a plane wave. Frame K' is moving with velocity $\boldsymbol{v} = (0.8c, 0, 0)$ relative to frame K. A plane wave is moving so that the ray is along the line having angle $\theta = 30.0$ degrees in K and $\theta' = 77.6$ degrees in K', given by $\tan\theta'/2 = \sqrt{(1+\beta)/(1-\beta)}\,\tan\theta/2$.

Using the formulas for the addition of velocities, we get the velocity vector as observed in K. Before doing the calculations, we note that the speed of light in K would be the same c, and we know that its z-component would be zero (because u'_z is zero). Therefore, the velocity vector must have the following form:

$$\boldsymbol{u} = \left(u_x, u_y, u_z\right) = (c \cos\theta, c \sin\theta, 0). \tag{3.71}$$

It is now easy to use the formulas for the addition of velocity and obtain the following formulas:

$$c \cos\theta = \frac{c \cos\theta' + v}{1 + \beta \cos\theta'}, \quad \beta := \frac{v}{c}, \tag{3.72}$$

$$c \sin\theta = \frac{1}{\gamma(v)} \cdot \frac{c \sin\theta'}{1 + \beta \cos\theta'}. \tag{3.73}$$

These two formulas can be set in a single simple formula:

$$\tan\frac{\theta}{2} = \frac{\sin\theta}{1 + \cos\theta} = \frac{\sin\theta'}{\gamma(v)\,(1 + \beta \cos\theta')} \cdot \frac{1}{1 + \dfrac{\cos\theta' + \beta}{1 + \beta \cos\theta'}} \tag{3.74}$$

$$= \frac{1}{\gamma(v)} \cdot \frac{\sin\theta'}{1 + \beta \cos\theta' + \cos\theta' + \beta} \tag{3.75}$$

$$= \frac{\sqrt{1 - \beta^2}}{1 + \beta} \cdot \frac{\sin\theta'}{1 + \cos\theta'} \tag{3.76}$$

$$= \sqrt{\frac{1 - \beta}{1 + \beta}} \tan\frac{\theta'}{2}. \tag{3.77}$$

For an example see Fig. 3.5.

Problem 15. For $\beta = 0.1$, compare the result of the previous formula with the Galilean case of Problem 4 on page 15 for $\theta = 5.00$ degrees, 45.0 degrees, 85.0 degrees, and 90.0 degrees.

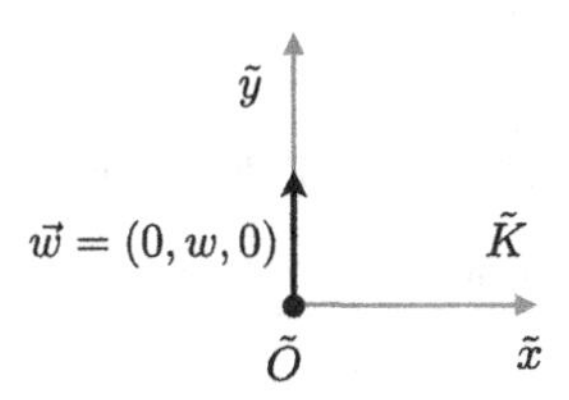

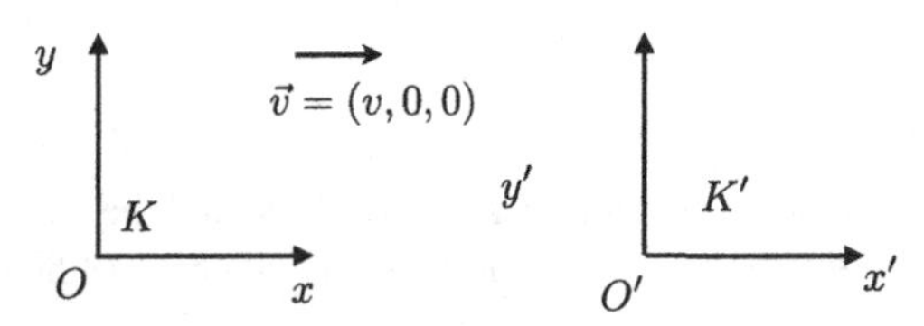

FIGURE 3.6 Pure boost to the rest frame of a particle moving with velocity $(v, w/\gamma(v), 0)$.

Frame K' is moving with velocity $(v, 0, 0)$ relative to K. Particle $\tilde{O}$ is moving with velocity $(0, w, 0)$ relative to K'. The velocity of $\tilde{O}$ with respect to K is $(v, w/\gamma(v), 0)$. Frame $\tilde{K}$ is moving with this velocity relative to K and its axes are parallel to those of K. This Lorentz transformation is given by (3.86).

3.11 Thomas rotation and precession

Notation. *In this section we set $c = 1$. Also we use X for the column matrix with entries t, x, y, and z. Thus for example*

$$X'' = \begin{pmatrix} t'' \\ x'' \\ y'' \\ z'' \end{pmatrix}. \tag{3.78}$$

Consider the following situation. As shown in Fig. 3.6, frame K' is moving with velocity $(v, 0, 0)$ relative to frame K. A particle, located at $\tilde{O}$, is moving with velocity $(0, w, 0)$ relative to K'. By the rules of addition of velocities, the velocity of the particle with respect to frame K is

$$u_x = v, \quad u_y = \frac{w}{\gamma(v)}, \quad u_z = 0, \tag{3.79}$$

and we have

$$\frac{1}{\gamma^2(\boldsymbol{u})} = 1 - \boldsymbol{u} \cdot \boldsymbol{u} = 1 - v^2 - \left(1 - v^2\right) w^2 \tag{3.80}$$

$$= \left(1 - v^2\right)\left(1 - w^2\right), \tag{3.81}$$

or

$$\gamma(\boldsymbol{u}) = \gamma(v)\,\gamma(w). \tag{3.82}$$

We introduce the notation

$$u_1 = v, \tag{3.83}$$

$$u_2 = \frac{w}{\gamma(v)}, \tag{3.84}$$

$$\gamma = \gamma(u). \tag{3.85}$$

Now, if frame $\tilde{K}$ is moving with velocity $\boldsymbol{u}$ relative to K and the axes of $\tilde{K}$ are parallel to those of K, the Lorentz boost relating K to $\tilde{K}$, as given by (3.56), reads

$$\begin{pmatrix} t \\ x \\ y \\ z \end{pmatrix} = \begin{pmatrix} \gamma & \gamma\, u_1 & \gamma\, u_2 & 0 \\ \gamma\, u_1 & 1+\frac{\gamma-1}{u^2}u_1^2 & \frac{\gamma-1}{u^2}u_1\,u_2 & 0 \\ \gamma\, u_2 & \frac{\gamma-1}{u^2}u_1\,u_2 & 1+\frac{\gamma-1}{u^2}u_2^2 & 0 \\ 0 & 0 & 0 & 1 \end{pmatrix} \begin{pmatrix} \tilde{t} \\ \tilde{x} \\ \tilde{y} \\ \tilde{z} \end{pmatrix}. \tag{3.86}$$

Note that this matrix is symmetric. As an example, if $v = 0.6$ and $w = 0.6$, we have

$$u_1 = 0.6, \quad u_2 = \frac{0.6}{1.25} = 0.48, \quad \gamma(u) = 1.563. \tag{3.87}$$

We write the above matrix equation as

$$X = M\,\tilde{X}, \tag{3.88}$$

where

$$M = \begin{pmatrix} 1.563 & 0.938 & 0.750 & 0.000 \\ 0.938 & 1.344 & 0.275 & 0.000 \\ 0.750 & 0.275 & 1.220 & 0.00 \\ 0.000 & 0.000 & 0.000 & 1.000 \end{pmatrix}. \tag{3.89}$$

Now consider the situation of Fig. 3.7, where K' is moving with velocity $(v, 0, 0)$ relative to K and K'' is moving with velocity $(0, w, 0)$ relative to K'. Writing the corresponding Lorentz transformations, we have

$$\begin{pmatrix} t \\ x \\ y \\ z \end{pmatrix} = \begin{pmatrix} \gamma(v) & \gamma(v)\,v & 0 & 0 \\ \gamma(v)\,v & \gamma(v) & 0 & 0 \\ 0 & 0 & 1 & 0 \\ 0 & 0 & 0 & 1 \end{pmatrix} \begin{pmatrix} t' \\ x' \\ y' \\ z' \end{pmatrix}, \tag{3.90}$$

$$\begin{pmatrix} t' \\ x' \\ y' \\ z' \end{pmatrix} = \begin{pmatrix} \gamma(w) & 0 & \gamma(w)\,w & 0 \\ 0 & 1 & 0 & 0 \\ \gamma(w)\,w & 0 & \gamma(w) & 0 \\ 0 & 0 & 0 & 1 \end{pmatrix} \begin{pmatrix} t'' \\ x'' \\ y'' \\ z'' \end{pmatrix}. \tag{3.91}$$

Combining these two matrix relations we get

$$\begin{pmatrix} t \\ x \\ y \\ z \end{pmatrix} = \begin{pmatrix} \gamma(v)\,\gamma(w) & \gamma(v)\,v & \gamma(v)\,\gamma(w)\,w & 0 \\ \gamma(v)\,\gamma(w)\,v & \gamma(v) & \gamma(v)\,\gamma(w)\,v\,w & 0 \\ \gamma(w)\,w & 0 & \gamma(w) & 0 \\ 0 & 0 & 0 & 1 \end{pmatrix} \begin{pmatrix} t'' \\ x'' \\ y'' \\ z'' \end{pmatrix}. \tag{3.92}$$

This matrix is not the same as (3.86). Note that this matrix, being asymmetric, is not a pure boost! For our example of $v = w = 0.6$, this matrix equation reads

$$X = N\,X'', \tag{3.93}$$

where

$$N = \begin{pmatrix} 1.563 & 0.750 & 0.938 & 0.000 \\ 0.938 & 1.250 & 0.563 & 0.000 \\ 0.750 & 0.000 & 1.250 & 0.000 \\ 0.000 & 0.000 & 0.000 & 1.000 \end{pmatrix}. \tag{3.94}$$

We then have

$$X = M\,\tilde{X}, \qquad X = N\,X'', \tag{3.95}$$

which means $M\,\tilde{X} = N\,X''$ or

$$\tilde{X} = M^{-1}\,N\,X''. \tag{3.96}$$

Calculating the matrix $M^{-1}\,N$ for the general values of the velocities is straightforward but tedious. What is important is that R is not the unit matrix. It is a rotation! For our numerical example

$$M^{-1}\,N = \begin{pmatrix} 1.000 & 0.000 & 0.000 & 0.000 \\ 0.000 & 0.976 & 0.219 & 0.000 \\ 0.000 & -0.219 & 0.976 & 0.000 \\ 0.000 & 0.000 & 0.000 & 1.000 \end{pmatrix} = \left[\begin{array}{c|c} 1 & 0 \\ \hline 0 & R \end{array}\right], \tag{3.97}$$

where R is the 3×3 matrix of rotation around the z-axis by -12.7 degrees.

Remark. The reader is strongly recommended to review the rotation of coordinate axes as given in Appendix B, especially Eqs. (B.24)–(B.26).

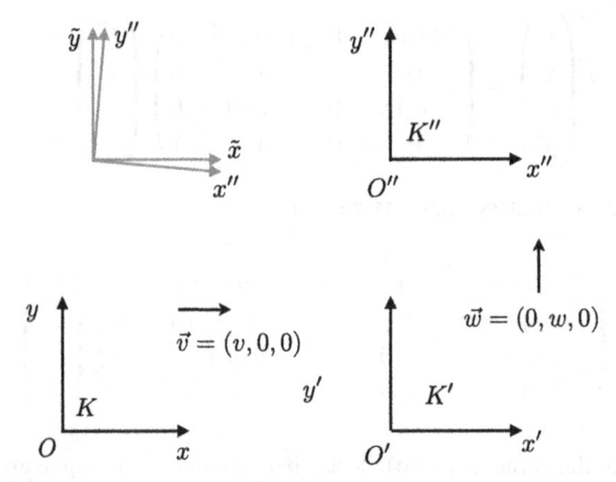

FIGURE 3.7 Combination of two orthogonal boosts.

Frame K' is moving relative to K with velocity v along the x-axis. Frame K'' is moving relative to K' with velocity w along the y'-axis.

In summary, the combination of two boosts in two different directions is not a pure boost but is equal to the combination of a pure boost and a rotation of the coordinate system! This rotation is called Thomas[1] rotation.

Thomas precession. Consider a particle moving in space, perhaps under the influence of various forces. The motion need not be in one direction. Let $\boldsymbol{u}$ be the velocity and $\boldsymbol{a}$ the acceleration of the particle in the inertial frame K. We would like to study the notion of the rest frame of the particle. A simple and intuitive way to ascribe a rest frame for the particle is by the Lorentz boost formula

$$X' = L(\boldsymbol{u})\,X, \tag{3.98}$$

where X is the column matrix with entries (t, x, y, z), defined in the inertial frame K, X' is the column matrix with entries (t', x', y', z'), defined in the rest frame of the particle, and $L(\boldsymbol{u})$ is the pure boost given by (3.56). If $\boldsymbol{u}$ is the velocity of the particle at time t and $\boldsymbol{u} + \delta\boldsymbol{u}$ is the velocity of the particle at time $t + \delta t$ (measured in the inertial frame K), and if we denote the coordinates of the rest frame at time t by subscript t, we get

$$X'_t = L(\boldsymbol{u})\,X, \tag{3.99}$$

$$X'_{t+\delta t} = L(\boldsymbol{u} + \delta\boldsymbol{u})\,X. \tag{3.100}$$

From these two equations we get

$$X'_{t+\delta t} = L^{-1}(\boldsymbol{u} + \delta\boldsymbol{u})\,L(\boldsymbol{u})\,X'_t. \tag{3.101}$$

[1] After Llewellyn Hilleth Thomas (1903–1992).

So, by the passing of time, the rest frame of the particle changes by the transformation

$$T = L^{-1}(\boldsymbol{u} + \delta \boldsymbol{u})\, L(\boldsymbol{u}). \tag{3.102}$$

It can be shown that T is the combination of a boost of $\delta \boldsymbol{u}$ and a rotation of

$$\Delta \boldsymbol{\Omega} = -\frac{\gamma^2(u)}{\gamma(u)+1} \boldsymbol{u} \times \delta \boldsymbol{u}. \tag{3.103}$$

The direction of this vector gives the axis of rotation, and its absolute value gives the angle of rotation. Dividing by δt we get *the angular frequency of the Thomas precession of the rest frame of the particle*

$$\boldsymbol{\omega}_{\mathrm{T}} = \frac{\gamma^2(u)}{\gamma(u)+1} \frac{\boldsymbol{a} \times \boldsymbol{u}}{c^2}. \tag{3.104}$$

Problem 16. A particle is moving on a circle of radius R with constant velocity u and angular frequency ω. Simplify the ratio $\omega_{\mathrm{T}}/\omega$ for two limiting cases: (1) the nonrelativistic limit $u \ll c$ and (2) the ultrarelativistic limit $u \simeq c$, $\gamma \gg 1$.

Guide to Literature. For a more detailed exposition of the Thomas rotation and precession see Sexl and Urbantke (1992, pp. 40–43) or Jackson (1999, pp. 548–553). For a detailed educational discussion of various aspects see Costella et al. (2001).

3.12 Temperature of moving bodies

If a sphere is at rest in K' and has temperature T_0, it radiates as a black body, and the radiation is isotropic. If K' (that is, the sphere) is moving with velocity v relative to frame K, for the observer in K the shape of the object is not a sphere, it is an oblate spheroid, and the radiation is not isotropic. The oblateness is due to the length contraction; the anisotropy is due to the aberration of light rays (see Fig. 3.8).

What could we say about the temperature of a moving object? Assigning a single temperature to a moving body is not trivial. The question is more than a century old, and still there is no consensus on the subject. The difficulty is how to generalize temperature to be applicable to moving objects. Remember that the experimental basis of defining temperature is the so-called zeroth law of thermodynamics, which states that if systems A and B are each in thermal equilibrium with a third system C, then A and B are in thermal equilibrium. Verifying this for systems in relative motion is not trivial.

Guide to Literature. As early as 1907, it was argued by Planck and Einstein that the temperature of a moving body is $T = T_0/\gamma(\beta)$, where T_0 is the temperature of the body in its rest frame and $\beta = v/c$ is its velocity. A line of reasoning to the same conclusion could be found in Pathria (1972, pp. 152–157). Later, by 1963, it was argued by H. Ott that $T = T_0\,\gamma(\beta)$. For a review and yet another transformation, we refer the reader to Farías et al. (2017).

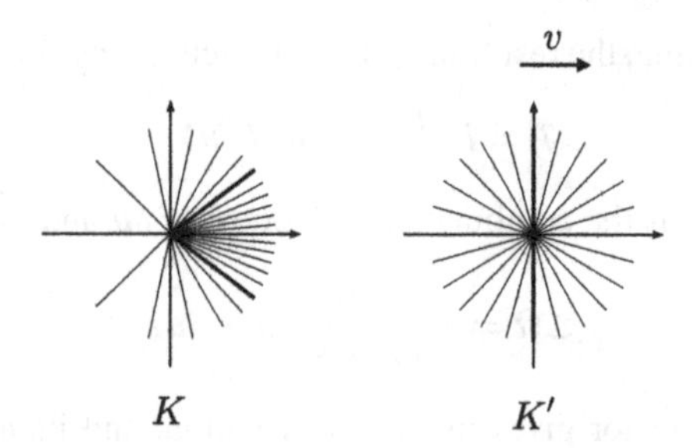

FIGURE 3.8 Light from a moving star.

The sphere in K' (small, at the center) is at rest and at temperature T so that it radiates isotropically. The same body as seen from frame K is not a sphere (because of length contraction) and its radiation is not isotropic. The rays in thick lines are those that in K' have $\theta' = 90$ degrees.

3.13 Transformation of accelerations

The acceleration of a particle, in any frame, has the usual definition—the derivative of the velocity with respect to time. In terms of formulas,

$$a_x := \frac{du_x}{dt}, \quad a_y := \frac{du_y}{dt}, \quad a_z := \frac{du_z}{dt}, \tag{3.105}$$

$$a'_x := \frac{du'_x}{dt'}, \quad a'_y := \frac{du'_y}{dt'}, \quad a'_z := \frac{du'_z}{dt'}. \tag{3.106}$$

Using the rules for addition of velocities, it is straightforward to get

$$du_x = \frac{1}{\gamma^2(v)} \cdot \frac{du'_x}{\left(1 + \frac{v}{c^2} u'_x\right)^2}, \tag{3.107}$$

$$du_y = \frac{1}{\gamma(v)} \cdot \frac{du'_y}{1 + \frac{v}{c^2} u'_x} - \frac{v}{c^2 \gamma(v)} \cdot \frac{u'_y \, du'_x}{\left(1 + \frac{v}{c^2} u'_x\right)^2}, \tag{3.108}$$

$$du_z = \frac{1}{\gamma(v)} \cdot \frac{du'_z}{1 + \frac{v}{c^2} u'_x} - \frac{v}{c^2 \gamma(v)} \cdot \frac{u'_z \, du'_x}{\left(1 + \frac{v}{c^2} u'_x\right)^2}. \tag{3.109}$$

And we remember that $dt = \gamma(v) \, dt' \left(1 + \frac{v}{c^2} u_x\right)$. Therefore we get

$$a_x = \frac{a'_x}{\gamma^3(v) \left(1 + \frac{v}{c^2} u'_x\right)^3}, \tag{3.110}$$

$$a_y = \frac{a'_y}{\gamma^2(v)\left(1+\frac{v}{c^2}u'_x\right)^2} - \frac{v\,u'_y\,a'_x}{c^2\,\gamma(v)\left(1+\frac{v}{c^2}u'_x\right)^3}, \tag{3.111}$$

$$a_z = \frac{a'_z}{\gamma^2(v)\left(1+\frac{v}{c^2}u'_x\right)^2} - \frac{v\,u'_z\,a'_x}{c^2\,\gamma(v)\left(1+\frac{v}{c^2}u'_x\right)^3}. \tag{3.112}$$

3.14 Proper acceleration

Consider a moving particle which is not necessarily moving with constant velocity. Suppose at time t the particle is moving with velocity $\boldsymbol{u}(t)$ and acceleration $\boldsymbol{a}(t)$. We assume these two functions are smooth. Therefore, in a sufficiently small interval around t_0, say $(t_0-\epsilon, t_0+\epsilon)$, we can write $\boldsymbol{u}(t)=\boldsymbol{u}(t_0)+\boldsymbol{a}(t_0)\,\mathrm{d}t$, where $\mathrm{d}t = t-t_0$. Now consider a system K' moving with constant velocity $\boldsymbol{u}(t_0)$. This system is called *the instantaneous rest frame of the particle at time t_0*. Of course, in this frame (K') the particle is at rest only momentarily—either before that moment or after that, the particle is not at rest due to the acceleration of the particle. The acceleration of the particle measured in K' is called *the proper acceleration* of the particle.

In K', at time $t=t_0$, the velocity of the particle is zero, and if ϵ is small enough, the velocity of the particle in time interval $(t_0-\epsilon, t_0+\epsilon)$ is small (compared to c). Therefore, in this frame nonrelativistic dynamics is valid because we know from experiments that for small velocities, Newton's nonrelativistic dynamics is valid. Let us assume $u_y=u_z=0$, which means that the particle, at the moment we are considering, is moving along the x-axis. From the theorem of addition of velocities we have

$$u'_x = \frac{u_x - v}{1-\frac{v}{c^2}u_x}, \quad u'_y = \frac{u_y}{\gamma(v)\left(1-\frac{v}{c^2}u_x\right)}, \quad u'_z = \frac{u_z}{\gamma(v)\left(1-\frac{v}{c^2}u_x\right)}. \tag{3.113}$$

Since $u_y=u_z=0$, we have $u'_y=u'_z=0$, and then from the transformation of accelerations we get

$$a_x = \frac{a'_x}{\gamma^3(v)}, \quad a_y = \frac{a'_y}{\gamma^2(v)}, \quad a_y = \frac{a'_z}{\gamma^2(v)}. \tag{3.114}$$

Now we note that $u=u_x=v$—the equality of $u=u_x$ is obvious, because $u_y=u_z=0$; the equality of $u=v$ results from the definition of the rest frame K': In K' the particle is at rest, and K' is moving with velocity v relative to K. We can write these equations in the following form:

$$a'_x=\gamma^3(u)\,a_x, \quad a'_y=\gamma^2(u)\,a_y, \quad a'_z=\gamma^2(u)\,a_z. \tag{3.115}$$

Here (a_x, a_y, a_z) is the acceleration vector in K and (a'_x, a'_y, a'_z) is the acceleration vector in the rest frame K'—the so-called *proper acceleration*.

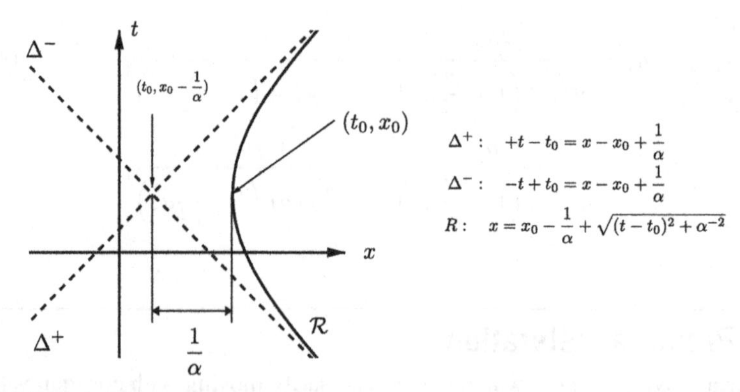

FIGURE 3.9 Hyperbolic motion.

(In this figure, $c = 1$.) $\mathcal{R}$ is the world-line of a particle P, with constant positive proper acceleration α. The initial condition is that at t_0, the particle is at rest at tpoint x_0. $\Delta^{\pm}$ are the asymptotes to the world-line. Events to the left ("north-west") of Δ^+ cannot be seen by P. The particle P cannot influence events to the left ("south-west") of Δ^-. The set $\Delta^+ \cup \Delta^-$ is called the event horizon for the observer $\mathcal{R}$.

3.15 Hyperbolic motion

Consider a particle moving with constant proper acceleration α on the x-axis with initial conditions $x(t_0) = x_0$, $u(t_0) = 0$. From the previous section we know that $\alpha = \gamma^3(u)\, a$, where $a = \mathrm{d}u/\mathrm{d}t$ is the acceleration. Now, it is easy to see that

$$\frac{\mathrm{d}}{\mathrm{d}t}[\gamma(u)\, u] = \gamma(u)\frac{\mathrm{d}u}{\mathrm{d}t} + u\frac{\mathrm{d}}{\mathrm{d}t}\left(1 - \frac{u^2}{c^2}\right)^{-1/2} = a\,\gamma^3(u) = \alpha, \tag{3.116}$$

from which it follows that for such a particle $\gamma(u)\, u = \alpha\, t + \text{constant}$. The constant is $-\alpha\, t_0$, because of the initial condition $u(t_0) = 0$. Therefore we get $u\,\gamma(u) = \alpha\,(t - t_0)$. Squaring and solving for u, we get

$$u = \pm\frac{\alpha\, c\,(t - t_0)}{\sqrt{c^2 + \alpha^2\,(t - t_0)^2}}. \tag{3.117}$$

For the positive sign, we get

$$\frac{\mathrm{d}x}{\mathrm{d}t} = \frac{\alpha\, c\,(t - t_0)}{\sqrt{c^2 + \alpha^2\,(t - t_0)^2}}. \tag{3.118}$$

The solution of this differential equation is

$$x = x_0 - \frac{c^2}{|\alpha|} + \mathrm{sgn}(\alpha)\sqrt{c^2\,(t - t_0)^2 + \frac{c^4}{\alpha^2}}, \tag{3.119}$$

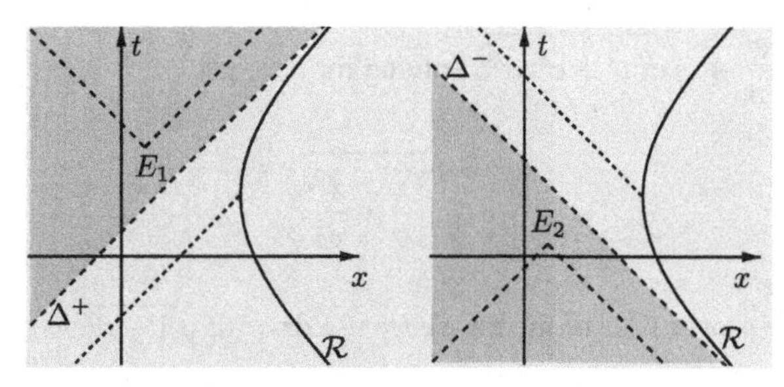

FIGURE 3.10 Horizons for the hyperbolic motion.

(a) The event E_1 is to the left of Δ^+ (the future event horizon). The future-pointing light-cone on this event does not intersect the world-line $\mathcal{R}$, and therefore the accelerated particle cannot see E_1. (b) The event E_2 is to the left of Δ^- (the past event horizon). The past-pointing light-cone on this event does not intersect the world-line $\mathcal{R}$, and therefore the accelerated particle cannot affect E_2, that is, it cannot see any light signal emitted by the particle.

which could be written in the following form:

$$\left(x - x_0 + \frac{c^2}{|\alpha|}\right)^2 - c^2 (t - t_0)^2 = \frac{c^4}{\alpha^2}. \tag{3.120}$$

This is a hyperbola with asymptotes

$$x - x_0 + \frac{c^2}{|\alpha|} = \pm c\,(t - t_0). \tag{3.121}$$

Horizon. The world-line (3.119) is drawn in Fig. 3.9. The asymptotes, referred to as the *horizons*, have important physical properties. As is obvious from this figure and Fig. 3.10, if an event is to the left of Δ^+, a light signal emitted from this event will never reach the observer. Hence Δ^+ is called the future event horizon. Also, a light signal emitted by the particle will never cause an effect on the events to the left of Δ^-. Hence Δ^- is called the past event horizon.

Problem 17. Consider a particle which is stationary at the point x_0 in the interval $-\infty < t \leq t_0$ and then moves with constant proper acceleration α in the interval $t_0 \leq t < \infty$. Draw the world-line. Show that this particle has a future particle horizon Δ^+, but no past particle horizon.

Worked Example 18. Consider a particle experiencing constant proper acceleration α, starting motion from rest ($u_0 = 0$). If the stationary observer K measures that after time t the particle is moved a distance x (measured by K), from (3.119) on page 70 we have $x = -\frac{c^2}{\alpha} + \sqrt{c^2 t^2 + c^4/\alpha^2}$. Solving for t we get

$$t = \sqrt{\frac{2x}{\alpha} + \frac{x^2}{c^2}}. \tag{3.122}$$

To calculate the proper time of the travel, we use $\mathrm{d}\tau = \mathrm{d}t\sqrt{1-\beta^2}$, where

$$\beta = \frac{u}{c}, \quad u = \frac{\mathrm{d}x}{\mathrm{d}t} = \frac{\mathrm{d}}{\mathrm{d}t}\left(c^2 t^2 + \frac{c^4}{\alpha^2}\right)^{1/2} = \frac{c^2 t}{\sqrt{c^2 t^2 + c^4/\alpha^2}}. \tag{3.123}$$

From this we get

$$\mathrm{d}\tau = \frac{c^2}{\alpha} \cdot \frac{\mathrm{d}t}{\sqrt{c^2 t^2 + \frac{c^4}{\alpha^2}}}. \tag{3.124}$$

Integrating this differential equation, we get

$$\tau = \frac{c}{\alpha}\sinh^{-1}\frac{\alpha t}{c} \quad \text{or} \quad t = \frac{c}{\alpha}\sinh\frac{\alpha\tau}{c}, \tag{3.125}$$

$$\beta = \tanh\left(\frac{\alpha\tau}{c}\right). \tag{3.126}$$

If a spaceship is traveling with $\alpha = 10\,\mathrm{m\,s^{-2}}$ for a duration of proper time (measured by the clocks in the ship) of $10^8\,\mathrm{s} \simeq 3.1$ years, the time passed for the stationary observer K would be

$$t = \frac{c}{\alpha}\sinh\frac{\alpha\tau}{c} = 4.2 \times 10^8\,\mathrm{s} \simeq 13.3\,\text{years}. \tag{3.127}$$

The terminal velocity would be

$$\beta = \tanh\left(\frac{10\,\mathrm{m\,s^{-2}} \times 10^8\,\mathrm{s}}{3 \times 10^8\,\mathrm{m\,s^{-1}}}\right) = \tanh(3.3) = 0.997. \tag{3.128}$$

The distance traveled by the particle (as measured by the stationary observer) would be

$$x = -\frac{c^2}{\alpha} + \sqrt{c^2 t^2 + \frac{c^4}{\alpha^2}} = 2.2 \times 10^{16}\,\mathrm{m} \simeq 2.6\,\mathrm{ly}. \tag{3.129}$$

Later we will see that the motion of a charged particle in a constant electric field is exactly such a motion, with quite high proper acceleration and quite short time of travel—see Worked Example 24 on page 99.

References

Clément, G., 1980. Does the Fizeau experiment really test special relativity? American Journal of Physics 48 (12), 1059–1062.

Costella, J.P., McKellar, B.H.J., Rawlinson, A.A., 2001. The Thomas rotation. American Journal of Physics 69 (8), 837–846.

Debs, T.A., Redhead, M.L.G., 1996. The twin "paradox" and the conventionality of simultaneity. American Journal of Physics 64 (4), 384–392.

Farías, C., Pinto, V.A., Moya, P.S., 2017. What is the temperature of a moving body? Scientific Reports 7 (1), 17657.

Grøn, Ø., Hervik, S., 2007. Einstein's General Theory of Relativity. Springer, Berlin.

Hawking, S.W., Ellis, G.F.R., 1973. The Large Scale Structure of Space-Time. Cambridge University Press, Cambridge.

Jackson, J.D., 1999. Classical Electrodynamics, 3rd edition. John Wiley & Sons, New York.

Pathria, R.K., 1972. Statistical Mechanics. Pergamon Press, Oxford.

Rindler, W., 1961. Length contraction paradox. American Journal of Physics 29 (6), 365–366.

Sexl, R.U., Urbantke, H.K., 1992. Relativity, Groups, Particles. Special Relativity and Relativistic Symmetry in Field and Particle Physics. Springer-Verlag, Wien.

CHAPTER

Development of the formalism

4

4.1 4-vectors

Any set of four quantities which under a Lorentz boost transform as four quantities $x^\mu = (t, x, y, z)$ are said to be the components of a 4-vector. The simplest example is x^μ itself, and after that the following four quantities:

$$U^0 := \gamma(u) = \frac{\mathrm{d}t}{\mathrm{d}\tau}, \tag{4.1}$$

$$U^1 := \gamma(u)\, u_x = \frac{\mathrm{d}x}{\mathrm{d}\tau} = \frac{\mathrm{d}t}{\mathrm{d}\tau} \cdot \frac{\mathrm{d}x}{\mathrm{d}t}, \tag{4.2}$$

$$U^2 := \gamma(u)\, u_y = \frac{\mathrm{d}x}{\mathrm{d}\tau} = \frac{\mathrm{d}t}{\mathrm{d}\tau} \cdot \frac{\mathrm{d}y}{\mathrm{d}t}, \tag{4.3}$$

$$U^3 := \gamma(u)\, u_z = \frac{\mathrm{d}x}{\mathrm{d}\tau} = \frac{\mathrm{d}t}{\mathrm{d}\tau} \cdot \frac{\mathrm{d}z}{\mathrm{d}t}. \tag{4.4}$$

The proof is trivial once we recall that $\mathrm{d}\tau$ is invariant, that is, it has the same value in both K and K'. Therefore, the transformation law of these four quantities is exactly the same as the transformation law of $\mathrm{d}x^\mu$s. Although this proof is rigorous and complete, I think it is better if we see another one. From the relativistic rules of addition of velocities (page 60) we have the following relations:

$$\gamma(u) = \gamma(v)\,\gamma(u')\left(1 + \frac{v}{c^2}u'_x\right) = \gamma(v)\left[\gamma(u') + \frac{v}{c^2}\gamma(u')\,u'_x\right], \tag{4.5}$$

$$u_x = \frac{u'_x + v}{1 + \frac{v}{c^2}u'_x}, \tag{4.6}$$

$$u_y = \frac{1}{\gamma(v)} \cdot \frac{u'_y}{1 + \frac{v}{c^2}u'_x}, \qquad u_z = \frac{1}{\gamma(v)} \cdot \frac{u'_z}{1 + \frac{v}{c^2}u'_x}. \tag{4.7}$$

The first line reads

$$U^0 = \gamma(v)\left(U'^0 + \frac{v}{c^2}\,U'^1\right) \tag{4.8}$$

and shows that $U^0 = \gamma(u)$ transforms as t. To prove that U^1, U^2, and U^3 transform as x, y, and z, we simply multiply the equations in the second and third lines by that

A Mathematical Approach to Special Relativity. https://doi.org/10.1016/B978-0-32-399708-9.00013-0

in the first line to obtain

$$\gamma(u)\,u_x = \gamma(v)\Big[\gamma(u')\,u'_x + v\,\gamma(u')\Big], \tag{4.9}$$

$$\gamma(u)\,u_y = \gamma(u')\,u'_y, \tag{4.10}$$

$$\gamma(u)\,u_z = \gamma(u')\,u'_z, \tag{4.11}$$

reading

$$U^1 = \gamma(v)\left(U'^1 + v\,U'^0\right), \tag{4.12}$$

$$U^2 = U'^2, \tag{4.13}$$

$$U^3 = U'^3. \tag{4.14}$$

4.2 Index notation

We use the Greek alphabet to denote indices with values 0, 1, 2, and 3. We use the Latin alphabet to denote indices with values 1, 2, and 3. We define

$$x^0 = t, \quad x^1 = x, \quad x^2 = y, \quad x^3 = z. \tag{4.15}$$

We also write the matrix of the Lorentz boost L (in the x^1-direction) as follows:

$$\left[L^\mu{}_\nu\right] = \begin{bmatrix} L^0{}_0 & L^0{}_1 & L^0{}_2 & L^0{}_3 \\ L^1{}_0 & L^1{}_1 & L^1{}_2 & L^1{}_3 \\ L^2{}_0 & L^2{}_1 & L^2{}_2 & L^2{}_3 \\ L^3{}_0 & L^3{}_1 & L^3{}_2 & L^3{}_3 \end{bmatrix} = \begin{bmatrix} \gamma(v) & \gamma(v)\dfrac{v}{c^2} & 0 & 0 \\ \gamma(v)v & \gamma(v) & 0 & 0 \\ 0 & 0 & 1 & 0 \\ 0 & 0 & 0 & 1 \end{bmatrix}. \tag{4.16}$$

With this notation, the Lorentz boost reads as follows:

$$x^\mu = \sum_{\nu=0}^{3} L^\mu{}_\nu\, x'^\nu. \tag{4.17}$$

Definition. Any set of four quantities which under a Lorentz boost transform as the (t, x, y, z) set is called an upper-index 4-vector. In other words,

$$A^\mu = \sum_{\nu=0}^{3} L^\mu{}_\nu\, A'^\nu. \tag{4.18}$$

Note that if $A = (A^0, A^1, A^2, A^3)$ is a 4-vector, then, necessarily, A^1/A^0 must have the dimensions of velocity.

Let c be the velocity of light in a vacuum. We call the following symmetric matrix the Minkowski[1] metric:

$$H = [\eta_{\mu\nu}] = \begin{bmatrix} \eta_{00} & \eta_{01} & \eta_{02} & \eta_{03} \\ \eta_{10} & \eta_{11} & \eta_{12} & \eta_{13} \\ \eta_{20} & \eta_{21} & \eta_{22} & \eta_{23} \\ \eta_{30} & \eta_{31} & \eta_{32} & \eta_{33} \end{bmatrix} = \begin{bmatrix} -\varsigma\, c^2 & 0 & 0 & 0 \\ 0 & \varsigma & 0 & 0 \\ 0 & 0 & \varsigma & 0 \\ 0 & 0 & 0 & \varsigma \end{bmatrix}, \tag{4.19}$$

where:

- $\varsigma = +1$ corresponds to the space-like convention, as in Weinberg (1972),
- $\varsigma = -1$ corresponds to the time-like convention, as in Landau and Lifshitz (1975).

Note that $\varsigma^2 = 1$. The inverse of this H is the following matrix:

$$H^{-1} = [\eta^{\mu\nu}] = \begin{bmatrix} \eta^{00} & \eta^{01} & \eta^{02} & \eta^{03} \\ \eta^{10} & \eta^{11} & \eta^{12} & \eta^{13} \\ \eta^{20} & \eta^{21} & \eta^{22} & \eta^{23} \\ \eta^{30} & \eta^{31} & \eta^{32} & \eta^{33} \end{bmatrix} = \begin{bmatrix} -\varsigma\, c^{-2} & 0 & 0 & 0 \\ 0 & \varsigma & 0 & 0 \\ 0 & 0 & \varsigma & 0 \\ 0 & 0 & 0 & \varsigma \end{bmatrix}. \tag{4.20}$$

Since these two symmetric matrices are inverse of each other, we have

$$\sum_{\alpha=0}^{3} \eta_{\mu\alpha}\, \eta^{\alpha\nu} = \sum_{\alpha=0}^{3} \eta_{\mu\alpha}\, \eta^{\nu\alpha} = \delta_\mu^\nu. \tag{4.21}$$

Raising and lowering indices. If A^μ is a 4-vector with upper indices, we define four quantities A_μ as follows:

$$A_\mu = \sum_{\alpha=0}^{3} \eta_{\mu\alpha}\, A^\alpha. \tag{4.22}$$

This means that if $A^\mu = (a, \boldsymbol{b})$, then $A_\mu = (-\varsigma\, c^2\, a, \varsigma\, \boldsymbol{b})$. Similarly, if A_μ is a lower-index 4-vector, we define

$$A^\mu = \sum_{\alpha=0}^{3} \eta^{\mu\alpha}\, A_\alpha. \tag{4.23}$$

These two operations of lowering and raising indices are consistent because

$$A_\mu = \sum_{\alpha=0}^{3} \eta_{\mu\alpha}\, A^\alpha = \sum_{\alpha=0}^{3} \eta_{\mu\alpha} \left(\sum_{\beta=0}^{3} \eta^{\alpha\beta}\, A_\beta \right) \tag{4.24}$$

[1] After Hermann Minkowski (1864–1909).

$$= \sum_{\beta=0}^{3} \left\{ \sum_{\alpha=0}^{3} \eta_{\mu\alpha}\, \eta^{\alpha\beta} \right\} A_\beta = \sum_{\beta=0}^{3} \delta_\mu^\beta\, A_\beta = A_\mu. \tag{4.25}$$

Definition. The Minkowski inner product of two 4-vectors is defined by

$$A \cdot B = \sum_{\mu=0}^{3} \sum_{\nu=0}^{3} \eta_{\mu\nu}\, A^\mu\, B^\nu. \tag{4.26}$$

What makes this definition useful is that it is invariant under a Lorentz transformation because

$$\sum_{\alpha,\beta} \eta_{\alpha\beta} L^\alpha_{\ \mu} L^\beta_{\ \nu} = \eta_{\mu\nu} \tag{4.27}$$

and we have

$$A \cdot B = \sum_{\alpha=0}^{3} \sum_{\beta=0}^{3} \eta_{\alpha\beta}\, A^\alpha\, B^\beta = \sum_{\alpha,\beta,\mu\,\nu=0}^{3} \eta_{\alpha\beta}\, L^\alpha_{\ \mu}\, A'^\mu\, L^\beta_{\ \nu}\, A'^\nu \tag{4.28}$$

$$= \sum_{\mu,\nu} \left\{ \sum_{\alpha,\beta} \eta_{\alpha\beta} L^\alpha_{\ \mu}\, L^\beta_{\ \nu} \right\} A'^\mu\, A'^\nu = \sum_{\mu,\nu} \eta_{\mu\nu}\, A'^\mu\, A'^\nu \tag{4.29}$$

$$= A' \cdot B'. \tag{4.30}$$

Einstein's summation convention. In the formulas we are facing, the summation symbol, $\sum$, becomes more and more annoying. However, in almost all cases, the summation indices have a simple structure—they occur once as a subscript and once as a superscript. Only rarely do we encounter a summation not obeying this rule or a repetition with no summation involved. Therefore, we use Einstein's summation convention (Einstein, 1916), which says that a repeated index, once as a subscript and once as a superscript, is to be summed over unless stated explicitly. Thus, for example

$$A^\mu\, B_{\mu\nu} = \sum_{\mu=0}^{3} A^\mu\, B_{\mu\nu} = A^0\, B_{0\nu} + A^1\, B_{1\nu} + A^2\, B_{2\nu} + A^3\, B_{3\nu}. \tag{4.31}$$

Transformation of lower-index 4-vectors. Recall that $A_\alpha = \eta_{\alpha\mu}\, A^\mu$ and $A^\mu = L^\mu_{\ \nu}\, A'^\nu$, from which it follows that

$$A_\alpha = \eta_{\alpha\mu}\, A^\mu = \eta_{\alpha\mu}\, L^\mu_{\ \nu}\, \eta^{\nu\beta}\, A'_\beta \tag{4.32}$$

$$= \left(\eta_{\alpha\mu}\, L^\mu_{\ \nu}\, \eta^{\nu\beta}\right) A'_\beta \tag{4.33}$$

$$= L_\alpha^{\ \beta}\, A'_\beta, \tag{4.34}$$

where

$$L_{\alpha}^{\ \beta} = \eta_{\alpha\mu} L^{\mu}_{\ \ \nu} \eta^{\nu\beta} = \left[H\, L\, H^{-1} \right]_{\alpha}^{\ \beta}. \tag{4.35}$$

Using the explicit form of the matrices H, H^{-1}, and L, we see that

$$\left[L_{\alpha}^{\ \beta} \right] = \begin{bmatrix} \gamma(v) & -\gamma(v)\, v & 0 & 0 \\ -\gamma(v)\frac{v}{c^2} & \gamma(v) & 0 & 0 \\ 0 & 0 & 1 & 0 \\ 0 & 0 & 0 & 1 \end{bmatrix}, \tag{4.36}$$

which is nothing but the transposed inverse of L (change v to $-v$ in L and transpose the result to see this). In index notation, we can write this result as

$$L_{\mu}^{\ \beta} L^{\nu}_{\ \ \beta} = \delta^{\nu}_{\mu}, \qquad L_{\mu}^{\ \beta} L^{\mu}_{\ \ \alpha} = \delta^{\beta}_{\alpha}. \tag{4.37}$$

Now see the following calculation:

$$L^{\alpha}_{\ \ \nu} A_{\alpha} = L^{\alpha}_{\ \ \nu} \left(L_{\alpha}^{\ \beta} A'_{\beta} \right) \tag{4.38}$$

$$= \left(L^{\alpha}_{\ \ \nu} L_{\alpha}^{\ \beta} \right) A'_{\beta} = \delta^{\beta}_{\nu} A'_{\beta}, \tag{4.39}$$

$$L^{\alpha}_{\ \ \nu} A_{\alpha} = A'_{\nu}. \tag{4.40}$$

In summary, we have the following rules for transforming lower-index and upper-index 4-vectors:

$$A^{\mu} = L^{\mu}_{\ \ \nu} A'^{\nu}, \quad \text{upper index}, \tag{4.41}$$

$$A'_{\nu} = L_{\mu}^{\ \nu} A_{\mu}, \quad \text{lower index}. \tag{4.42}$$

The prototype of an upper-index 4-vector is $\mathrm{d}x^{\mu}$. Now we look at the prototype of a lower-index 4-vector. From calculus we know that

$$\frac{\partial f}{\partial x'^{\nu}} = \frac{\partial x^{\mu}}{\partial x'^{\nu}} \cdot \frac{\partial f}{\partial x^{\mu}}. \tag{4.43}$$

From $x^{\mu} = L^{\mu}_{\ \ \nu} x'^{\nu}$ we see that

$$\frac{\partial x^{\mu}}{\partial x'^{\nu}} = L^{\mu}_{\ \ \nu}. \tag{4.44}$$

We therefore see that the four quantities $\frac{\partial f}{\partial x^{\mu}}$ form a lower-index 4-vector, for which we use the more convenient notation

$$\partial_{\mu} f := \frac{\partial f}{\partial x^{\mu}}. \tag{4.45}$$

It is customary to omit the function f from the above equation and think of ∂_{μ} as both a vector field and an operator.

For later use, we write the explicit form of the transformation of ∂_μ for a boost along the x-axis:

$$\frac{\partial}{\partial t}=\gamma(v)\frac{\partial}{\partial t'}-\gamma(v)\,v\frac{\partial}{\partial x'}, \tag{4.46}$$

$$\frac{\partial}{\partial x}=-\gamma(v)\frac{v}{c^2}\frac{\partial}{\partial t'}+\gamma(v)\frac{\partial}{\partial x'}, \tag{4.47}$$

$$\frac{\partial}{\partial y}=\frac{\partial}{\partial y'}, \tag{4.48}$$

$$\frac{\partial}{\partial z}=\frac{\partial}{\partial z'}. \tag{4.49}$$

4.3 Quadratic scalars

When we say that a quantity is scalar, we mean that the value of the quantity is invariant under a Lorentz transformation. The following quadratic forms are scalars:

1. The line element. We have

$$\mathrm{d}s^2 := \eta_{\mu\nu}\,\mathrm{d}x^\mu\,\mathrm{d}x^\nu=\varsigma\left(-c^2\,\mathrm{d}t^2+\mathrm{d}x^2+\mathrm{d}y^2+\mathrm{d}z^2\right), \tag{4.50}$$

   where $\mathrm{d}s^2$ is called the "square" of the line element. It should be noted that it is not always the square of a real number because $\mathrm{d}s^2$ could be negative.
2. The "square" of the gradient. If f is a function of x^μ, the scalar

$$\mathrm{d}f=\mathrm{d}x^\mu\frac{\partial f}{\partial x^\mu}=\mathrm{d}x^\mu\,\partial_\mu f \tag{4.51}$$

   is simply the differential of f, which means the incremental change of f if the argument is changed by the 4-vector $\mathrm{d}x^\mu$. The lower-index 4-vector $\partial_\mu f$ is simply the components of the gradient of f. Sometimes we denote this gradient by ∂f, whose square is

$$(\partial f)^2=\eta^{\mu\nu}\left(\partial_\mu f\right)(\partial_\nu f)=\varsigma\left[-\frac{1}{c^2}\left(\frac{\partial f}{\partial t}\right)^2+\nabla f\cdot\nabla f\right]. \tag{4.52}$$

3. d'Alembert's operator. The "square" or the gradient operator ∂ is

$$\Box := \partial\cdot\partial=\eta^{\mu\nu}\partial_\mu\,\partial_\nu=\varsigma\left[-\frac{1}{c^2}\frac{\partial^2}{\partial t^2}+\frac{\partial^2}{\partial x^2}+\frac{\partial^2}{\partial y^2}+\frac{\partial^2}{\partial z^2}\right]. \tag{4.53}$$

   This is a second-order differential operator called d'Alembert's operator. The wave equation, with wave velocity c for the function f, reads $\Box f=g$, where g is called the source.

4.4 The 4-current

In frame K, we denote the charge density by ρ and the current density by $\boldsymbol{J}$. As we saw in §1.18.1, in this frame the conservation of charge is expressed by the differential equation

$$\frac{\partial \rho}{\partial t}+\frac{\partial J_x}{\partial x}+\frac{\partial J_y}{\partial y}+\frac{\partial J_z}{\partial z}=0. \tag{4.54}$$

Using the transformation of the partial derivatives we see that

$$0=\left(\gamma\frac{\partial}{\partial t'}-\gamma v\frac{\partial}{\partial x'}\right)\rho+\left(\gamma\frac{\partial}{\partial x'}-\gamma\frac{v}{c^2}\frac{\partial}{\partial t'}\right)J_x+\frac{\partial J_y}{\partial y'}+\frac{\partial J_z}{\partial z'} \tag{4.55}$$

$$=\frac{\partial}{\partial t'}\left[\gamma\left(\rho-\frac{v}{c^2}J_x\right)\right]+\frac{\partial}{\partial x'}\left[\gamma\left(J_x-v\rho\right)\right]+\frac{\partial J_y}{\partial y'}+\frac{\partial J_z}{\partial z'}. \tag{4.56}$$

The Galilean principle of relativity states that the conservation of charge, being a law of physics, must be valid in frame K' as well, which means that in K' it must be true that

$$\frac{\partial \rho'}{\partial t'}+\frac{\partial J_x'}{\partial x'}+\frac{\partial J_y'}{\partial y'}+\frac{\partial J_z'}{\partial z'}=0. \tag{4.57}$$

These two differential equations must be equivalent, that is, there must be an "if and only if" relation between the two. In other words, assuming the validity of one, the other must be deduced from algebra. This is so if we have

$$\rho'=\gamma(v)\left(\rho-\frac{v}{c^2}J_x\right), \tag{4.58}$$

$$J_x'=\gamma(v)\left(J_x-v\,\rho\right), \tag{4.59}$$

$$J_y'=J_y, \tag{4.60}$$

$$J_z'=J_z. \tag{4.61}$$

Inverting this transformation we get

$$\rho=\gamma(v)\left(\rho'+\frac{v}{c^2}J_x'\right), \tag{4.62}$$

$$J_x=\gamma(v)\left(J_x'+v\,\rho'\right), \tag{4.63}$$

$$J_y=J_y', \tag{4.64}$$

$$J_z=J_z', \tag{4.65}$$

which simply means that the four quantities

$$J^\mu=(\rho,\boldsymbol{J}) \tag{4.66}$$

form a 4-vector, which we call the 4-current.

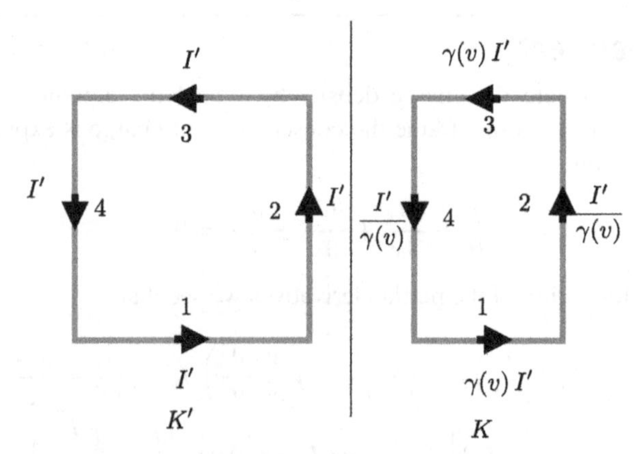

FIGURE 4.1 Charge and current density in a moving current loop.

In frame K' (left) the rectangular loop carries current I'. As seen by the observer in the lab, the rectangle is moving, and its length (segments 1 and 3) is contracted by a factor of $\gamma(v)$. The cross-section areas of segments 2 and 3 are also contracted by the same factor. Segments 1 and 3, as seen in K, are charged with densities $\pm\gamma(v)\,v\,I'/(c^2 A')$, where A' is the cross-section area of the wire (in K').

Worked Example 19. Consider a loop of wire, in the shape of a rectangle, with current I' passing through it. The wire is stationary in frame K' and lies in the (x', y')-plane. Frame K' is moving with velocity v relative to frame K (the lab frame). There is no net charge on the wire, so that $\rho' = 0$. The magnitude of the current density is $J' = I'/A'$, where A' is the cross-section area of the wire (in K'). Referring to Fig. 4.1, the current density 4-vector is given by

$$\begin{aligned}
&\text{segment 1:} \quad \rho' = 0, \quad J'_x = J', \quad J'_y = 0, \quad J'_z = 0,\\
&\text{segment 2:} \quad \rho' = 0, \quad J'_x = 0, \quad J'_y = J', \quad J'_z = 0,\\
&\text{segment 3:} \quad \rho' = 0, \quad J'_x = -J', \quad J'_y = 0, \quad J'_z = 0,\\
&\text{segment 4:} \quad \rho' = 0, \quad J'_x = 0, \quad J'_y = -J', \quad J'_z = 0.
\end{aligned} \tag{4.67}$$

Applying the Lorentz transformation (4.62) we see that in frame K the 4-current is given by

$$\begin{aligned}
&\text{segment 1:} \quad \rho = \gamma(v)\,v\,J'/c^2, \quad J_x = \gamma(v)\,J', \quad J_y = 0, \quad J_z = 0,\\
&\text{segment 2:} \quad \rho = 0, \quad J_x = 0, \quad J_y = J', \quad J_z = 0,\\
&\text{segment 3:} \quad \rho = -\gamma(v)\,v\,J'/c^2, \quad J_x = -\gamma(v)\,J', \quad J_y = 0, \quad J_z = 0,\\
&\text{segment 4:} \quad \rho = 0, \quad J_x = 0, \quad J_y = -J', \quad J_z = 0.
\end{aligned} \tag{4.68}$$

Since a Lorentz transformation does not affect lengths normal to the direction of motion, the cross-section area of the wire in segments 1 and 3 is the same in both

K' and K—equal to A'. Therefore, the current in segments 1 and 3 is $\gamma(v)\,J'\,A' = \gamma(v)\,I'$. However, due to the length contraction, the cross-section area of segments 2 and 4 is $A'/\gamma(v)$. Therefore, the current in segments 2 and 4 (in frame K) is given by $J'\,A'/\gamma(v) = I'/\gamma(v)$. Note also that segments 1 and 3 are charged (in frame K), with charge densities

$$\rho = \pm\frac{\gamma(v)\,v}{c^2}\cdot\frac{I'}{A'}. \tag{4.69}$$

An important consequence of this charge, observed in K, is that *a moving current loop produces a magnetic* and *an electric field.*

4.5 The 4-potential

Of the four Maxwell equations, two are homogeneous and two are inhomogeneous. The two homogeneous equations say that there exists a scalar and a vector potential. The reasoning is as follows.

From $\nabla\cdot\boldsymbol{B} = 0$, we conclude that there exists a vector field $\boldsymbol{A}$, called the vector potential, such that $\boldsymbol{B} = \nabla\times\boldsymbol{A}$. Now if we insert this form of $\boldsymbol{B}$ in $\nabla\times\boldsymbol{E} + \partial\boldsymbol{B}/\partial t = 0$, we see that $\nabla\times(\boldsymbol{E} + \partial\boldsymbol{A}/\partial t) = 0$, from which it follows that there exists a scalar function ϕ such that $\boldsymbol{E} + \partial\boldsymbol{A}/\partial t = -\nabla\phi$. In summary, from the homogeneous Maxwell equations we conclude that a vector field $\boldsymbol{A}$ and a scalar field ϕ exist, such that

$$\boldsymbol{B} = \nabla\times\boldsymbol{A}, \tag{4.70}$$

$$\boldsymbol{E} = -\nabla\phi - \frac{\partial\boldsymbol{A}}{\partial t}. \tag{4.71}$$

We now insert these forms into the inhomogeneous equations:

$$\nabla\cdot\boldsymbol{E} = \nabla\cdot\left(-\nabla\phi - \frac{\partial\boldsymbol{A}}{\partial t}\right) = -\nabla^2\phi - \frac{\partial}{\partial t}\nabla\cdot\boldsymbol{A} \tag{4.72}$$

$$= \frac{1}{c^2}\frac{\partial^2\phi}{\partial t^2} - \nabla^2\phi - \frac{1}{c^2}\frac{\partial^2\phi}{\partial t^2} - \frac{\partial}{\partial t}\nabla\cdot\boldsymbol{A} \tag{4.73}$$

$$= -\varsigma\,\Box\phi - \frac{\partial}{\partial t}\left[\frac{1}{c^2}\frac{\partial\phi}{\partial t} + \nabla\cdot\boldsymbol{A}\right] = \frac{\rho}{\epsilon_0} = \frac{\mu_0\,\rho}{\mu_0\,\epsilon_0} = \mu_0\,c^2\,\rho, \tag{4.74}$$

$$\nabla\times\boldsymbol{B} - \frac{1}{c^2}\frac{\partial\boldsymbol{E}}{\partial t} = \nabla\times(\nabla\times\boldsymbol{A}) - \frac{1}{c^2}\frac{\partial}{\partial t}\left(-\frac{\partial\boldsymbol{A}}{\partial t} - \nabla\phi\right) \tag{4.75}$$

$$= \nabla\nabla\cdot\boldsymbol{A} - \nabla^2\boldsymbol{A} + \frac{1}{c^2}\frac{\partial^2\boldsymbol{A}}{\partial t^2} + \nabla\left(\frac{1}{c^2}\frac{\partial\phi}{\partial t}\right) \tag{4.76}$$

$$= -\varsigma\,\Box\boldsymbol{A} + \nabla\left(\frac{1}{c^2}\frac{\partial\phi}{\partial t} + \nabla\cdot\boldsymbol{A}\right) = \mu_0\,\boldsymbol{J}. \tag{4.77}$$

Let us summarize:

$$-\varsigma\,\Box\phi - \frac{\partial\chi}{\partial t} = \mu_0\,c^2\,\rho, \tag{4.78}$$

$$-\varsigma\,\Box\boldsymbol{A} + \nabla\chi = \mu_0\,\boldsymbol{J}, \tag{4.79}$$

where

$$\chi = \frac{1}{c^2}\frac{\partial\phi}{\partial t} + \nabla\cdot\boldsymbol{A} \tag{4.80}$$

and $\Box = \partial\cdot\partial$ is the d'Alembert operator.

For simplicity, suppose $\chi = 0$, which is called the Lorenz condition.[2] It can be shown that one can always find ϕ and $\boldsymbol{A}$ such that this condition is fulfilled.[3] If the Lorenz condition is satisfied, then the inhomogeneous Maxwell equations, in terms of the scalar and vector potentials, have the following form:

$$\frac{1}{c^2}\Box\phi = -\varsigma\,\mu_0\,\rho, \tag{4.81}$$

$$\Box\boldsymbol{A} = -\varsigma\,\mu_0\,\boldsymbol{J}. \tag{4.82}$$

Since, under a Lorentz boost, the d'Alembert operator $\Box$ is invariant, we see that the inhomogeneous Maxwell equations in both K and K' frames have the same form if (1) the Lorenz condition is satisfied in both of them and (2) four quantities

$$A^\mu = \left(\frac{\phi}{c^2}, \boldsymbol{A}\right) \tag{4.83}$$

transform as a 4-vector. For if that is the case, then the transformations of A^μ and J^μ at the two sides of this equation cancel each other. The Lorenz condition can be written as

$$\partial\cdot A = \partial_\mu A^\mu = 0. \tag{4.84}$$

It is then obvious that if this condition is satisfied in K, then it is also true in K', and vice versa.

The 4-potential, as a lower-index 4-vector, has the form

$$A_\mu = (-\varsigma\,\phi, \varsigma\,\boldsymbol{A}), \tag{4.85}$$

so that we have

$$A_\mu\,\mathrm{d}x^\mu = -\varsigma\,\phi\,\mathrm{d}t + \varsigma\,\boldsymbol{A}\cdot\mathrm{d}\boldsymbol{r}. \tag{4.86}$$

[2] This condition is named after Ludvig Valentin Lorenz (1829–1891), while the Lorentz transformations are named after Hendrik Antoon Lorentz (1853–1928)—see Nevels and Shin (2001). Note the spelling of these two names; one has an extra t!

[3] The proof is based on the gauge freedom in defining the potentials; see Jackson (1999, p. 240, §6.3).

4.6 Electromagnetic field tensor

The two equations $\boldsymbol{B}=\nabla\times\boldsymbol{A}$ and $\boldsymbol{E}=-\nabla\phi-\partial_t\boldsymbol{A}$ could be written in a compact form. To see this, remember that $A^\mu=\left(c^{-2}\phi,\boldsymbol{A}\right)$, which means

$$A_\mu=(-\varsigma\,\phi,\varsigma\,\boldsymbol{A}), \tag{4.87}$$

in which ς is the sign convention of metric (see (4.19) on page 77) defined by

$$\mathrm{d}s^2=\varsigma\left[-c^2\,\mathrm{d}t^2+\mathrm{d}x^2+\mathrm{d}y^2+\mathrm{d}z^2\right]. \tag{4.88}$$

As usual, we define

$$F_{\mu\nu}:=\partial_\mu A_\nu-\partial_\nu A_\mu. \tag{4.89}$$

Let us see what is this antisymmetric matrix:

$$\begin{aligned}F_{01}=\partial_0A_1-\partial_1A_0&=\varsigma\left[\frac{\partial}{\partial t}(A_x)-\frac{\partial(-\phi)}{\partial x}\right]=-\varsigma\left(\frac{\partial A_x}{\partial t}+\frac{\partial\phi}{\partial x}\right)\\&=\varsigma\,E_x,\end{aligned} \tag{4.90}$$

$$\begin{aligned}F_{12}=\varsigma\left[\partial_1A_2-\partial_2A_1\right]&=\varsigma\left[\frac{\partial A_x}{\partial y}-\frac{\partial A_y}{\partial x}\right]\\&=-\varsigma B_z.\end{aligned} \tag{4.91}$$

Calculating other entries this way we get

$$\left[F_{\mu\nu}\right]=\varsigma\begin{bmatrix}0&E_x&E_y&E_z\\-E_x&0&-B_z&B_y\\-E_y&B_z&-0&-B_x\\-E_z&-B_y&B_x&0\end{bmatrix}. \tag{4.92}$$

This quantity, which we write in the form of a matrix $\left[F_{\mu\nu}\right]$, is a *tensor of rank 2*, by which we mean it has the following transformation (under a Lorentz transformation):

$$F'_{\mu\nu}(x'^\sigma)=L^\alpha{}_\mu\,L^\beta{}_\nu F_{\alpha\beta}\left(L^\sigma{}_\kappa x^\kappa\right). \tag{4.93}$$

The proof is simple, but let us first clarify the notation about the argument: The argument of the left-hand side is $x'^\sigma=(t',x',y',z')$. We know that under a Lorentz transformation we have $x'^\sigma=L^\sigma{}_\kappa x^\kappa$, which means that the four quantities (t',x',y',z') are functions of (t,x,y,z). This is explicitly written in the above formula. Usually in textbooks, people do not write this. It is understood that the arguments must be transformed appropriately, that is, in the abovementioned way. Now let us go through the proof. We know that $\partial'_\mu=L^\alpha{}_\mu\partial_\alpha$ and $A'_\nu=L^\beta{}_\nu A_\beta$. From these two relations it follows that

$$F'_{\mu\nu}=\partial'_\mu A'_\nu-\partial'_\nu A'_\mu \tag{4.94}$$

$$
\begin{aligned}
&= L^{\alpha}{}_{\mu} L^{\beta}{}_{\nu} \partial_{\alpha} A_{\beta} - L^{\beta}{}_{\nu} L^{\alpha}{}_{\mu} \partial_{\beta} A_{\alpha} && (4.95)\\
&= L^{\alpha}{}_{\mu} L^{\beta}{}_{\nu} \left(\partial_{\alpha} A_{\beta} - \partial_{\beta} A_{\alpha}\right) && (4.96)\\
&= L^{\alpha}{}_{\mu} L^{\beta}{}_{\nu} F_{\alpha\beta}. && (4.97)
\end{aligned}
$$

Now define

$$G_{\alpha\mu\nu} := \partial_{\alpha} F_{\mu\nu} + \partial_{\mu} F_{\nu\alpha} + \partial_{\nu} F_{\alpha\mu}. \tag{4.98}$$

Note that the successive terms are obtained from the first term by two cyclic reshufflings of the indices $\alpha \to \mu \to \nu \to \alpha$.

Since $F_{\mu\nu}$ is antisymmetric in the indices, we can prove that $G_{\alpha\mu\nu}$ is also antisymmetric in any of its pairs of indices, that is,

$$G_{\alpha\mu\nu} = -G_{\mu\alpha\nu} = -G_{\nu\mu\alpha} = -G_{\alpha\nu\mu}. \tag{4.99}$$

The proof is straightforward. We write the proof that $G_{\mu\alpha\nu} = -G_{\alpha\mu\nu}$:

$$
\begin{aligned}
G_{\mu\alpha\nu} &= +\partial_{\mu} F_{\alpha\nu} + \partial_{\nu} F_{\mu\alpha} + \partial_{\alpha} F_{\nu\mu}\\
&= -\partial_{\mu} F_{\nu\alpha} - \partial_{\nu} F_{\alpha\mu} - \partial_{\alpha} F_{\mu\nu}\\
&= -\left(\partial_{\alpha} F_{\mu\nu} + \partial_{\mu} F_{\nu\alpha} + \partial_{\nu} F_{\alpha\mu}\right) = -G_{\alpha\mu\nu}. && (4.100)
\end{aligned}
$$

Since $G_{\alpha\mu\nu}$ is totally antisymmetric, we conclude that it has only four independent components: G_{012}, G_{013}, G_{023}, and G_{123}. Any other component is zero, one of these, or minus one of these. Let us list these independent components:

$$
\begin{aligned}
G_{021} &= \partial_0 F_{21} + \partial_2 F_{10} + \partial_1 F_{02} = \varsigma \left(-\frac{\partial B_z}{\partial t} + \frac{\partial E_x}{\partial y} - \frac{\partial E_y}{\partial x}\right), && (4.101)\\
G_{013} &= \partial_0 F_{13} + \partial_3 F_{01} + \partial_1 F_{30} = \varsigma \left(-\frac{\partial B_y}{\partial t} - \frac{\partial E_x}{\partial y} + \frac{\partial E_y}{\partial x}\right), && (4.102)\\
G_{032} &= \partial_0 F_{32} + \partial_2 F_{03} + \partial_3 F_{20} = \varsigma \left(-\frac{\partial B_x}{\partial t} - \frac{\partial E_z}{\partial y} + \frac{\partial E_y}{\partial z}\right), && (4.103)\\
G_{132} &= \partial_1 F_{32} + \partial_2 F_{13} + \partial_3 F_{21} = \varsigma \left(-\frac{\partial B_x}{\partial x} - \frac{\partial B_y}{\partial y} - \frac{\partial B_z}{\partial z}\right). && (4.104)
\end{aligned}
$$

These independent components vanish, because of the homogeneous Maxwell equations. If we write $F_{\mu\nu}$ in terms of the potentials, $F_{\mu\nu} = \partial_{\mu} A_{\nu} - \partial_{\nu} A_{\mu}$, then the independent components of $G_{\alpha\mu\nu}$ vanish identically:

$$
\begin{aligned}
G_{\alpha\mu\nu} &= \partial_{\alpha}\left(\partial_{\mu} A_{\nu} - \partial_{\nu} A_{\mu}\right) + \partial_{\mu}\left(\partial_{\nu} A_{\alpha} - \partial_{\alpha} A_{\nu}\right) + \partial_{\nu}\left(\partial_{\alpha} A_{\mu} - \partial_{\mu} A_{\alpha}\right)\\
&= \partial_{\alpha}\partial_{\mu} A_{\nu} - \partial_{\alpha}\partial_{\nu} A_{\mu} + \partial_{\mu}\partial_{\nu} A_{\alpha} - \partial_{\mu}\partial_{\alpha} A_{\nu} + \partial_{\nu}\partial_{\alpha} A_{\mu} - \partial_{\nu}\partial_{\mu} A_{\alpha}\\
&= 0. && (4.105)
\end{aligned}
$$

Raising the indices of $F_{\mu\nu}$ we get

$$F^{\alpha\beta} = \eta^{\alpha\mu} F_{\mu\nu} \eta^{\nu\beta} = \varsigma \begin{bmatrix} 0 & c^{-2}E_x & c^{-2}E_y & c^{-2}E_z \\ -c^{-2}E_x & 0 & B_z & -B_y \\ -c^{-2}E_y & -B_z & 0 & B_x \\ -c^{-2}E_z & B_y & -B_x & 0 \end{bmatrix}. \tag{4.106}$$

Let us see what is the transformation law for $F^{\mu\nu}$. In compact form they are $F^{\mu\nu} = L^{\mu}{}_{\alpha} L^{\nu}{}_{\beta} F'^{\alpha\beta}$. We want to find the explicit form in terms of the electric and magnetic fields E and B. This is simply a matrix product of three 4×4 matrices. We do this only for E_x, leaving the derivation of the other relations to the reader. We have

$$\begin{aligned}
\varsigma \frac{E_x}{c^2} &= F^{01} = L^{0}{}_{\alpha} L^{1}{}_{\beta} F'^{\alpha\beta} \\
&= L^{0}{}_{0} L^{1}{}_{1} F'^{01} + L^{0}{}_{1} L^{1}{}_{0} F'^{10} \\
&+ L^{0}{}_{1} L^{1}{}_{2} F'^{12} + L^{0}{}_{2} L^{1}{}_{1} F'^{21} \\
&+ L^{0}{}_{1} L^{1}{}_{3} F'^{13} + L^{0}{}_{3} L^{1}{}_{1} F'^{31} \\
&+ L^{0}{}_{2} L^{1}{}_{3} F'^{23} + L^{0}{}_{3} L^{1}{}_{2} F'^{31} \\
&= \left(L^{0}{}_{0} L^{1}{}_{1} - L^{0}{}_{1} L^{1}{}_{0}\right) F'^{01} + \left(L^{0}{}_{1} L^{1}{}_{2} - L^{0}{}_{2} L^{1}{}_{1}\right) F'^{12} \\
&+ \left(L^{0}{}_{1} L^{1}{}_{3} - L^{0}{}_{3} L^{1}{}_{1}\right) F'^{13} + \left(L^{0}{}_{2} L^{1}{}_{3} - L^{0}{}_{3} L^{1}{}_{2}\right) F'^{23}, \\
\varsigma \frac{E_x}{c^2} &= \varsigma \frac{E'_x}{c^2} \left(\gamma^2 - \gamma^2 \frac{v^2}{c^2}\right) = \varsigma \frac{E'_x}{c^2}.
\end{aligned} \tag{4.107}$$

For later reference, here we write all the transformation formulas:

$$\begin{array}{l|l}
E_x = E'_x, & B_x = B'_x, \\
E_y = \gamma(v)\left(E'_y + v\, B'_z\right), & B_y = \gamma(v)\left(B'_y - \frac{v}{c^2} E'_z\right), \\
E_z = \gamma(v)\left(E'_z - v\, B'_y\right), & B_z = \gamma(v)\left(B'_z + \frac{v}{c^2} E_y\right).
\end{array} \tag{4.108}$$

Now let us calculate $\varsigma\, \partial_\beta F^{\alpha\beta}$:

$$\varsigma\, \partial_\beta F^{0\beta} = \frac{1}{c^2}\left(\frac{\partial E_x}{\partial x} + \frac{\partial E_y}{\partial y} + \frac{\partial E_z}{\partial z}\right) = \mu_0\, \rho, \tag{4.109}$$

$$\varsigma\, \partial_\beta F^{1\beta} = -\frac{1}{c^2}\frac{\partial E_x}{\partial t} + \frac{\partial B_z}{\partial y} - \frac{\partial B_y}{\partial z} = \mu_0\, J_x, \tag{4.110}$$

$$\varsigma\, \partial_\beta F^{2\beta} = -\frac{1}{c^2}\frac{\partial E_y}{\partial t} - \frac{\partial B_z}{\partial x} + \frac{\partial B_x}{\partial z} = \mu_0\, J_y, \tag{4.111}$$

$$\varsigma\, \partial_\beta F^{3\beta} = -\frac{1}{c^2}\frac{\partial E_z}{\partial t} + \frac{\partial B_y}{\partial x} - \frac{\partial B_x}{\partial y} = \mu_0\, J_z. \tag{4.112}$$

Recalling

$$J^\mu = (\rho, \boldsymbol{J}), \qquad c^2 = \frac{1}{\mu_0\, \epsilon_0}, \tag{4.113}$$

the Maxwell equations can be written in the following two compact forms:

$$G_{\alpha\mu\nu} := \partial_\alpha F_{\mu\nu} + \partial_\mu F_{\nu\alpha} + \partial_\nu F_{\alpha\mu} = 0, \tag{4.114}$$

$$\partial_\nu F^{\mu\nu} = \varsigma\, \mu_0\, J^\mu. \tag{4.115}$$

From the homogeneous equations, $G_{\alpha\mu\nu} = 0$, it follows that there exists a 4-potential $A_\mu = (-\varsigma\, \phi, \varsigma\, \boldsymbol{A})$, such that $F_{\mu\nu} = \partial_\mu A_\nu - \partial_\nu A_\mu$. Inserting this in the inhomogeneous equations we get

$$\partial_\nu \left(\partial^\mu A^\nu - \partial^\nu A^\mu\right) = \partial^\mu (\partial \cdot A) - \Box A^\mu, \tag{4.116}$$

where $\partial \cdot A = \partial_\nu A^\nu$. Now, if we impose the so-called Lorenz condition, $\partial \cdot A = 0$, we get the usual wave equations for the 4-potential:

$$\Box A^\alpha = -\varsigma\, \mu_0\, J^\alpha. \tag{4.117}$$

4.7 6-vectors

Let A and B be two 4-vectors. Mimicking the definition of the cross-product in the 3D space, we define the antisymmetric matrix (tensor) $C = A \times B$, where $C^{\mu\nu} := A^\mu B^\nu - A^\nu B^\mu$. This matrix, being antisymmetric, has only six independent entries—C^{01}, C^{02}, C^{03}, C^{12}, C^{13}, and C^{23}. We can organize them as follows:

$$C^{\mu\nu}(\boldsymbol{a}, \boldsymbol{b}) = \begin{bmatrix} 0 & a_x & a_y & a_z \\ -a_x & 0 & b_z & -b_y \\ -a_y & -b_z & 0 & b_x \\ -a_z & b_y & -b_z & 0 \end{bmatrix}. \tag{4.118}$$

Note that, because of our conventions, the space components of 4-vectors (for example, x) do not have the same dimensions as the time components (the t). For $C^{\mu\nu}$ this means that

$$[b_i] = [c\, a_i]. \tag{4.119}$$

The Lorentz transformation of this antisymmetric tensor is obviously

$$C'^{\mu\nu} = A'^\mu B'^\nu - A'^\nu B'^\mu \tag{4.120}$$

$$= L^\mu{}_\alpha A^\alpha L^\nu{}_\beta B^\beta - L^\mu{}_\alpha B^\alpha L^\nu{}_\beta A^\beta \tag{4.121}$$

$$= L^{\mu}{}_{\alpha} L^{\nu}{}_{\beta} \left(A^{\alpha} B^{\beta} - B^{\alpha} A^{\beta}\right) \tag{4.122}$$

$$= L^{\mu}{}_{\alpha} L^{\nu}{}_{\beta} C^{\alpha\beta}, \tag{4.123}$$

or, in matrix notation,

$$C' = L^T C L. \tag{4.124}$$

It is an easy exercise to see that, in terms of $\boldsymbol{a}$ and $\boldsymbol{b}$, defined above, the Lorentz transformation reads

$$a'_x = a_x, \qquad b'_x = b_x, \tag{4.125}$$

$$a'_y = \gamma\left(a_y - c^{-2} v\, b_z\right), \qquad b'_y = \gamma\left(b_y + v\, a_z\right), \tag{4.126}$$

$$a'_z = \gamma\left(a_z + c^{-2} v\, b_y\right), \qquad b'_z = \gamma\left(b_z - v\, a_y\right). \tag{4.127}$$

Any set of six quantities $(\boldsymbol{a}, \boldsymbol{b})$ which under a Lorentz transformation transform as above are called the six components of a 6-vector. The above defined *cross*-product of two 4-vectors is only one example of a 6-vector.

Problem 20. By straightforward calculation, show that if $(\boldsymbol{a}, \boldsymbol{b})$ is a 6-vector, then both $\boldsymbol{a} \cdot \boldsymbol{b}$ and $c^2\, \boldsymbol{a} \cdot \boldsymbol{a} - \boldsymbol{b} \cdot \boldsymbol{b}$ are scalars, *i.e.*, invariant under a Lorentz transformation. This means that

$$\boldsymbol{a}' \cdot \boldsymbol{b}' = \boldsymbol{a} \cdot \boldsymbol{b}, \tag{4.128}$$

$$c^2\, |\boldsymbol{a}'|^2 - |\boldsymbol{b}'|^2 = c^2\, |\boldsymbol{a}|^2 - |\boldsymbol{b}|^2. \tag{4.129}$$

Note also that these two quantities are invariant under rotations of space—because the inner product is so.

Remark. The term "6-vector" is very old (Cunningham, 1914), but it is not used frequently in modern literature. The modern terminology is antisymmetric tensor of rank 2.

References

Cunningham, E., 1914. The Principle of Relativity. Cambridge University Press.
Einstein, A., 1916. Die grundlage der allgemeinen relativitätstheorie. Annalen der Physics 49, 769–822.
Jackson, J.D., 1999. Classical Electrodynamics, 3rd edition. John Wiley & Sons, New York.
Landau, L.D., Lifshitz, E.M., 1975. The Classical Theory of Fields, 4th revised English edition. Landau and Lifshitz Course of Theoretical Physics, vol. 2. Pergamon Press, Oxford.
Nevels, R., Shin, C.-S., 2001. Lorenz, Lorentz, and the gauge. IEEE Antennas and Propagation Magazine 43 (3), 70–71.
Weinberg, S., 1972. Gravitation and Cosmology. Principles and Applications of the General Theory of Relativity. John Wiley & Sons, New York.

CHAPTER

Relativistic dynamics of particles

5

5.1 Momentum and energy of a particle

In nonrelativistic mechanics a particle with mass m and velocity $\boldsymbol{u}$ has the momentum $\boldsymbol{P} = m\,\boldsymbol{u}$. Newton's law is that $\boldsymbol{F} = d\boldsymbol{P}/dt$. Work is defined as the integral $W = \int \boldsymbol{F}\cdot d\boldsymbol{r}$ on the path of the particle. It can be shown that the work done on the particle by the total force is equal to the change of the kinetic energy $K = \frac{1}{2}\,m\,u^2 = p^2/(2\,m)$. If a constant force is applied to a particle, it will have constant acceleration, which eventually leads to faster than light velocity.

Now let us consider the quantity $\boldsymbol{P} := m\,\gamma(u)\,\boldsymbol{u}$, in which m is a constant called the rest mass of the particle. This quantity is mass times the three spatial components of U^μ. Let us postulate that Newton's second law has the same form, but for this *relativistic* momentum.

Postulate. *In any inertial frame, the equation of motion of a point particle is* $d\boldsymbol{P}/dt = \boldsymbol{F}$, *where* $\boldsymbol{P} = m\,\gamma(\boldsymbol{u})\,\boldsymbol{u}$ *is the momentum of the particle,* $\boldsymbol{u}$ *is its velocity,* m *is a parameter known as the rest mass, and* $\boldsymbol{F}$ *is the force acting on the particle.*

Before proceeding, we notice that if $\boldsymbol{F} = F\,\hat{\boldsymbol{x}}$ is constant and the particle is moving along the x-axis, then the equation of motion reads

$$\frac{\mathrm{d}}{\mathrm{d}t}\gamma(u)\,u = \alpha = \frac{F}{m}, \tag{5.1}$$

which is exactly the one with constant proper acceleration $\alpha = F/m$ studied previously in §3.15. Now let us consider the case when the force is constant, but the particle has an initial transverse velocity. We will see that in this case, there is a deceleration, transverse to the direction of force.

Worked Example 21. A particle of mass m is subject to the constant force $\boldsymbol{F} = F\,\hat{\boldsymbol{x}}$. Initially, that is, at $t = 0$, it is at the origin, moving with velocity $\boldsymbol{u}(0) = (0, v_0, 0)$. Let us find the velocity and the equations of motion.

Setting $c = 1$ for simplicity of calculations, let $\boldsymbol{u} = (u, v, w)$ be the velocity vector and let $\gamma(\boldsymbol{u})$ be the Lorentz factor. The equations of motion read

$$m\,\frac{\mathrm{d}}{\mathrm{d}t}\gamma(\boldsymbol{u})\,u = F, \tag{5.2}$$

A Mathematical Approach to Special Relativity. https://doi.org/10.1016/B978-0-32-399708-9.00014-2

$$m\,\frac{\mathrm{d}}{\mathrm{d}t}\gamma(\boldsymbol{u})\,v=0, \tag{5.3}$$
$$m\,\frac{\mathrm{d}}{\mathrm{d}t}\gamma(\boldsymbol{u})\,w=0. \tag{5.4}$$

From the last one we see that $\gamma(\boldsymbol{u})\,w$ is constant, and since $w(0)=0$, we get $\gamma(\boldsymbol{u}(t))\,w(t)=\gamma(\boldsymbol{u}(0))\,w(0)=0$, from which it follows that

$$w(t)=0. \tag{5.5}$$

Similarly, from (5.3) we find that $\gamma(\boldsymbol{u}(t))\,v(t)=\gamma(\boldsymbol{u}(0))\,v_0$, and since $\boldsymbol{u}(0)=(0,v_0,0)$, we have $\gamma(\boldsymbol{u}(0))=\gamma(v_0)$, and therefore

$$\gamma(\boldsymbol{u}(t))\,v(t)=\gamma(v_0)\,v_0. \tag{5.6}$$

Squaring this equation we get $\frac{v^2}{1-u^2-v^2}=\frac{v_0^2}{1-v_0^2}$, from which we get

$$v^2=(1-u^2)\,v_0^2. \tag{5.7}$$

Now we define $a:=F/m$, and we solve Eq. (5.2) to get $\gamma(\boldsymbol{u})\,u=a\,t$, from which it follows that

$$\frac{u^2}{1-u^2-v^2}=a^2\,t^2. \tag{5.8}$$

From Eq. (5.7) we get

$$1-u^2-v^2=1-u^2-(1-u^2)\,v_0^2=\left(1-u^2\right)\left(1-v_0^2\right). \tag{5.9}$$

Using this in (5.8) we get

$$u(t)=\frac{\sqrt{1-v_0^2}\,a\,t}{\sqrt{1+\left(1-v_0^2\right)\,a^2\,t^2}} \qquad c=1. \tag{5.10}$$

Also from this equation we find

$$1-u^2=\frac{1}{1+\left(1-v_0^2\right)\,a^2\,t^2}. \tag{5.11}$$

Now, using this and (5.7) we find

$$v(t)=\frac{v_0}{\sqrt{1+\left(1-v_0^2\right)\,a^2\,t^2}} \qquad c=1. \tag{5.12}$$

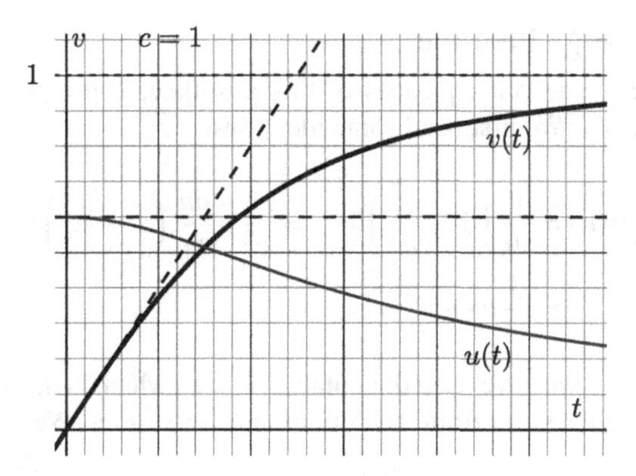

FIGURE 5.1

The x-component (Eq. (5.10), thick) and the y-component (Eq. (5.12), thin) of a particle subject to a constant force $\boldsymbol{F} = F\,\hat{x}$ along the x-axis, with $a = 1$ and with initial transverse velocity $v_0 = 0.6c$ along the y-axis. Dashed lines are the Galilean limits. Note the transverse deceleration. Horizontal dotted line indicates the scale of velocity, $c = 1$.

Finding equations of motion is a matter of elementary integration:

$$y(t) = \int_0^t v(t')\,dt' = v_0 \int_0^t \frac{dt}{\sqrt{1+(1-v_0^2)\,a^2 t^2}} \tag{5.13}$$

$$= \frac{v_0}{a\sqrt{1-v_0^2}} \sinh^{-1}\left(\sqrt{1-v_0^2}\,a\,t\right) \qquad c = 1 \tag{5.14}$$

$$\simeq v_0\,t, \qquad t \sim 0, \tag{5.15}$$

$$x(t) = \int_0^t u(t')\,dt' \tag{5.16}$$

$$= \int_0^t \frac{\sqrt{1-v_0^2}\,a\,s\,ds}{\sqrt{1+(1-v_0^2)\,a^2 s^2}} \tag{5.17}$$

$$= \frac{1}{a\sqrt{1-v_0^2}}\left[\left(1+(1-v_0^2)\,a^2 t^2\right)^{\frac{1}{2}} - 1\right] \tag{5.18}$$

$$\simeq \frac{1}{2}\sqrt{1-v_0^2}\,a\,t^2, \qquad t \sim 0. \tag{5.19}$$

Fig. 5.1 shows $x(t)$ and $y(t)$. Note the behavior for $t \sim 0$, and note the transverse deceleration.

Problem 22. Prove that the trajectory of the previous worked example is a catenary. Draw a graph comparing the relativistic trajectory with the Newtonian one.

5.1.1 Work-energy theorem

We now investigate the consequences of this postulate; consequences that can be tested experimentally. We first need some identities:

$$\frac{\mathrm{d}}{\mathrm{d}t}\gamma(u) = \frac{\mathrm{d}}{\mathrm{d}t}\left(1-\frac{u^2}{c^2}\right)^{-1/2} = \frac{1}{c^2}u\frac{\mathrm{d}u}{\mathrm{d}t}\left(1-\frac{u^2}{c^2}\right)^{-3/2} \tag{5.20}$$

$$= \frac{1}{c^2}\boldsymbol{u}\cdot\boldsymbol{a}\,\gamma^3(u). \tag{5.21}$$

In the last line above we have used the notation $\dot{\boldsymbol{u}} = \boldsymbol{a}$, which is the acceleration, and $u\,\mathrm{d}u = \boldsymbol{u}\cdot\mathrm{d}\boldsymbol{u}$, which follows from the definition of $u^2 = \boldsymbol{u}\cdot\boldsymbol{u}$. We also note that

$$\boldsymbol{u}\cdot\frac{\mathrm{d}}{\mathrm{d}t}\gamma(u)\,\boldsymbol{u} = \boldsymbol{u}\cdot\left(\frac{1}{c^2}\gamma^3(u)\,(\boldsymbol{u}\cdot\boldsymbol{a})\,\boldsymbol{u} + \gamma(u)\,\boldsymbol{a}\right) \tag{5.22}$$

$$= \gamma(u)\,(\boldsymbol{u}\cdot\boldsymbol{a})\left(\gamma^2(u)\frac{u^2}{c^2}+1\right) \tag{5.23}$$

$$= \gamma^3(u)\,\boldsymbol{u}\cdot\boldsymbol{a}, \tag{5.24}$$

where we have used

$$\gamma^2(u)\frac{u^2}{c^2}+1 = \frac{u^2/c^2}{1-u^2/c^2}+1 = \frac{1}{1-u^2/c^2} = \gamma^2(u). \tag{5.25}$$

Defining as usual $\mathrm{d}W = \boldsymbol{F}\cdot\mathrm{d}\boldsymbol{r}$, it is now easy to calculate the work done on the particle:

$$W = \int_i^f \frac{\mathrm{d}\boldsymbol{P}}{\mathrm{d}t}\cdot\boldsymbol{u}\,\mathrm{d}t = m\int_i^f \mathrm{d}t\,\boldsymbol{u}\cdot\frac{\mathrm{d}}{\mathrm{d}t}\left(\gamma(u)\,\boldsymbol{u}\right) \tag{5.26}$$

$$= m\int_i^f \gamma^3(u)\,\boldsymbol{u}\cdot\boldsymbol{a}\,\mathrm{d}t \tag{5.27}$$

$$= m\,c^2\int_i^f \left(\frac{\mathrm{d}}{\mathrm{d}t}\gamma(u)\right)\mathrm{d}t \tag{5.28}$$

$$= m\,c^2\,\Delta\gamma(u), \tag{5.29}$$

where by $\Delta\gamma(u)$ we mean $\gamma(u_f)-\gamma(u_i)$. If the particle is initially at rest, $u_i = 0$, we have $\gamma(u_i) = 1$ and $W = m\,c^2\,(\gamma(u)-1)$, where u is the final velocity of the particle. We therefore define the following.

Definition. The kinetic energy of a particle of mass m and velocity $\boldsymbol{u}$ is

$$K := m\,c^2\,(\gamma(u)-1)\,. \tag{5.30}$$

This "kinetic energy" has the following properties:

1. The kinetic energy of a particle at rest is zero.

2. It has the expansion

$$K = \frac{1}{2} m u^2 + \frac{3}{8} m \frac{u^4}{c^2} + \frac{5}{16} m \frac{u^6}{c^4} + \ldots, \tag{5.31}$$

so that for $u \ll c$ we have the nonrelativistic kinetic energy.

We can summarize the above calculation as follows.

Theorem. *The work done on the particle by the total force is equal to the change in kinetic energy of the particle.*

Definition. $m c^2$ is called the rest energy of the particle.

Since $\Delta\gamma(u) = \Delta(\gamma(u) - 1)$, we can add $m c^2$ to the kinetic energy, and we get *the* energy

$$E := \gamma(u) m c^2 = m c^2 + K. \tag{5.32}$$

Worked Example 23. An electron is ejected from an accelerator with kinetic energy 200 keV. We know that the rest mass of the electron is $m_e = 511\,\text{keV}/c^2$. Using $K = (\gamma(\beta) - 1) m_e c^2$, we get $\gamma(\beta) = 1 + 200/511 = 1.391$, from which we get $\beta = 0.67$.

5.1.2 The energy-momentum 4-vector

Now let us recall the definition of the 4-velocity: $U^\mu = \gamma(u)(1, \boldsymbol{u})$. If we multiply this 4-vector by the scalar m, we get a 4-vector called the energy-momentum 4-vector:

$$P^\mu = m U^\mu = \Big(m\gamma(u),\ m\gamma(u)\boldsymbol{u}\Big) = \left(\frac{E}{c^2},\ \boldsymbol{P}\right), \tag{5.33}$$

$$P_\mu = \eta_{\mu\nu} P^\nu = (-\varsigma E,\ \varsigma \boldsymbol{P}). \tag{5.34}$$

From the definition of P^μ it is obvious that

$$\boldsymbol{u} = \frac{m\gamma(u)\boldsymbol{u}}{m\gamma(u)} = \frac{\boldsymbol{P}}{E/c^2} = \frac{c^2 \boldsymbol{P}}{E}, \tag{5.35}$$

which we write in the following more symmetric form:

$$\boldsymbol{\beta} := \frac{\boldsymbol{u}}{c} = \frac{c\boldsymbol{P}}{E}. \tag{5.36}$$

This last form shows that if a particle moves with the speed of light, then E should be equal to $c P$.

Since the square of the 4-velocity is $U_\mu U^\mu = \varsigma c^2$, the square of the energy-momentum 4-vector is $\varsigma m^2 c^2$; therefore, whatever the signature of the metric is $(\varsigma = \pm 1)$, we have

$$\frac{E^2}{c^2} - P^2 = m^2 c^2 \qquad \text{or} \qquad E^2 - c^2 P^2 = m^2 c^4. \tag{5.37}$$

This relation between E, P, and m is called *the mass shell* relation. Note that in the 4D space of (E, P_x, P_y, P_z) it is the defining relation of a hyperboloid of revolution, whose asymptote is the cone $E^2/c^2 - P^2 = 0$. This asymptote would result if the rest mass of the particle is zero, in which case $E = c\,P$, and the particle always moves with the speed of light. Therefore particles of light have rest mass zero.

Note that the mass shell relation, being quadratic in both E and P, says nothing about the sign of E, but from the defining relation $E = \gamma(u)\,m\,c^2$, the energy of a particle is positive. The mass shell relation, $E^2/c^2 - P^2 = m^2\,c^2$, defines a two-sheet hyperboloid of revolution—a positive sheet for which E is positive and a negative sheet for which E is negative. At this moment we cannot interpret the negative part—it is related to the so-called *antiparticles*.

5.1.3 Relativistic cyclotron frequency

Consider a particle of charge q moving in a constant magnetic field $\boldsymbol{B}$. The force exerted on the particle is $q\,\boldsymbol{u} \times \boldsymbol{B}$, so the equation of motion is

$$\frac{\mathrm{d}\boldsymbol{P}}{\mathrm{d}t} = q\,\boldsymbol{u} \times \boldsymbol{B}. \tag{5.38}$$

For $\boldsymbol{P} = m\,\gamma(u)\,\boldsymbol{u}$ we get

$$m\,\frac{\mathrm{d}}{\mathrm{d}t}\,(\gamma(u)\boldsymbol{u}) = q\,\boldsymbol{u} \times \boldsymbol{B}. \tag{5.39}$$

Taking the inner product of this equation with $\boldsymbol{u}$, noting $\boldsymbol{u} \cdot \boldsymbol{u} \times \boldsymbol{B} = 0$, and using (5.24), we get

$$\boldsymbol{u} \cdot \frac{\mathrm{d}}{\mathrm{d}t}\,((\gamma(u)\,\boldsymbol{u}) = \gamma^3(u)\,\boldsymbol{u} \cdot \boldsymbol{a} = 0. \tag{5.40}$$

Note also that $\boldsymbol{u} \cdot \boldsymbol{a} = \frac{1}{2}\frac{\mathrm{d}}{\mathrm{d}t}\,(\boldsymbol{u} \cdot \boldsymbol{u})$, so the above equation means $\mathrm{d}u^2/\mathrm{d}t = 0$. Therefore u and $\gamma(u)$ are constant. Now, writing Eq. (5.39) as

$$m\,\gamma(u)\,\frac{\mathrm{d}\boldsymbol{u}}{\mathrm{d}t} = q\,\boldsymbol{u} \times \boldsymbol{B} \tag{5.41}$$

and introducing

$$\boldsymbol{n} = \frac{\boldsymbol{u}}{u}, \tag{5.42}$$

$$\boldsymbol{\omega} = \frac{q\,\boldsymbol{B}}{\gamma(u)\,m}, \qquad \text{relativistic cyclotron frequency,} \tag{5.43}$$

we get

$$\frac{\mathrm{d}\boldsymbol{n}}{\mathrm{d}t} = -\boldsymbol{\omega} \times \boldsymbol{n}. \tag{5.44}$$

This equation says that the direction of $\boldsymbol{u}$ is rotating (or precessing) with angular velocity $\boldsymbol{\omega}$. The quantity $\omega_0 = q\,B/m$ is the nonrelativistic cyclotron frequency; $\sqrt{1-\beta^2}\,\omega_0$ is the relativistic cyclotron frequency. When designing cyclotrons, engineers have to consider the $1/\gamma(u)$ term; otherwise their device will not function properly. This is an experimental verification of the $\boldsymbol{P} = m\,\gamma(u)\,\boldsymbol{u}$.

5.2 Lagrangian of a particle

The material in this section requires knowledge of Lagrangian and Hamiltonian mechanics, which are usually covered in textbooks on classical mechanics such as Goldstein (1980) or Symon (1971).

Earlier we postulated that Newton's second law is valid provided that we define the momentum as follows: $\boldsymbol{P} = m\,\gamma(u)\,\boldsymbol{u}$. Consider now the following Lagrangian:

$$L = -m\,c^2\sqrt{1-\frac{\dot{x}^2+\dot{y}^2+\dot{z}^2}{c^2}} = -\frac{m\,c^2}{\gamma(u)}. \tag{5.45}$$

The action $S = \int L\,dt$ now has a simple interpretation:

$$S = -m\,c^2\int_i^f\sqrt{1-\frac{u^2}{c^2}}\,\mathrm{d}t = -m\,c^2\int_i^f \mathrm{d}\tau = -m\,c^2\left(\tau_f-\tau_i\right). \tag{5.46}$$

In words, it is a constant multipole of the proper time lapse!

Let us now calculate the momenta. We have

$$P_x := \frac{\partial L}{\partial \dot{x}} = \frac{m\,\dot{x}}{\sqrt{1-u^2/c^2}} = m\,\gamma(u)\,u_x \tag{5.47}$$

and similar formulas for P_y and P_z, which we can write in the following compact vector form:

$$\boldsymbol{P} = m\,\gamma(u)\,\boldsymbol{u}. \tag{5.48}$$

This is exactly the spatial components of $P^\mu = m\,U^\mu$. Therefore postulating $\boldsymbol{F} = \dot{\boldsymbol{P}}$ for $\boldsymbol{P} = m\,\gamma(u)\,\boldsymbol{u}$ is equivalent to postulating $L = -m\,c^2\sqrt{1-\beta^2}$.

Now let us calculate the *energy*, by which we mean the following quantity:

$$\begin{aligned} E &:= \dot{x}\,P_x + \dot{y}\,P_y + \dot{z}\,P_z - L \\ &= m\,\gamma(u)\left(\dot{x}^2+\dot{y}^2+\dot{z}^2\right) - \frac{m\,c^2}{\gamma(u)} \\ &= m\,\gamma(u)\,u^2 + \frac{m\,c^2}{\gamma(u)} = \frac{m\,c^2}{\gamma(u)}\left[1+\gamma^2(u)\frac{u^2}{c^2}\right] \\ &= m\,\gamma(u)\,c^2. \end{aligned} \tag{5.49}$$

If we add a potential $V(x, y, z)$ to the Lagrangian, we get the following equations of motion:

$$\frac{\mathrm{d}\boldsymbol{P}}{\mathrm{d}t} = -\nabla V. \tag{5.50}$$

5.3 Interaction of a particle with an electromagnetic field

Let $A^\mu = (c^{-2}\phi, \boldsymbol{A})$ be a 4-vector field. By this we mean that all its components depend on t and $\boldsymbol{r}$, and they transform as (t, x, y, z). Now let us investigate the effect of adding

$$S_\mathrm{I} = -\varsigma \int q\, A_\mu\, dx^\mu \tag{5.51}$$

to the action (see (4.83), (4.85), and (4.86)). In this expression, q is a constant—the charge—and $A_\mu = \eta_{\mu\nu} A^\nu = (\varsigma\,\phi, -\varsigma\,\boldsymbol{A})$. Now we have

$$\begin{aligned} S &= -m\,c^2 \int d\tau - \varsigma\, q \int A_\mu\, dx^\mu \\ &= \int \left(-m\,c^2 \sqrt{1 - \frac{u^2}{c^2}} - \varsigma\, q\, A_\mu \frac{\mathrm{d}x^\mu}{\mathrm{d}t} \right) dt \\ &= \int \left(-m\,c^2 \sqrt{1 - \frac{u^2}{c^2}} - q\,\phi + q\,\boldsymbol{A}\cdot\boldsymbol{u} \right) dt. \end{aligned} \tag{5.52}$$

We see that the Lagrangian is now

$$L = -m\,c^2 \sqrt{1 - \frac{u^2}{c^2}} - q\,\phi + q\,\boldsymbol{A}\cdot\boldsymbol{u}. \tag{5.53}$$

From this it follows that the *canonical* momenta are

$$\boldsymbol{p} := \frac{\partial L}{\partial \dot{\boldsymbol{r}}} = m\,\gamma(\boldsymbol{u})\,\boldsymbol{u} + q\,\boldsymbol{A} = \boldsymbol{\pi} + q\,\boldsymbol{A}, \tag{5.54}$$

where the *kinematic* momenta are, by definition,

$$\boldsymbol{\pi} = m\,\gamma(\boldsymbol{u})\,\boldsymbol{u}. \tag{5.55}$$

Now let us investigate the equations of motion. We have

$$\begin{aligned} 0 &= \frac{\mathrm{d}}{\mathrm{d}t}\frac{\partial L}{\partial \dot{x}} - \frac{\partial L}{\partial x} \\ &= \frac{\mathrm{d}}{\mathrm{d}t}(m\,\gamma\,u_x + q\,A_x) + \frac{\partial}{\partial x}(q\,\phi - q\,\boldsymbol{u}\cdot\boldsymbol{A}), \end{aligned} \tag{5.56}$$

$$\frac{d\pi_x}{dt} = q\left\{-\frac{dA_x}{dt} - \frac{\partial\phi}{\partial x} + \frac{\partial}{\partial x}\boldsymbol{u}\cdot\boldsymbol{A}\right\}$$
$$= q\left\{-\frac{\partial A_x}{\partial t} - \frac{\partial\phi}{\partial x} + u_y\left(\frac{\partial A_y}{\partial x} - \frac{\partial A_x}{\partial y}\right) + u_z\left(\frac{\partial A_z}{\partial x} - \frac{\partial A_x}{\partial z}\right)\right\}$$
$$= q\,E_x + q\,(\boldsymbol{u}\times\boldsymbol{B})_x, \tag{5.57}$$

where

$$\boldsymbol{B} := \nabla\times\boldsymbol{A}, \tag{5.58}$$
$$\boldsymbol{E} := -\nabla\phi - \frac{\partial\boldsymbol{A}}{\partial t}. \tag{5.59}$$

Thus, the equations of motion read

$$\frac{d\boldsymbol{\pi}}{dt} = q\,(\boldsymbol{E} + \boldsymbol{u}\times\boldsymbol{B}). \tag{5.60}$$

Worked Example 24. Let $\boldsymbol{A} = 0$ and $\phi = -\mathcal{E}\,x$ for a constant $\mathcal{E}$. We then get $\boldsymbol{B} = \nabla\times\boldsymbol{A} = 0$ and $\boldsymbol{E} = -\nabla\phi = \mathcal{E}\,\hat{\boldsymbol{x}}$. Since $\boldsymbol{A} = 0$, we get $\boldsymbol{\pi} = \boldsymbol{p}$, and therefore equations of motion read $\frac{d\boldsymbol{p}}{dt} = q\,\mathcal{E}\,\hat{\boldsymbol{x}}$. Remembering $\boldsymbol{p} = m\,\gamma(u)\,\boldsymbol{u}$ and assuming motion along the x-axis, we get

$$\frac{d}{dt}(\gamma(u)\,u) = \alpha := \frac{q\,\mathcal{E}}{m}. \tag{5.61}$$

As we saw in §3.15, $\frac{d}{dt}(\gamma(u)\,u) = a\,\gamma^3(u) = \alpha$. Here $a = \frac{du}{dt}$ is the *acceleration* and α is the *proper acceleration*. We therefore see that a charged particle in a constant electric field experiences a constant force, and its motion is a motion with constant proper acceleration (a hyperbolic motion).

Consider for example an accelerator with length 10.0 m with constant electric field 100 MV m^{-1}. If an electron with $m_e = 511\,\text{keV}\,c^{-2}$ and $q = -\text{e}$ is being accelerated in this field, it acquires a kinetic energy of $1000\,\text{MeV} = 1\,\text{e}\times 100\,\text{MV}\,\text{m}^{-1}\times 10\,\text{m}$. The proper acceleration of the electron would be $\alpha = \text{e}\,\text{E}/m_e = 19.6\,c^2\,\text{m}^{-1} = 1.76\times 10^{18}\,\text{m}\,\text{s}^{-2} \simeq 1.8\times 10^{17}\,\text{g}$. Now we can find the time of flight (from when the electron starts its motion in the accelerator until when it exits the accelerator) exactly as we did in Worked Example 18 on page 71. It is easy to show that—if we use $c = 3.00\times 10^8\,\text{m}\,\text{s}^{-1}$—we get $t = 33.5$ ns, while $\tau = 1$ ns. The time of flight of a photon for the same distance of 10.0 m is 33.3 ns.

5.4 Angular momentum 6-vector of a particle

For a point particle with 4-momentum p we define $L = x \times p$, where x is the position 4-vector and p is the momentum 4-vector. In terms of components

$$L^{\mu\nu} := x^{\mu} p^{\nu} - x^{\nu} p^{\mu}. \tag{5.62}$$

For a free particle ($\mathrm{d}p^{\mu}/\mathrm{d}\tau = 0$), all the components of this 6-vector are constant:

$$\begin{aligned}\frac{\mathrm{d}L}{\mathrm{d}t} &= \frac{\mathrm{d}\tau}{\mathrm{d}t} \cdot \frac{\mathrm{d}L}{\mathrm{d}\tau} = \frac{1}{\gamma} \frac{\mathrm{d}}{\mathrm{d}\tau} \left(x^{\mu} p^{\nu} - x^{\nu} p^{\mu}\right) \\ &= \frac{1}{\gamma} \left(U^{\mu} p^{\nu} + x^{\mu} \frac{\mathrm{d}p^{\nu}}{\mathrm{d}\tau} - U^{\nu} p^{\mu} - x^{\nu} \frac{\mathrm{d}p^{\mu}}{\mathrm{d}\tau}\right) \\ &= \frac{1}{\gamma} \left(U^{\mu} m U^{\nu} + 0 - U^{\nu} m U^{\mu} - 0\right) = 0.\end{aligned} \tag{5.63}$$

The time-space components of this tensor are $L^{0i} = x^{i} p^{0} - x^{0} p^{i}$, which are the Cartesian components of $c^{-2} E \boldsymbol{r} - t \boldsymbol{P}$. To see what this vector is, we should remember the equation of motion of a free particle: $\boldsymbol{r}(t) = \boldsymbol{r}(0) + t \boldsymbol{u}$. Since $\boldsymbol{u} = c^2 \boldsymbol{P}/E$, we have

$$(L^{0i}) = c^{-2} E \boldsymbol{r}(t) - t \boldsymbol{P} = c^{-2} E \boldsymbol{r}(0). \tag{5.64}$$

Therefore, L^{0i} is $c^{-2} E$ times the x^{i}-component of the initial position vector. The conservation of these quantities follows from the conservation of the energy E and the initial position. Note that the value of $\boldsymbol{r}(0)$ does not change by the passing of time!

Now let us look at the space-space components of L:

$$L^{12} = x P_y - y P_x, \qquad L^{23} = y P_z - z P_y, \qquad L^{31} = z P_x - x P_z. \tag{5.65}$$

These are simply the components of the orbital angular momentum $\boldsymbol{L} = \boldsymbol{r} \times \boldsymbol{P}$, so that we have

$$L = \left[L^{\mu\nu}\right] = \begin{bmatrix} 0 & c^{-2} E\, x(0) & c^{-2} E\, y(0) & c^{-2} E\, z(0) \\ -c^{-2} E\, x(0) & 0 & L_z & -L_y \\ -c^{-2} E\, y(0) & -L_z & 0 & L_x \\ -c^{-2} E\, z(0) & L_y & -L_x & 0 \end{bmatrix}. \tag{5.66}$$

The tensor L is called the orbital angular momentum tensor. It is important to note that the nonvanishing components of this tensor depend on the choice of the origins of the t-, x-, y-, and z-axes.

From elementary mechanics we know that for a classical, nonrelativistic point particle, $\boldsymbol{L}$ is the total angular momentum. It is also well known that for a system composed of more than one classical point particle, *e.g.*, a solid body with mass M, the total angular momentum vector is the sum of two parts—the angular momentum in the center of mass system, $\boldsymbol{S}$, and the orbital angular momentum of a point body with the same mass M comoving with the center of mass, $\boldsymbol{L}$. That is, the total angular momentum of such a system is $\boldsymbol{J} = \boldsymbol{S} + \boldsymbol{L}$ (see, for example, Goldstein, 1980, pp. 7–9).

5.5 Inertia

Inertia or mass is a property of matter which deserves a detailed study. It first appears in Newton's second law:

$$\boldsymbol{F} = m\,\boldsymbol{a} = m\,\frac{\mathrm{d}\boldsymbol{u}}{\mathrm{d}t} = \frac{\mathrm{d}}{\mathrm{d}t}\,(m\,\boldsymbol{u})\,. \tag{5.67}$$

This equation says that when we want to change the momentum of a particle $(m\,\boldsymbol{u})$, somehow, it resists. If there is no force, the momentum is constant; if there is a force, the more the inertia (m) of the particle, the more it resists. Let us see why we call this resistance. Suppose a force $\boldsymbol{F}$ is applied on the particle with mass m. In time interval Δt the change in the velocity of the particle would be

$$\Delta\boldsymbol{u} = \frac{1}{m}\boldsymbol{F}\,\Delta t. \tag{5.68}$$

The larger the mass (inertia), the smaller would be the change in the velocity.

Inertia is an observable. Measurement of inertia depends on the size of it. To measure the mass of an electron, or an apple, or a star, we use completely different methods, and we must be aware of this when developing the theory. At the moment we are not going into the details of experimental methods to measure the mass in such a huge range of values. Here, we would like to see what could be deduced from Newton's laws about measurement of inertia. Here is the logic of assigning inertial mass to bodies: Choose a rigid body (a platinum cylinder, for example) to be the unit of mass. Then put that on a frictionless horizontal table. Attach to that a very light spring, to which the body whose mass we want to measure is attached. Push the bodies to each other and release them. The two bodies will exert forces on each other, which according to Newton's third law are the same. Therefore according to Newton's second law, their accelerations are proportional to the inverse of the masses. A measurement of the accelerations would yield the mass of the test body. In summary:

1. We accept Newton's laws.
2. We prepare an interaction between the particle whose mass we want to measure—the test particle—and a particle whose mass we know, such that after a while, the two particles are not interacting any more.
3. We then calculate their momenta before and after this interaction (which is called a collision), and from that information we deduce the unknown mass.

The special theory of relativity changes the definition of momentum as follows:

$$\boldsymbol{p} = m\,\gamma(u)\,\boldsymbol{u}. \tag{5.69}$$

Here m is called *the rest mass* of the particle, which is a Lorentz invariant. The rest mass is Lorentz invariant because, by its definition, it must be measured when the particle is at rest, and therefore all observers agree on its value—any observer who wants to measure the rest mass of the electron must do that for an electron which is at rest relative to him/her. If different inertial observers find different values for the rest mass of the electron, then we have a method to distinguish between them and find the absolute rest frame: the frame in which the electron's rest mass is the least (or greatest) is the distinguished frame.

5.5.1 Parallel and transverse inertia

Let us see what could be said about the inertia of a moving particle, based on the validity of special relativity, that is, based on the validity of the Lorentz transformations. Consider a particle with rest mass m, moving with velocity $\boldsymbol{u}$. The force exerted on the particle is, by definition,

$$\boldsymbol{F} = \frac{\mathrm{d}\boldsymbol{p}}{\mathrm{d}t} = \frac{\mathrm{d}}{\mathrm{d}t}\left(m\,\gamma(u)\,\boldsymbol{u}\right). \tag{5.70}$$

Any vector, including the force exerted on this particle and the acceleration, could be decomposed into a sum of a vector parallel to $\boldsymbol{u}$ and a vector perpendicular to $\boldsymbol{u}$:

$$\boldsymbol{F} = \boldsymbol{F}_{||} + \boldsymbol{F}_{\perp}, \tag{5.71}$$

$$\boldsymbol{a} = \boldsymbol{a}_{||} + \boldsymbol{a}_{\perp}. \tag{5.72}$$

We already know that (5.20)

$$\frac{\mathrm{d}\gamma}{\mathrm{d}t} = \gamma^3(u)\,c^{-2}\,\boldsymbol{a}\cdot\boldsymbol{u}, \tag{5.73}$$

from which it follows that

$$\boldsymbol{u}\cdot\boldsymbol{F} = m\,\gamma(u)\,\boldsymbol{a}\cdot\boldsymbol{u} + m\,\gamma^3(u)\,c^{-2}\,u^2\,\boldsymbol{a}\cdot\boldsymbol{u} \tag{5.74}$$

$$= m\,\gamma(u)\,\boldsymbol{a}\cdot\boldsymbol{u}\left(1 + \frac{u^2/c^2}{1-u^2/c^2}\right) \tag{5.75}$$

$$= m\,\gamma^3(u)\,\boldsymbol{a}\cdot\boldsymbol{u}. \tag{5.76}$$

Therefore

$$\boldsymbol{F} = m\,\gamma(u)\,\boldsymbol{a} + c^{-2}\,\boldsymbol{u}\,(\boldsymbol{F}\cdot\boldsymbol{u}), \tag{5.77}$$

$$\boldsymbol{F}\cdot\frac{\boldsymbol{u}}{u} = m\,\gamma(u)\,\frac{\boldsymbol{a}\cdot\boldsymbol{u}}{u} + \frac{1}{c^2}\,\boldsymbol{u}\cdot\boldsymbol{u}\,\frac{\boldsymbol{F}\cdot\boldsymbol{u}}{u}, \tag{5.78}$$

$$F_{\|} = m\,\gamma(u)\,a_{\|} + \frac{u^2}{c^2}\,F_{\|}, \tag{5.79}$$

$$\boldsymbol{F}_{\|} = m\,\gamma^3(u)\,\boldsymbol{a}_{\|}. \tag{5.80}$$

Now let us find $\boldsymbol{F}_{\perp}$:

$$\boldsymbol{F}_{\perp} = \boldsymbol{F} - \boldsymbol{F}_{\|} \tag{5.81}$$

$$= m\,\frac{\mathrm{d}}{\mathrm{d}t}\,(\gamma(u)\,\boldsymbol{u}) - \boldsymbol{F}_{\|} \tag{5.82}$$

$$= m\,\gamma(u)\,\frac{\mathrm{d}\boldsymbol{u}}{\mathrm{d}t} + m\,\frac{\mathrm{d}\gamma}{\mathrm{d}t}\,\boldsymbol{u} - \boldsymbol{F}_{\|} \tag{5.83}$$

$$= m\,\gamma(u)\,\boldsymbol{a}_{\perp} + \left\{ m\,\gamma(u)\,\boldsymbol{a}_{\|} + m\,\frac{\mathrm{d}\gamma}{\mathrm{d}t}\boldsymbol{u} - \boldsymbol{F}_{\|} \right\}. \tag{5.84}$$

Since the terms in { } are all parallel to $\boldsymbol{u}$, we have

$$\boldsymbol{F}_{\perp} = m\,\gamma(u)\,\boldsymbol{a}_{\perp}. \tag{5.85}$$

Therefore, we can make the following statement.

Statement. Let $\boldsymbol{F}$ be the total force applied on a body with rest mass m, which is moving with velocity $\boldsymbol{u}$. If the acceleration is parallel to the velocity $\boldsymbol{u}$, the inertia of the particle is $m\,\gamma^3(u)$, and if the acceleration is perpendicular to the velocity $\boldsymbol{u}$, the inertia of the particle is $m\,\gamma(u)$:

$$m\,\gamma^3(u) = \quad \text{parallel inertia}, \tag{5.86}$$

$$m\,\gamma(u) = \quad \text{transverse inertia}. \tag{5.87}$$

5.6 Inertia and the energy content

Consider a system with internal structure. At the atomic scale, this means something like a nucleus or an atom. At the macroscopic scale, any system has internal structure. It is one of the most important and famous discoveries of Einstein that the inertia of such a system depends on its energy content—the famous $E = m\,c^2$ equation, which is usually misinterpreted.

Here is a beautiful proof of this statement by Einstein himself (Einstein, 1946): Consider the following process, which increases the internal energy of the system S with inertia M which is at rest in frame K. Two photons, that is, two plane electromagnetic waves, each with energy $E/2$ and momentum $E/(2\,c)$, are absorbed by the body simultaneously (see Fig. 5.2). The two photons were moving in opposite directions, x and $-x$ say, such that the total momentum of the system before the absorption of photons is zero. Because of the law of conservation of momentum, the momentum of the system after the collision must also be zero. Therefore in this process the body whose inertia before the absorption was M must remain at rest.

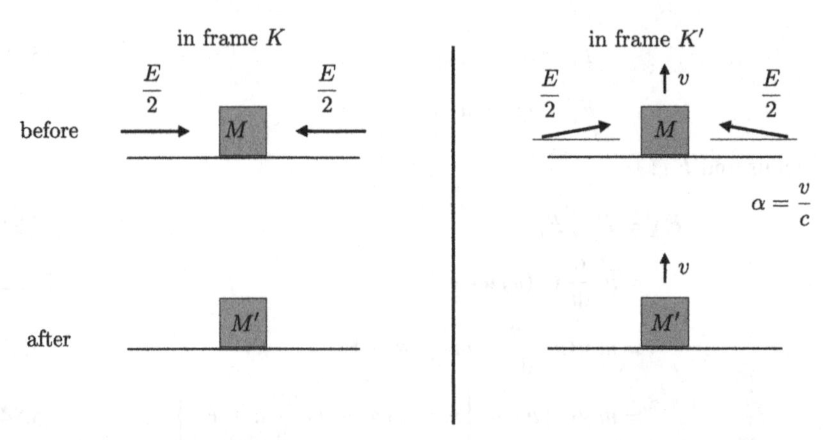

FIGURE 5.2 Mass-energy equivalence.

Einstein's thought experiment demonstrating $\Delta E = \Delta M\, c^2$. Frame K' is moving downward with respect to frame K.

Now consider an observer who is moving in the $-y$-direction with a very small velocity v—we call this the frame K'. This observer sees that the body with inertia M was moving with velocity v in the $+y$ direction, and two photons were approaching it almost in the x- and $-x$-directions. Almost, not completely, since, because of the aberration, the photons have a small angle $\alpha = v/c$ with respect to the x-axis. This means that the two photons have, before the absorption, a total momentum in the $+y$ direction equal to

$$2\left(\frac{E}{2\,c}\right)\sin\frac{v}{c} \simeq \frac{E\,v}{c^2}. \tag{5.88}$$

Now the total momentum of this system (in the $+y$-direction), before the absorption of the photons, is

$$M\,v + \frac{E}{c^2}\,v, \tag{5.89}$$

while its total momentum after the absorption of the photons must be equal to the inertia of the system (after the absorption) times its velocity. The velocity of the system cannot change, because it did not change in K (the rest frame of the system), and therefore it must not change in K' either. So the velocity after the absorption must be the same as its velocity before the absorption. That is, it must be $M'\,v$, so we must have

$$M'\,v = \left(M + \frac{E}{c^2}\right)\,v. \tag{5.90}$$

Canceling v from both sides, we get

$$\Delta M = M' - M = \frac{E}{c^2}. \tag{5.91}$$

This proof depends on the following assumptions:

1. Conservation of linear momentum.
2. That a plane electromagnetic wave with energy E has linear momentum E/c.
3. That for a small velocity, the angle of aberration is v/c.
4. That for a small velocity, the momentum of a system is $M\,v$, where M is its inertia and v is its velocity.
5. That all these laws are valid in all inertial frames—the Galilean principle of relativity.

Note that there are very solid experimental confirmations for all these assertions.

Binding energy. This equivalence of mass and energy is very important in atomic and subatomic systems. When an atom absorbs radiation (that is, energy in the form of electromagnetic waves), its mass increases by E/c^2. Vice versa, if an atom emits radiation, its mass is decreased by E/c^2.

Consider for example a hydrogen atom, which is a bound state of a proton of mass m_p and charge $+e$ and an electron of mass m_e and charge $-e$. The total energy of the system is

$$E_\mathrm{H} = m_\mathrm{H}\,c^2 = m_\mathrm{p}\,c^2 + m_\mathrm{e}\,c^2 + K_\mathrm{p} + K_\mathrm{e} + V, \tag{5.92}$$

where K_p and K_e are the kinetic energy of the proton and the electron and V is the potential energy of the interaction. To a very good approximation,

$$K_\mathrm{p} + K_\mathrm{e} + V = -\frac{1}{2}\,m_\mathrm{e}\,c^2\,\alpha^2\,\frac{1}{n^2}, \qquad \alpha \sim \frac{1}{137}, \tag{5.93}$$

where $n \geq 1$ is an integer—the so-called principal quantum number of the hydrogen atom. Since this is negative, we see that the rest mass of the hydrogen atom is smaller than the sum of the rest masses of the proton and the electron.

The quantity

$$B = (m_\mathrm{H} - m_\mathrm{p} - m_\mathrm{e})c^2 \tag{5.94}$$

is called the binding energy of the hydrogen atom. Using this definition we can say that the rest mass of the hydrogen atom is equal to the sum of the rest masses of its constituents *minus* the mass equivalent of the binding energy. The same is true for any bound state of a set of particles.

5.7 Kinematics of decays and collisions

Decays and collisions of atomic and subatomic particles are among the most important tools we have to investigate atomic and subatomic physics. Special relativity offers valuable tools to this field and has been very successful in this respect. In this section we study the implications of special relativity for the decay and collision of particles.

A free point particle has very few physical properties—its mass m, its spin s, its momentum $\boldsymbol{P}$, its energy E, and its electric charge q.[1] Energy, momentum, and mass are not all independent quantities. The 4-velocity of a point particle is, by definition,

$$U = (U^\mu) = \frac{\mathrm{d}x^\mu}{\mathrm{d}\tau} = (\gamma(u), \gamma(u)\,\boldsymbol{u})\,. \tag{5.95}$$

The invariant Minkowski product $U \cdot U$ is the constant number $-\varsigma\, c^2$:

$$U \cdot U = \eta_{\mu\nu} \frac{\mathrm{d}x^\mu}{\mathrm{d}\tau} \frac{\mathrm{d}x^\nu}{\mathrm{d}\tau} = \frac{(\mathrm{d}s)^2}{(\mathrm{d}\tau)^2} = -\varsigma\, c^2. \tag{5.96}$$

The 4-momentum of the particle is denoted by $p = (p^\mu)$ and is, by definition, $m\,U$, which means that

$$p = (p^\mu) = \left(c^{-2}\, E, \boldsymbol{P}\right), \qquad E = m\, c^2\, \gamma(u), \qquad \boldsymbol{P} = m\, \gamma(u)\, \boldsymbol{u}. \tag{5.97}$$

We therefore have $U \cdot U = -\varsigma\, c^2$ and $p \cdot p = -\varsigma\, m^2\, c^2$, which could be written

$$c^{-2}\, E^2 - P^2 = m^2\, c^2. \tag{5.98}$$

The $c = 1$ convention. The formulas we are going to present look simpler if we adopt the convention of setting $c = 1$, that is, if we use a system of units in which the value of c is 1. We can always restore the powers of c by dimensional analysis of the formulas. In what follows, we use this convention in calculations, but we restore the powers of c in the resulting formulas. Although this seems to be an inconsistent notation, I think that what is most important for anyone working in this field is to have the ability to switch back and forth between these two conventions.
The mass shell relation. We have

$$E = \gamma(u)\, m, \qquad \boldsymbol{P} = m\, \gamma(u)\, \boldsymbol{u}, \qquad E^2 - P^2 = m^2. \tag{5.99}$$

Since $P^2 = P_x^2 + P_y^2 + P_z^2$, the equation $E^2 - P^2 = m^2$ defines a 3D hyperboloid of revolution in $\mathbb{R}^4$. Given the momentum, P (a nonnegative quantity), we can find the energy, E. The equation $E^2 - P^2 = m^2$ has two roots, $E = \pm\sqrt{(m^2 + P^2)}$. Classically, only the positive root is acceptable, but quantum field theoretically both roots are acceptable. For the moment let us concentrate on the positive root $E = \sqrt{(m^2 + P^2)}$.

---

[1] Particles such as quarks and leptons do possess other kinds of charges, such as "color" or "lepton number."

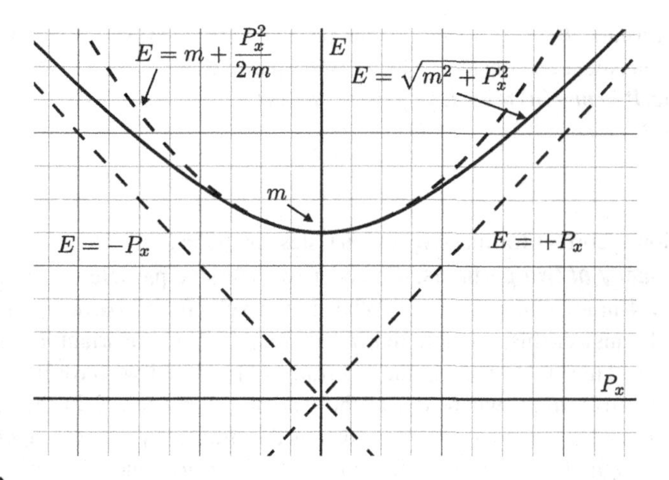

FIGURE 5.3

Mass shell diagram for a particle moving along the x-axis ($c = 1$). Note that the relativistic expression (the hyperbola, solid line) lies below the nonrelativistic expression (the parabola, dashed line). In words, for a high-energy particle with given momentum P, the energy E is *less* than the value predicted by the nonrelativistic expression; and conversely, for a given energy E, the true relativistic momentum is *more* than the value that the nonrelativistic expression predicts. Note also that as $m \to 0$, the hyperbola approaches its asymptotes, representing the mass shell of a photon.

If $P < m$ (which means $P < m\,c$), we can expand E in powers of P^2, getting

$$E = m\,c^2 + \frac{P^2}{2m} - \frac{P^4}{8\,m^3\,c^2} + \frac{P^6}{16\,m^5\,c^4} + \ldots. \tag{5.100}$$

The first term is the rest energy of the particle. The second term, $P^2/(2m)$, is the nonrelativistic kinetic energy. The functions $\sqrt{(m^2 + P^2)}$ and $m + P^2/(2m)$ are drawn in Fig. 5.3.

For $m \to 0$, the hyperbola becomes a cone, with vertex at the origin ($E = 0$, $\boldsymbol{P} = 0$). Restricted to the (E, P_x) plane, this is the union of two straight lines $E = \pm P_x$ (that is, $E = \pm c\,P_x$). The interpretation is obvious: Since these lines are the asymptotes, for any particle, if the momentum is high enough ($P \gg m\,c$), the relation between energy and momentum is approximately $E = c\,P$.

Plane electromagnetic waves carry energy and momentum. In vacuum, if a plane wave is absorbed completely by an object, the object's energy and momentum will increase. If the increase of energy is E, the increase of the momentum of the object would be $c^{-1}\,E$ (see, for example, Jackson (1999, pp. 258–262) or Landau and Lifshitz (1975, pp. 110–112)). Therefore if the plane electromagnetic wave is to be considered as a stream of particles (photons), the relation between the energy and the momentum of these particles must be $E = c\,P$. This means that we can use the same mass shell formula $E^2 - P^2 = m^2$ for the photons, provided we set $m = 0$. In other words, photons must be considered as *massless* particles.

Dividing $\boldsymbol{P} = m\,\gamma(u)\,\boldsymbol{u}$ by $E = \gamma(u)\,m$, we get

$$\frac{\boldsymbol{u}}{c} = \frac{c\,\boldsymbol{P}}{E}. \tag{5.101}$$

This equation is correct for massive and massless particles.

Relative velocity of two particles. Consider two massive particles with masses m_1 and m_2 and 4-momenta p_1 and p_2. What is the velocity of particle 2 relative to particle 1? To answer this, we note that $p_1 \cdot p_2$ is a Lorentz invariant, *i.e.*, it has the same value in all inertial frames. In the rest frame of particle 1 we have $p_1 = m_1(1, 0)$ and $p_2 = m_2\,\gamma_{21}(1, \boldsymbol{u}_{21})$, where $\boldsymbol{u}_{21}$ is the velocity of particle 2 in the rest frame of particle 1, and $\gamma_{21} = \gamma(u_{12})$. Therefore in this frame $p_1 \cdot p_2 = -\varsigma\, m_1\, m_2\, \gamma_{21}$. Now, in the original lab frame we have $p_1 = m_1\,\gamma_1(1, \boldsymbol{u}_1)$ and $p_2 = m_2\,\gamma_2(1, \boldsymbol{u}_2)$, from which it follows that $p_1 \cdot p_2 = -\varsigma\, m_1\, m_2\, \gamma_1\, \gamma_2(1 - \boldsymbol{u}_1 \cdot \boldsymbol{u}_2)$. Equating this with $p_1 \cdot p_2 = -\varsigma\, m_1\, m_2\, \gamma_{21}$ we get

$$\gamma(u_{21}) = \gamma_1\,\gamma_2\left(1 - \frac{\boldsymbol{u}_1 \cdot \boldsymbol{u}_2}{c^2}\right). \tag{5.102}$$

Remembering $\gamma = E/m$ and $\boldsymbol{u} = \boldsymbol{P}/E$, we can write this expression in terms of the energies and momenta. Thus, the energy of particle 2 in the rest frame of particle 1 is

$$E_2' = \frac{E_1\,E_2}{m_1\,c^2}\left(1 - c^2\,\frac{\boldsymbol{P}_1 \cdot \boldsymbol{P}_2}{E_1\,E_2}\right). \tag{5.103}$$

Center of momentum and the invariant mass. Consider a system S, consisting of N *free* particles. Let us label the particles by an index $i \in \{1, \ldots N\}$. Particle i has mass m_i, velocity $\boldsymbol{u}_i$, momentum $\boldsymbol{P}_i$, and energy E_i. The 4-momentum of particle i is $p_i = (E_i, \boldsymbol{P}_i)$. We define the total energy-momentum 4-vector as follows:

$$p = \sum_{i=1}^{N} p_i = (E, \boldsymbol{P}), \tag{5.104}$$

where $E = \sum_{i=1}^{N} E_i$ and $\boldsymbol{P} = \sum_{i=1}^{N} \boldsymbol{P}_i$ are called the total energy and the total momenta of the system S. We now prove an important property of p.

Theorem. *We have $E^2 - P^2 \geq 0$, and equality holds only for the special case where all particles are massless with momenta all in the same direction.*

Proof. The proof is simple:

$$E^2 = \Big(\sum_i E_i\Big)^2 = \sum_{i,j} E_i\,E_j = \sum_i E_i^2 + \sum_i \sum_{j \neq i} E_i\,E_j, \tag{5.105}$$

$$P^2 = \left(\sum_i P_i\right)^2 = \sum_{i,j} P_i \cdot P_j = \sum_i P_i^2 + \sum_i \sum_{j \neq i} P_i \cdot P_j, \tag{5.106}$$

$$E^2 - P^2 = \sum_i \left(E_i^2 - P_i^2\right) + \sum_i \sum_{j \neq i} \left(E_i E_j - P_i \cdot P_j\right) \tag{5.107}$$

$$= \sum_i m_i^2 + A, \tag{5.108}$$

where $A := \sum' (E_i E_j - \boldsymbol{P}_i \cdot \boldsymbol{P}_j)$ and the prime on $\sum$ means that there is no sum for $i = j$. Using $\boldsymbol{P}_i = E_i \boldsymbol{u}_i$ (no sum on i) we see that

$$A = \sum{}' E_i E_j \left(1 - \boldsymbol{u}_i \cdot \boldsymbol{u}_j\right). \tag{5.109}$$

Since $\boldsymbol{u}_i$ is a vector with length ≤ 1, all terms in parentheses are nonnegative. Therefore A is nonnegative so that $E^2 - P^2$ is nonnegative. The equality $E^2 = P^2$ holds only if all terms $1 - \boldsymbol{u}_i \cdot \boldsymbol{u}_j$ are identically zero. This is the case only if all $\boldsymbol{u}_i$s are in the same direction and all particles are massless—because for a massive particle $|\boldsymbol{u}_i| < 1$. □

The inequality $E^2 - P^2 \geq 0$, which means $E > |\boldsymbol{P}|$, is the key to assign two important observables to the system S consisting of N free particles—its mass $M = \surd\left(E^2 - P^2\right)$ and its velocity $\boldsymbol{u} = \boldsymbol{P}/E$. This mass is zero only if all particles are massless *and* all are moving in the same direction. An idealized laser pulse (if its divergence is zero) is such a system. Except this special case, all systems with more than one particle are massive, that is, $M > 0$ and $|\boldsymbol{u}| < 1 = c$.

The quantity $E^2 - P^2$, being the inner product of the 4-vector p with itself, is a Lorentz invariant. This means that it has the same value in any inertial frame. Hence $M = \surd\left(E^2 - P^2\right)$ is called *the invariant mass* of system S.

Center of momentum frame. For an observer moving with velocity $\boldsymbol{u} = \boldsymbol{P}/E$, the total momentum of the system is zero. This frame is called the center of momentum (COM) frame. If other laws of nature like the conservation of electric charge permit, *kinematically* it is possible that all the particles of the system annihilate and a single particle with mass M is being created. The resulting particle would move with velocity $\boldsymbol{u} = \boldsymbol{P}/E$.

Conservation of 4-momentum. The most general interaction we are considering in this chapter is something like

$$A_1 + A_2 + \cdots + A_n \to B_1 + B_2 + \cdots + B_m, \tag{5.110}$$

where A_is are the initial particles and B_js are the final particles. It is assumed that initially the particles are far enough away such that there is no interaction between them. The particles approach a small region of space, get near to each other, and interact, and then, because of the interactions between them, a set of m final particles emerge. At sufficiently long times, these particles are again far enough apart that there is no interaction between them. There is a set of conservation laws in any such interaction—which we call scattering. Note that a decay of a particle is the special case $n = 1$ of this general scattering. We do not know the details of the interaction between the particles. We assume that whatever these interactions are, they preserve the following quantities: total energy, total momentum, and total angular momentum. In total, these are $1+3+3=7$ quantities. These conservation laws could be deduced from other assumptions—that the space-time has certain symmetries. Here we want to know how these laws are formulated in the general scattering problem.

Conservation of total energy and total linear momenta is contained in the equation

$$\sum_{i=1}^{n} p_i = \sum_{j=1}^{m} p'_j, \tag{5.111}$$

where p_i is the 4-momentum of particle A_i and p'_j is the 4-momentum of particle B_j. In words, the total energy and momentum of the system before the interaction are the same as these quantities after the interaction. In other words, the interactions cannot change the total energy and total linear momenta of the system.

Worked Example 25. Consider an ultrahigh-energy photon (γ) of energy $E_1 = 5.5 \times 10^{14}$ eV moving along the x-axis in the intergalactic space. It is possible that this photon collides (that is, interacts) with a photon of the cosmic background radiation (γ_{CMB}) of energy $E_2 = 6.6 \times 10^{-4}$ eV, which is moving in the $-\hat{x}$-direction. Let us calculate the invariant mass and the momentum of this system:

$$M^2 c^4 = (E_1 + E_2)^2 - c^2 \left(\frac{E_1}{c} - \frac{E_2}{c} \right)^2 = 4\, E_1\, E_2, \tag{5.112}$$

$$M c^2 = \sqrt{4\, E_1\, E_2} \simeq 1.2\,\text{MeV}. \tag{5.113}$$

Since this mass is greater than twice the mass of an electron, *kinematically*[2] it is possible that such a collision happens, resulting in the creation of an electron–positron pair.

Problem 26. Show that if $E_1 < E_c = 9.89 \times 10^{13}$ eV, then it is impossible for the interaction $\gamma + \gamma_{\text{CMB}} \to e^+ + e^-$ to happen.

Problem 27 (The GZK limit). Almost 90% of the cosmic rays are protons. A high-energy proton could interact with a photon of the cosmic microwave background (of energy $\sim 6 \times 10^{-4}$ eV) according to

$$p + \gamma_{\text{CMB}} \to p + \pi^0, \tag{5.114}$$

where π^0 is a neutral particle of mass $m_{\pi^0} = 135\,\text{MeV}/c^2$. The mass of a proton is $m_p = 938\,\text{MeV}/c^2$. Find the threshold energy of the proton for this interaction to be

[2] By this, we mean this is allowed by the conservation of energy and momentum. One should also check other conservation laws, such as the conservation of charge, etc. They are satisfied in this case.

possible. This threshold energy is called the GZK limit.[3] Because of this interaction, we expect there is a sudden decrease in the cosmic rays at this energy. There are observations confirming this; see Bahcall and Waxman (2003) and Weinberg (2008, p. 108).

Decay of a particle into two particles. Consider a particle (or system) with invariant mass M. If, by some internal forces, the system is divided into two parts with invariant masses m_1 and m_2, the energy and hence the magnitude of the momenta of these two particles (systems) are determined uniquely. To see this, let p, p_1, and p_2 denote the 4-momenta of particles M, m_1, and m_2, respectively. Because of the conservation of 4-momentum, we have $p = p_1 + p_2$, or $p_2 = p - p_1$. Squaring this identity we get

$$m_2^2 = M^2 + m_1^2 + 2\varsigma\, p \cdot p_1. \tag{5.115}$$

In the rest frame of M we have $p = (M, 0)$ and $p_1 = (E_1, \boldsymbol{P}_1)$, and hence

$$p \cdot p_1 = -\varsigma\, M\, E_1. \tag{5.116}$$

So in the rest frame of the decaying particle with mass M we have

$$m_2^2 = M^2 + m_1^2 - 2\, M\, E_1, \tag{5.117}$$

from which we get

$$E_1 = \frac{M^2 + m_1^2 - m_2^2}{2\, M}, \tag{5.118}$$

and a similar expression for E_2 (just change $1 \leftrightarrow 2$).

An important example of this is the emission of a photon from an excited atom or nucleus. Recall that the mass (inertia) of an atom is not just the sum of the masses of its constituents (see §5.6 on page 103). One has to add the potential energy, which for a bound state is negative. Let M be the mass of the atom *after* emission (that is, m_2 in the above formula) and let $M + \Delta E$ be the mass of the atom before emission. The rest mass of the photon is zero ($m_1 = 0$).

Problem 28. Show that the energy of the emitted photon—valid for $\Delta E \ll M$—is

$$E_\gamma \simeq \Delta E \left(1 - \frac{\Delta E}{M\, c^2}\right). \tag{5.119}$$

We see that the energy of the emitted photon is a little less than ΔE. The amount $(\Delta E)^2/(M\, c^2)$ is the recoil energy.

---

[3] After Kenneth Ingvard Greisen (1918–2007), Georgiy Timofeyevich Zatsepin (1917–2010), and Vadim Alekseyevich Kuzmin (1937–2015).

Decay of a particle into three particles. Now let us see what happens when a particle with mass M, initially at rest, decays into more than two particles. For the case of three secondary particles we have $p = p_1 + p_2 + p_3$, which we can write as $p - p_1 = p_2 + p_3$. Squaring this we get

$$M^2 + m_1^2 - 2 M E_1 = m_2^2 + m_3^2 + 2 E_2 E_3 (1 - \boldsymbol{u}_2 \cdot \boldsymbol{u}_3) . \tag{5.120}$$

Solving this for E_1 we get

$$E_1 = \frac{M^2 + m_1^2 - m_2^2 - m_3^2}{2 M} - \frac{E_2 E_3}{M} (1 - \boldsymbol{u}_2 \cdot \boldsymbol{u}_3) . \tag{5.121}$$

Compton scattering. In 1927 Arthur H. Compton won the Nobel Prize in Physics for a discovery that now bears his name—Compton scattering. When a beam of X-ray is incident on matter, it is scattered. Compton discovered that this scattering could be analyzed by the following assumptions.

1. The X-ray is made up of a stream of massless particles each having momentum h/λ, where h is the Planck constant and λ is the wavelength of the X-ray.
2. These massless particles (photons) interact with electrons, which could be considered as free electrons at rest.[4]
3. The laws of conservation of energy and momentum are valid, exactly as given by the special theory of relativity, that is, the energy and momentum of a massive particle with mass m are $E = m \gamma(u) c^2$ and $\boldsymbol{P} = m \gamma(u) \boldsymbol{u}$, and those of a massless particle are related by $E = c |\boldsymbol{P}|$.

Let us denote the 4-momentum of the incident photon by q_1 and that of the scattered photon by q_2. Let p_1 and p_2 be, respectively, the 4-momentum of the initial and final electrons. We chose the coordinate system such that the incident photon is moving along the x-axis. The scattered photon's momentum will make an angle θ with the x-axis. We chose the y-axis such that this vector lies in the (x, y)-plane. Clearly we have

$$q_1 = (E, E \hat{x}), \qquad q_2 = (E', E' \hat{n}), \qquad \hat{n} \cdot \hat{x} = \cos\theta, \tag{5.122}$$

$$p_1 = (m, 0), \qquad p_2 = m(\gamma(u), \gamma(u) \boldsymbol{u}). \tag{5.123}$$

Conservations of energy and momentum are summarized in $q_1 + p_1 = q_2 + p_2$. Since usually we do not know p_2, we write this as $q_1 + p_1 - q_2 = p_2$ and square this equation. Remembering $q_1^2 = q_2^2 = 0$ and $p_1^2 = p_2^2 = m^2$, we get

$$m^2 = m^2 - 2 q_1 \cdot q_2 + 2 q_1 \cdot p_1 - 2 q_2 \cdot p_1 . \tag{5.124}$$

[4] Actually, for nonmetals, the electrons are bound to heavy nuclei, but this only causes a broadening of the scattered lines.

Obviously

$$q_1 \cdot q_2 = E\,E'(1-\cos\theta), \qquad q_1 \cdot p_1 = m\,E, \qquad q_2 \cdot p_1 = m\,E'. \tag{5.125}$$

Inserting these in the previous equation we get $\frac{1}{E} - \frac{1}{E'} = \frac{2}{m\,c^2}\sin^2\frac{\theta}{2}$. Now, using $E = h\,c/\lambda$ and $E' = h\,c/\lambda'$, we get the famous Compton relation,

$$\lambda - \lambda' = 2\,\lambda_c \sin^2\frac{\theta}{2}, \tag{5.126}$$

where

$$\lambda_c = \frac{h}{m\,c} \tag{5.127}$$

is called the Compton wavelength of the particle.

Inverse Compton scattering. Consider a charged particle, for example, an electron moving along the x-axis to the right, and a photon moving along the x-axis to the left—a head-on collision of a charged particle and a photon. After the collision, we would have a secondary photon and a secondary electron. Let the momentum of the scattered photon be in the direction of $\hat{\boldsymbol{n}}$, making an angle θ with the positive x-axis, *i.e.*, $\hat{x} \cdot \hat{\boldsymbol{n}} = \cos\theta$. What would be the energy of the scattered photon? Let p_1 and q_1 denote the 4-momenta of the incident particle and photon, and let p_2 and q_2 denote the 4-momenta of the scattered particle and photon. We have

$$p_1 = (m\,\gamma, m\,u\,\gamma\,\hat{x}), \qquad q_1 = (E, -E\,\hat{x}), \qquad q_2 = (E', E'\,\hat{\boldsymbol{n}}), \tag{5.128}$$

where u is the magnitude of the velocity of the incident particle and $\gamma = \gamma(u)$. We have $p_1 + q_1 = p_2 + q_2$, but since we do not know p_2 we write this as $p_2 = p_1 + q_1 - q_2$ and square it. Since $p_1^2 = p_2^2 = m^2$ and $q_1^2 = q_2^2 = 0$, we get $m^2 + 2\,p_1 \cdot q_1 - 2\,p_1 \cdot q_2 - 2\,q_1 \cdot q_2 = m^2$, which simplifies to $q_2 \cdot (p_1 + q_1) = p_1 \cdot q_1$. From this it is easy to find

$$E' = \frac{m\,\gamma(1+u)E}{m\,\gamma(1-u\,\cos\theta) + E(1+\cos\theta)} \tag{5.129}$$

$$= \frac{1+u}{1-u\,\cos\theta} \cdot \frac{E}{1 + \dfrac{1+\cos\theta}{1-u\,\cos\theta} \cdot \dfrac{E}{m\,c^2}}. \tag{5.130}$$

For $E \ll m\,c^2$, *i.e.*, for particles colliding head-on with low-energy photons, we can neglect the second term in the denominator of the second fraction and get

$$\frac{E'}{E} = f(u,\theta) = \frac{1+u}{1-u\,\cos\theta}. \tag{5.131}$$

For a fixed u, *i.e.*, a fixed value of the energy of the incident particle, $f(u,\theta)$ is a decreasing function of θ, for $0 \le \theta \le \pi$, with maximum at $\theta = 0$. It is an easy exercise to see that

$$f(u,0) \underset{u\to 1}{\sim} 4\gamma^2, \tag{5.132}$$

so that for very high-energy particles we have

$$\max E' = 4\left(\frac{U}{m\,c^2}\right)^2 E, \tag{5.133}$$

where E is the energy of the incident photon, U is the energy of the incident particle, and E' is the energy of the scattered photon.

Guide to Literature. Compton's original article (Compton, 1923) is quite readable. For applications of the Compton effect in astrophysics, see Bradt (2008, pp. 329–354).

Elastic scattering of two particles. Consider two particles A and B, with different masses $m_A \neq m_B$. In the scattering $A+B \to A+B$ the kinetic energy of each particle (and hence its velocity) does not change. To prove this we write the conservation of the 4-momentum, $p_A + p_B = p'_A + p'_B$, and square it to get $p_A \cdot p_B = p'_A \cdot p'_B$. In the COM frame $\boldsymbol{P}_B = -\boldsymbol{P}_A$ and $\boldsymbol{P}'_B = -\boldsymbol{P}'_A$, and we have $p_A = (E_A, \boldsymbol{P}_A)$, $p_B = (E_B, -\boldsymbol{P}_A)$, $p_A \cdot p_B = E_A\,E_B + |\boldsymbol{P}_A|^2$, and similar expressions for the primed quantities. Now we use $m_A^2 = E_A^2 - |\boldsymbol{P}_A|^2$ and write it as $|\boldsymbol{P}_A|^2 = E_A^2 - m_A^2$. Similarly $|\boldsymbol{P}'_A|^2 = (E'_A)^2 - m_A^2$. Using these relations to rewrite $p_A \cdot p_B = p'_A \cdot p'_B$, we get $E_A\,E_B + E_A^2 - m_A^2 = E'_A\,E'_B + {E'_A}^2 - m_A^2$, from which it follows that $E_A\,(E_A + E_B) = E'_A\,(E'_A + E'_B)$. Combining this last equation with $E_A + E_B = E'_A + E'_B$, we get $E_A = E'_A$. By a similar reasoning we get $E_B = E'_B$.
The Breit frame. Consider the scattering $A+B \to A+B$ in the COM frame. We know that the energy and therefore the magnitude of the momentum of particle A are the same before and after the collision. Let us define the scattering angle Θ, in the COM frame, as the angle between the momentum of particle A after the collision with that of the same particle A before the collision (see Fig. 5.4). From the figure, it is obvious that if one moves with velocity $v_A \cos\frac{\Theta}{2}$ in the positive x-direction, then this observer will see only a y-component for the momentum of particle A, both before and after the collision. In other words, in this frame we see that the momentum of particle A after the collision is just the negative of its momentum before the collision—as if it is hitting a rigid wall. This frame is called *the infinite momentum* frame or the *Breit* frame.[5] It is related to the lab frame by

$$V_{\text{Breit}} = v_A \cos\frac{\Theta}{2}, \qquad v_A = \frac{P_A}{E_A}. \tag{5.134}$$

[5] After Gregory Breit (1899–1981).

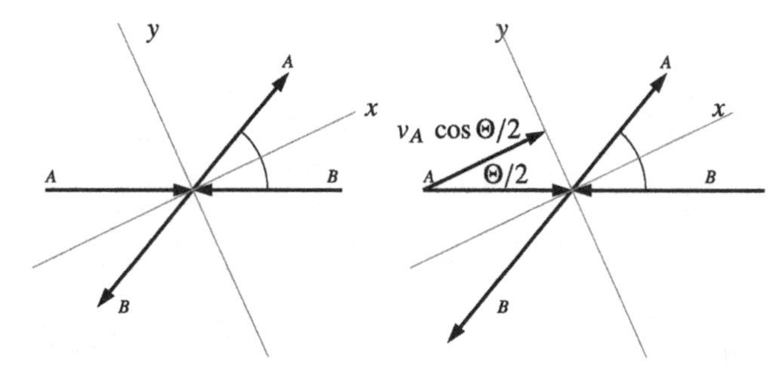

FIGURE 5.4 Breit frame.

The left figure shows the momenta, which are equal in the COM frame. The right figure, also in the COM frame, shows the velocities which are not equal. The velocity of each particle before and after the collision is the same. If one moves with velocity $v_A \cos\frac{\Theta}{2}$ in the positive x-direction, then this observer—using coordinates (x', y')—will see only a y'-component of the momentum of particle A, both before and after the collision. Before the collision, particle A is moving in the $-y'$-direction, and after the collision, particle A is moving in the $+y'$-direction. This *moving* frame is the Breit frame.

References

Bahcall, J.N., Waxman, E., 2003. Has the GZK suppression been discovered? Physics Letters B 556 (1), 1–6.

Bradt, H., 2008. Astrophysics Processes. Cambridge University Press, Cambridge.

Compton, A.H., 1923. A quantum theory of the scattering of x-rays by light elements. Physical Review 21 (5), 483–502.

Einstein, A., 1946. Elementary derivation of the equivalence of mass and energy. Technion Journal 5, 16–17.

Goldstein, H., 1980. Classical Mechanics, 2nd edition. Addison-Wesley, Reading, MA.

Jackson, J.D., 1999. Classical Electrodynamics, 3rd edition. John Wiley & Sons, New York.

Landau, L.D., Lifshitz, E.M., 1975. The Classical Theory of Fields, fourth revised English edition. Landau and Lifshitz Course of Theoretical Physics, vol. 2. Pergamon Press, Oxford.

Symon, K.R., 1971. Mechanics, 3rd edition. Addison-Wesley, Reading, MA.

Weinberg, S., 2008. Cosmology. Oxford University Press, Oxford.

CHAPTER

Electrodynamics in covariant form

6

6.1 Density and current density

General case. Consider a set of N particles and some observable like charge, mass, energy, etc., which is defined for any of the particles. We index particles by an index $a \in \{1, 2, \dots N\}$ and denote the quantity ω so that ω_a is the value of quantity ω for particle a. The path of particle a in space is specified by a vector $\boldsymbol{r}_a(t) := (x_a(t), y_a(t), z_a(t))$, and its velocity is the vector $\boldsymbol{u}_a(t) := (\dot{x}_a(t), \dot{y}_a(t), \dot{z}_a(t))$. We now define the following four quantities, a scalar and a vector called respectively *the density* and *the current density* of quantity ω:

$$\rho_\omega(t, \boldsymbol{r}) := \sum\nolimits_a \omega_a\, \delta^{(3)}\left(\boldsymbol{r} - \boldsymbol{r}_a(t)\right), \tag{6.1}$$

$$\boldsymbol{J}_\omega(t, \boldsymbol{r}) := \sum\nolimits_a \omega_a\, \boldsymbol{u}_a(t)\, \delta^{(3)}\left(\boldsymbol{r} - \boldsymbol{r}_a(t)\right). \tag{6.2}$$

These four quantities are fields, that is, they are functions of (t, x, y, z) defined on the whole of space-time. However, because of the Dirac delta distributions in the definitions above, these fields are almost everywhere zero. They have nonvanishing values only on the world-lines of the particles. For a continuous medium like a fluid, $(\rho_\omega, \boldsymbol{J}_\omega)$ are smooth functions.

The significance of $\rho_\omega(t, \boldsymbol{r})$ could be deduced from the following integral:

$$\int_V \mathrm{d}x\, \mathrm{d}y\, \mathrm{d}z\, \rho_\omega(t, \boldsymbol{r}) = \sum_{a \in V} \omega_a := \Omega(t, V). \tag{6.3}$$

In words, the integral of $\rho_\omega(t, \boldsymbol{r})$ over a region V is the sum of the values ω_a for those particles which are inside the volume V. Some notes:

1. To avoid ambiguity, we think of V such that any particle is either inside or outside V, but not on its surface S.
2. The integral is a *simultaneous* integral, that is, the values of the integrand must all be observed simultaneously and then summed.
3. This quantity $\Omega(t, V)$ is useful if the quantity ω is extensive (additive). To see this, consider a fluid being made up of macroscopic particles. By this, we mean the elements are large enough to contain a macroscopic number of molecules. For such a *macroscopic particle*, one can define both temperature and entropy. If ω is considered to be the entropy, then $\Omega(t, V)$ is the total entropy of volume V

A Mathematical Approach to Special Relativity. https://doi.org/10.1016/B978-0-32-399708-9.00015-4

at time t. If ω is considered to be the temperature, then $\Omega(t, V)$ is not a useful quantity, because temperature is not an extensive quantity.

Consider an infinitesimal surface with area $\mathrm{d}A$ normal to the unit vector $\hat{\boldsymbol{n}}$; $\boldsymbol{J}_\omega(t, \boldsymbol{r}) \cdot \hat{\boldsymbol{n}}\,\mathrm{d}A$ is the rate of passing of quantity ω across $\hat{\boldsymbol{n}}\,\mathrm{d}A$. From this, it follows that

$$\oint_S \boldsymbol{J}_\omega(t, \boldsymbol{r}) \cdot \hat{\boldsymbol{n}}\,\mathrm{d}A \tag{6.4}$$

is the net flux of quantity ω over the volume V—the boundary of V is the surface S (sometimes denoted by ∂V).

If the quantity ω is conserved, then it must be true that

$$\frac{\mathrm{d}\Omega}{\mathrm{d}t} = -\oint_S \boldsymbol{J}_\omega(t, \boldsymbol{r}) \cdot \hat{\boldsymbol{n}}\,\mathrm{d}A. \tag{6.5}$$

The use of the ordinary differentiation symbol instead of the partial differentiation symbol is customary. Note that Ω is a function of t and V, but we regard V as being constant.

As we have already seen in §1.18.1 (page 24), this integral identity is true for any volume V if and only if

$$\frac{\partial \rho_\omega}{\partial t} + \nabla \cdot \boldsymbol{J}_\omega = 0. \tag{6.6}$$

If we define the 4-current

$$J^\mu_\omega := (\rho_\omega, \boldsymbol{J}_\omega) \tag{6.7}$$

the continuity equation will have the following form:

$$\partial \cdot J_\omega = \partial_\mu J^\mu_\omega = 0. \tag{6.8}$$

Remark for the casual reader: The subscripts ω in the above formulas are not indices like μ; they remind us that we are talking about the quantity ω.

6.2 Electric charge density and its current density

As our first illustrative example, let us consider the electric charge as ω, which we prefer to denote by e. It is an experimental fact that charge is both conserved and additive. So from our above reasoning, $J^\mu_e = (\rho_e, \boldsymbol{J}_e)$ satisfies $\partial_\mu J^\mu_e = 0$. Since ∂_μ is a lower-index 4-vector, J^μ_e must be an upper-index 4-vector. Here is a subtlety: We know that $(1, \boldsymbol{u}_a)$ *is not* a 4-vector, but $\gamma(u_a)(1, \boldsymbol{u}_a)$ is. Looking at the defining formulas of ρ_e and $\boldsymbol{J}_e$, we note the presence of the Dirac delta distribution. This delta distribution does the trick. The point is that *the spatial Dirac delta distribution, $\delta^{(3)}(\boldsymbol{r} - \boldsymbol{r}')$, is not invariant under a Lorentz transformation*, or in other words, it is not a scalar! So let us prove that $(\rho_e, \boldsymbol{J}_e)$ is a 4-vector by another reasoning. To do this, let us note that

$$\int_{-\infty}^{\infty} \delta(t-t') f(t') \mathrm{d}t' = f(t). \tag{6.9}$$

Using this, let us rewrite the definition of J_e^μ as follows:

$$\rho_e(t,x,y,z) = \int_{-\infty}^{\infty} \sum_a e_a \delta^{(3)}(\boldsymbol{r}-\boldsymbol{r}_a(t))\, \delta(t-t')\, \mathrm{d}t', \tag{6.10}$$

$$J_e^i(t,x,y,z) = \int_{-\infty}^{\infty} \sum_a e_a \delta^{(3)}(\boldsymbol{r}-\boldsymbol{r}_a(t))\, \delta(t-t') \frac{\mathrm{d}x_a^i}{\mathrm{d}t'}\, \mathrm{d}t'. \tag{6.11}$$

Noting that for all particles we have $x_a^0(t) = t$ and changing the order of integration and summation, we see that these four equations could be written in the compact form

$$J_e^\mu(x) = \sum_a e_a \int_{-\infty}^{\infty} \delta^{(3)}(\boldsymbol{r}-\boldsymbol{r}_a(t))\, \delta(t-t') \frac{\mathrm{d}x_a^\mu(t')}{\mathrm{d}t'}\, \mathrm{d}t'. \tag{6.12}$$

This relation is the sum of N integrals. In the integral corresponding to particle a, we can change the variable of integration from t' to τ_a. The variable t' is called *the coordinate time*, while τ_a is called the proper time of particle a. The latter variable, τ_a, is a scalar (a Lorentz invariant). So we see that

$$J_e^\mu(x) = \sum_a e_a \int_{-\infty}^{\infty} \delta^{(3)}(\boldsymbol{r}-\boldsymbol{r}_a(\tau_a))\, \delta(t-\tau_a) \frac{\mathrm{d}x_a^\mu(\tau_a)}{\mathrm{d}\tau_a}\, \mathrm{d}\tau_a \tag{6.13}$$

$$= \sum_a e_a \int_{-\infty}^{\infty} \delta^{(4)}(x - x_a(\tau_a)) \frac{\mathrm{d}x_a^\mu}{\mathrm{d}\tau_a}\, \mathrm{d}\tau_a. \tag{6.14}$$

This last form proves that J_e^μ is a scalar once we recollect that *the 4D Dirac delta distribution*

$$\delta^{(4)}\left(x-x'\right) = \delta\left(t-t'\right) \delta\left(x-x'\right) \delta\left(y-y'\right) \delta\left(z-z'\right) \tag{6.15}$$

is a Lorentz invariant, because the determinant of the Lorentz transformation is equal to 1.

6.3 The Lorentz force

We know that if a point particle with charge e is moving with velocity $\boldsymbol{u}$ in the lab and there are an electric field $\boldsymbol{E}$ and a magnetic field $\boldsymbol{B}$, a force $\boldsymbol{F} = e(\boldsymbol{E} + \boldsymbol{u} \times \boldsymbol{B})$ acts on the particle. For an extended object, the charge is represented by the density ρ and the current $\boldsymbol{J}$. Now instead of the 3-force, we have a density of 3-force which has the form $\boldsymbol{f} = \rho \boldsymbol{E} + \boldsymbol{J} \times \boldsymbol{B}$. We have used the symbol $\boldsymbol{f}$ for this, because it has the dimensions of force per volume. This force, or force density, is called the Lorentz 3-force or 3-force density. As we already know, this 3-force, like any 3-force, is related to a 4-force $F^\mu = \gamma(u)\left(c^{-2}\boldsymbol{u}\cdot\boldsymbol{F},\ \boldsymbol{F}\right)$.

The observation of this section is that this Lorentz 4-force density could be written in the following form:

$$f^\mu = F^{\mu\nu}\, J_\nu = F^\mu_{\ \nu}\, J^\nu. \tag{6.16}$$

The proof is a simple matrix multiplication:

$$\begin{bmatrix} f^0 \\ f^1 \\ f^2 \\ f^3 \end{bmatrix} = \varsigma \begin{bmatrix} 0 & c^{-2}E_x & c^{-2}E_y & c^{-2}E_z \\ -c^{-2}E_x & 0 & B_z & -B_y \\ -c^{-2}E_y & -B_z & 0 & B_x \\ -c^{-2}E_z & B_y & -B_x & 0 \end{bmatrix} \begin{bmatrix} -\varsigma c^2 \rho \\ \varsigma J_x \\ \varsigma J_y \\ \varsigma J_z \end{bmatrix}, \tag{6.17}$$

which leads to

$$f^0 = c^{-2}\,\boldsymbol{E}\cdot\boldsymbol{J},$$

$$\left.\begin{aligned} f^1 &= \rho\, E_x + \left(J_y\, B_z - J_z\, B_y\right) \\ f^2 &= \rho\, E_y + \left(J_z\, B_x - J_x\, B_z\right) \\ f^3 &= \rho\, E_z + \left(J_x\, B_y - J_y\, B_x\right) \end{aligned}\right\} \quad \boldsymbol{f} = \rho\,\boldsymbol{E} + \boldsymbol{J}\times\boldsymbol{B}. \tag{6.18}$$

Remark. If $\boldsymbol{F}$ is the 3-force, the corresponding 4-force is $\gamma(u)\,(\boldsymbol{u}\cdot\boldsymbol{F},\boldsymbol{F})$. Note the presence of the $\gamma(u)$—the spatial components of the 4-force are $\gamma(u)\boldsymbol{F}$, not the $\boldsymbol{F}$ themselves. However, if $\boldsymbol{f}$ is a 3-force density, the spatial components of the corresponding 4-force density are $\boldsymbol{f}$ themselves—no factor of γ! To have a better understanding of this, we should recollect that a force density has the dimension of force per volume. The "per volume" dimension comes from something like a Dirac delta distribution, which is not invariant under a Lorentz boost. Because of length contraction, a volume $\mathrm{d}x'\,\mathrm{d}y'\,\mathrm{d}z'$ moving with velocity v in frame K has the volume $\frac{1}{\gamma(u)}\mathrm{d}x\,\mathrm{d}y\,\mathrm{d}z$. This transformation property of the volume leads to the transformation property of the Dirac delta distribution, from which the above statement follows.

6.4 Energy-momentum-stress tensor of matter

If a particle with rest mass m_a is moving with velocity u_a, its total energy is $E_a = m_a\,\gamma(u_a)\,c^2$. Note that this energy consists of the rest energy $m_a\,c^2$ and the kinetic energy, but not the potential energy. For the moment just forget about the potential energy. We also note that for particles we have $E_a = c^2\,P^0_a$ and $u^i_a = P^i_a/P^0_a$. Now we can define the following density and current density for energy:

$$\rho_E(t,\boldsymbol{r}) = \sum\nolimits_a E_a\,\delta^{(3)}\left(\boldsymbol{r}-\boldsymbol{r}_a(t)\right) = c^2\sum\nolimits_a P^0_a\,\delta^{(3)}\left(\boldsymbol{r}-\boldsymbol{r}_a(t)\right), \tag{6.19}$$

$$J^i_E(t,\boldsymbol{r}) = \sum\nolimits_a E_a\,u^i_a(t)\,\delta^{(3)}\left(\boldsymbol{r}-\boldsymbol{r}_a(t)\right) = c^2\sum\nolimits_a P^i_a\,\delta^{(3)}\left(\boldsymbol{r}-\boldsymbol{r}_a(t)\right). \tag{6.20}$$

These four quantities *are not* the components of a 4-vector because E_a is not a scalar, that is, it is not a Lorentz invariant. Anyhow, ρ_E is the density of energy and J^i_E is the i-th component of the energy current density.

The particle a with mass m_a and velocity u^i_a has three more quantities like energy—the three components of the momentum. The i-th component of the momentum has a corresponding density and a corresponding current density:

$$\rho_{Pi}(t, \boldsymbol{r}) = \sum\nolimits_a P^i_a \, \delta^{(3)}(\boldsymbol{r} - \boldsymbol{r}_a(t)), \tag{6.21}$$

$$J^j_{Pi}(t, \boldsymbol{r}) = \sum\nolimits_a P^i_a u^j_a \, \delta^{(3)}(\boldsymbol{r} - \boldsymbol{r}_a(t)) = \sum\nolimits_a \frac{P^i_a P^j_a}{P^0_a} \delta^{(3)}(\boldsymbol{r} - \boldsymbol{r}_a(t)). \tag{6.22}$$

Here, J^j_{Pi} is the j-th component of the current density of the i-th component of momentum.

We now define the following 4×4 matrix of 16 quantities, which we call the energy-momentum-stress tensor:

$$T^{\mu\nu}_{\text{mat}} = \sum\nolimits_a \frac{P^\mu_a P^\nu_a}{P^0_a} \delta^{(3)}(\boldsymbol{r} - \boldsymbol{r}_a(t)). \tag{6.23}$$

The index mat in $T^{\mu\nu}_{\text{mat}}$ should remind us of *Matter*, by which we mean the totality of the material particles of the system. This tensor has some properties which we list in a moment. Before that, we should justify our use of the word *tensor*.

When we say $T^{\mu\nu}$ is a tensor of rank 2, we mean that under a Lorentz transformation, it has the following transformation law:

$$T^{\mu\nu} = L^\mu{}_\alpha L^\nu{}_\beta T'^{\alpha\beta}. \tag{6.24}$$

To prove that this property holds, we again use $\int_{-\infty}^{\infty} dt' \delta(t - t') f(t') = f(t)$ to write $T^{\mu\nu}_{\text{mat}}$ as

$$T^{\mu\nu}_{\text{mat}} = \sum\nolimits_a \int_{-\infty}^{\infty} \delta(t - t') \, \delta^{(3)}(\boldsymbol{r} - \boldsymbol{r}_a(t')) \, \frac{P^\mu_a(t') P^\nu_a(t')}{P^0_a(t')}. \tag{6.25}$$

This is a sum of integrals over different particles. For particle a we change the integration variable from the coordinate time t' to the proper time τ_a and use $P^\mu_a / P^0_a = U^\mu_a = \dfrac{dx^\mu_a}{d\tau_a}$ to transform the above form to the following:

$$T^{\mu\nu}_{\text{mat}} = \sum\nolimits_a \int_{-\infty}^{\infty} d\tau_a \, \delta^{(4)}(x - x_a(\tau_a)) \, P^\mu_a(\tau_a) \frac{dx^\nu_a}{d\tau_a}. \tag{6.26}$$

Now, since $\delta^{(4)}(x - x')$ is invariant under a Lorentz transformation and both P^μ and dx^ν are 4-vectors, it follows that $T^{\mu\nu}_{\text{mat}}$ has the claimed transformation law.

Now let us list the properties of the energy-momentum-stress tensor.

1. It is symmetric in the two indices, that is, $T^{\mu\nu} = T^{\nu\mu}$. (Temporarily we drop the subscript mat.)
2. T^{00} is the density of the zeroth component of the 4-momentum, that is, the density of energy divided by c^2 which can be named the density of *mass*.
3. $T^{0i} = T^{i0}$ has two interpretations: (1) the density of the i-th component of the momentum and (2) the i-th component of the current density of the energy divided by c^2, that is, $c^{-2} J_E^i$.
4. T^{ij} has two interpretations: (1) the i-th component of the current of the j-th component of momentum and (2) the j-th component of the current of the i-th component of the momentum. Symmetry of T^{ij} says that these two are equal.

Let us define the energy-momentum-stress tensor of one single particle as follows:

$$T_a^{\mu\nu}(t, \boldsymbol{r}) = \frac{P_a^{\mu}(t)\, P_a^{\nu}(t)}{P_a^0(t)} \delta^{(3)}\left(\boldsymbol{r} - \boldsymbol{r}_a(t)\right). \tag{6.27}$$

We can prove the following identity:

$$\partial_\nu T_a^{\mu\nu} = \frac{\mathrm{d}P_a^{\mu}}{\mathrm{d}t} \delta^{(3)}\left(\boldsymbol{r} - \boldsymbol{r}_a(t)\right). \tag{6.28}$$

The right-hand side of this equation is simply the density of the 4-force. The proof is just taking derivatives:

$$\sum_{\nu=0}^{3} \partial_\nu T_a^{\mu\nu} = \partial_0 T_a^{\mu 0} + \sum_{i=1}^{3} \partial_i T_a^{\mu i} \tag{6.29}$$

$$\begin{aligned}
&= \frac{\partial}{\partial t}\left(P_a^{\mu}(t)\,\delta^{(3)}\left(\boldsymbol{r} - \boldsymbol{r}_a(t)\right)\right) \\
&\quad + \sum_{i=1}^{3} \frac{\partial}{\partial x^i}\left(\frac{P_a^{\mu}(t)\, P_a^{i}(t)}{P_a^0(t)} \delta^{(3)}\left(\boldsymbol{r} - \boldsymbol{r}_a(t)\right)\right) \\
&= \frac{\mathrm{d}P_a^{\mu}}{\mathrm{d}t} \delta^{(3)}\left(\boldsymbol{r} - \boldsymbol{r}_a(t)\right) + P_a^{\mu}(t) \frac{\partial}{\partial t} \delta^{(3)}\left(\boldsymbol{r} - \boldsymbol{r}_a(t)\right) \\
&\quad + \sum_{i=1}^{3} \frac{P_a^{\mu}(t)\, P_a^{i}(t)}{P_a^0(t)} \frac{\partial}{\partial x^i} \delta^{(3)}\left(\boldsymbol{r} - \boldsymbol{r}_a(t)\right).
\end{aligned} \tag{6.30}$$

Now we note that

$$\frac{\partial}{\partial t} \delta^{(3)}\left(\boldsymbol{r} - \boldsymbol{r}_a(t)\right) = \sum_{i=1}^{3} \left(\frac{\partial}{\partial x_a^i} \delta^{(3)}\left(\boldsymbol{r} - \boldsymbol{r}_a(t)\right)\right) \frac{\mathrm{d}x_a^i}{\mathrm{d}t}, \tag{6.31}$$

$$\frac{\partial}{\partial x_a^i} \delta^{(3)}\left(\boldsymbol{r} - \boldsymbol{r}_a(t)\right) = -\frac{\partial}{\partial x^i} \delta^{(3)}\left(\boldsymbol{r} - \boldsymbol{r}_a(t)\right), \tag{6.32}$$

$$\frac{P_a^i(t)}{P_a^0(t)} = u_a^i = \frac{\mathrm{d}x_a^i}{\mathrm{d}t}. \tag{6.33}$$

From these three identities it follows that

$$\sum_{i=1}^{3} \frac{P_a^\mu(t)\, P_a^i(t)}{P_a^0(t)} \frac{\partial}{\partial x^i} \delta^{(3)}(\boldsymbol{r} - \boldsymbol{r}_a(t)) = \sum_{i=1}^{3} P_a^\mu(t) \frac{\mathrm{d}x_a^i}{dt} \frac{\partial}{\partial x^i} \delta^{(3)}(\boldsymbol{r} - \boldsymbol{r}_a(t))$$
$$= -\sum_{i=1}^{3} P_a^\mu(t) \frac{\mathrm{d}x_a^i}{dt} \frac{\partial}{\partial x_a^i} \delta^{(3)}(\boldsymbol{r} - \boldsymbol{r}_a(t))$$
$$= -P_a^\mu(t) \frac{\partial}{\partial t} \delta^{(3)}(\boldsymbol{r} - \boldsymbol{r}_a(t)) . \tag{6.34}$$

Now if we insert this in (6.30), we get the proof. It is now obvious that for the energy-momentum-stress tensor $T_{\mathrm{mat}}^{\mu\nu}$ we have

$$\partial_\nu T_{\mathrm{mat}}^{\mu\nu}(t, \boldsymbol{r}) = \sum\nolimits_a \frac{\mathrm{d}P_a^\mu}{\mathrm{d}t} \delta^{(3)}(\boldsymbol{r} - \boldsymbol{r}_a(t)) , \tag{6.35}$$

where the right-hand side is the total 4-force density of the system.

6.5 The energy-momentum-stress tensor of the electromagnetic field

The density of force acting on the charged continuum is $f^\mu = F^{\mu\nu} J_\nu$, and because of the Maxwell equations, $\partial_\alpha F^{\alpha\beta} = -\varsigma\, \mu_0\, J^\beta$. Now let us calculate $\varsigma\, \mu_0\, f^\beta$:

$$\varsigma\, \mu_0\, f^\beta = \varsigma\, \mu_0 \left(\varsigma\, F^{\beta\lambda} J_\lambda\right) = \left(F^{\beta\lambda}\right)(\mu_0 J_\lambda) = \left(-F^{\lambda\beta}\right)\left(-\partial^\mu F_{\mu\lambda}\right) = F^{\lambda\beta}\, \partial^\mu F_{\mu\lambda}$$
$$= F^{\lambda\beta}\, \partial^\mu F_{\mu\lambda} + \frac{1}{2} F_{\mu\lambda} \underbrace{\overbrace{\left(\partial^\mu F^{\lambda\beta} + \partial^\lambda F^{\mu\beta}\right)}^{\text{symmetric in } \mu\lambda}}_{=0}$$
$$= F^{\lambda\beta}\, \partial^\mu F_{\mu\lambda} + \frac{1}{2} F_{\mu\lambda} \left(\partial^\mu F^{\lambda\beta} - \partial^\lambda F^{\beta\mu}\right)$$
$$= F^{\lambda\beta}\, \partial^\mu F_{\mu\lambda} + \frac{1}{2} F_{\mu\lambda} \big(\underbrace{\partial^\beta F^{\mu\lambda} + \partial^\mu F^{\lambda\beta} + \partial^\lambda F^{\beta\mu}}_{=0} - \partial^\lambda F^{\beta\mu} + \partial^\mu F^{\mu\beta}\big)$$
$$= F^{\lambda\beta}\, \partial^\mu F_{\mu\lambda} + \frac{1}{2} F_{\mu\lambda} \left(\partial^\beta F^{\mu\lambda} + 2\partial^\mu F^{\lambda\beta}\right)$$
$$= \frac{1}{2} F_{\mu\lambda} \partial^\beta F^{\mu\lambda} + F^{\lambda\beta}\, \partial^\mu F_{\mu\lambda} + F_{\mu\lambda} \partial^\mu F^{\lambda\beta}$$
$$= \frac{1}{4} \partial^\beta \left(F_{\mu\lambda} F^{\mu\lambda}\right) + \partial^\mu \left(F_{\mu\lambda} F^{\lambda\beta}\right)$$
$$= \partial_\alpha \left[\frac{1}{4} \eta^{\alpha\beta} F_{\mu\lambda} F^{\mu\lambda} + \eta^{\alpha\mu} F_{\mu\lambda} F^{\lambda\beta}\right] .$$

Defining

$$T_{\rm em}^{\alpha\beta} := -\frac{\varsigma}{\mu_0}\left[\frac{1}{4}\eta^{\alpha\beta}F_{\mu\lambda}F^{\mu\lambda} + \eta^{\alpha\mu}F_{\mu\lambda}F^{\lambda\beta}\right], \tag{6.36}$$

we have $-f^{\beta} = \partial_\alpha T_{\rm em}^{\alpha\beta}$. (The above calculation seems to be very tricky. To be honest, I calculated $\partial_\alpha T_{\rm em}^{\alpha\beta}$ and wrote the calculation backward!)

Let us define *the total energy-momentum-stress tensor*:

$$T^{\alpha\beta} := T_{\rm mat}^{\alpha\beta} + T_{\rm em}^{\alpha\beta}. \tag{6.37}$$

This tensor satisfies the following equation:

$$\partial_\alpha T^{\alpha\beta} = 0. \tag{6.38}$$

To find the explicit form of the different components of $T_{\rm em}^{\mu\nu}$, we first note that

$$F_{0i} = -\varsigma\, E_i, \qquad F^{0i} = \varsigma\, c^{-2} E_i, \tag{6.39}$$

$$F_{ij} = \varsigma\, \epsilon_{ijk} B_k, \qquad F^{ij} = \varsigma\, \epsilon_{ijk} B_k. \tag{6.40}$$

From these equations it follows that

$$F_{0i}F^{0i} = -\frac{1}{c^2}\boldsymbol{E}\cdot\boldsymbol{E} = -\frac{E^2}{c^2}, \tag{6.41}$$

$$F_{ij}F^{ij} = \epsilon_{ijm}\epsilon_{ijn}B_m B_n = 2\,\delta_{nm} B_n B_m = 2\,\boldsymbol{B}\cdot\boldsymbol{B} = 2\,B^2, \tag{6.42}$$

$$F_{\mu\nu}F^{\mu\nu} = F_{0i}F^{0i} + F_{i0}F^{i0} + F_{ij}F^{ij} = 2\left(B^2 - \frac{E^2}{c^2}\right). \tag{6.43}$$

Now, let us temporarily define $S^{\alpha\beta} := -\varsigma\,\eta^{\alpha\mu}F_{\mu\lambda}F^{\lambda\beta}$, which appears in the formula for $T_{\rm em}^{\alpha\beta}$:

$$S^{00} = -\varsigma\,\eta^{0\mu}F_{\mu\lambda}F^{\lambda 0} = -\varsigma\eta^{00}F_{0i}F^{i0} = -\varsigma\left(\frac{-\varsigma}{c^2}\right)\left(\frac{E^2}{c^2}\right) = \frac{E^2}{c^4}, \tag{6.44}$$

$$\begin{aligned} S^{0i} &= -\varsigma\,\eta^{0\mu}F_{\mu\lambda}F^{\lambda i} = -\varsigma\,\eta^{00}F_{0j}F^{ji} \\ &= -\varsigma\left(\frac{-\varsigma}{c^2}\right)(-\varsigma E_j)(\varsigma\,\epsilon_{jik}B_k) = \frac{1}{c^2}\epsilon_{ijk}E_j B_k \\ &= \frac{1}{c^2}(\boldsymbol{E}\times\boldsymbol{B})_i = S^{i0}, \end{aligned} \tag{6.45}$$

$$\begin{aligned} S^{ij} &= -\varsigma\,\eta^{im}F_{m\lambda}F^{\lambda j} = -\varsigma\,(\varsigma\,\delta_{im})F_{m0}F^{0j} - \varsigma\,(\varsigma\,\delta_{im})F_{mn}F^{nj} \\ &= -F_{i0}F^{0j} - F_{in}F^{nj} \\ &= -(\varsigma\,E_i)\left(\frac{\varsigma}{c^2}E_j\right) - (\varsigma\,\epsilon_{inm}B_m)(\varsigma\,\epsilon_{njk}B_k) \\ &= -\frac{1}{c^2}E_i E_j - \epsilon_{min}\,\epsilon_{jkn}B_m B_k \end{aligned}$$

$$= -\frac{1}{c^2} E_i\, E_j - \left(\delta_{mj}\,\delta_{ik} - \delta_{mk}\,\delta_{ij}\right) B_m\, B_k$$

$$= \frac{1}{c^2} E_i\, E_j - B_i\, B_j + B^2\,\delta_{ij} = S^{ji}. \tag{6.46}$$

From these equations it is now easy to show that

$$T^{00}_{\text{em}} = \left(\frac{1}{c^2}\right) \frac{1}{2\,\mu_0} \left(B^2 + \frac{E^2}{c^2}\right), \tag{6.47}$$

$$T^{0i}_{\text{em}} = T^{i0} = \left(\frac{1}{c^2}\right) \frac{1}{\mu_0} (\boldsymbol{E} \times \boldsymbol{B})_i\,, \tag{6.48}$$

$$T^{ij}_{\text{em}} = \frac{1}{\mu_0} \left[\frac{1}{2}\delta_{ij} \left(B^2 + \frac{E^2}{c^2}\right) - B_i\, B_j - \frac{1}{c^2} E_i\, E_j\right]. \tag{6.49}$$

Remark. The usual definition of the Poynting vector is $\boldsymbol{S} = \mu_0^{-1} \boldsymbol{E} \times \boldsymbol{B}$, that is, c^2 times our T^{0i}_{em}, and the usual definition of the density of momentum is $\boldsymbol{g} = c^{-2} \boldsymbol{S}$, which is our T^{0i}_{em}. The factor c^{-2} in T^{00}_{em} and T^{0i}_{em} is present, because by the conventions used in this book, $x^0 = t$ and $P^0 = m\,\gamma(u)$. That is, T^{00}_{em} is the density of mass, or better said, energy divided by c^2; T^{0i}_{em} is also the current density of mass. To convert these equations to cgs units, drop the c^{-2}, change μ_0 to 4π, and of course, change $\boldsymbol{E}/c$ to $\boldsymbol{E}$.

6.6 The conservation laws

We saw in §1.18.1 (page 24) that the conservation of a quantity ω, like mass or charge, is formulated by the partial differential equation $\partial_\mu J^\mu_\omega = 0$. The conservation of ω says that whenever we have a closed surface S, enclosing a volume V, the change of quantity ω inside V is only due to some quantity ω crossing the boundary S. In other words, ω is neither created nor destroyed anywhere. We now elaborate on this subject. To do that, it is better to first rederive the integral identity.

If the equation $\sum_{\mu=0}^{3} \partial_\mu J^\mu_\omega = 0$ is valid for the 4-current of the quantity ω, then we define

$$\Omega(t, V) = \int_V J^0_\omega(t, x, y, z)\, \mathrm{d}x\, \mathrm{d}y\, \mathrm{d}z. \tag{6.50}$$

Now we have

$$\frac{\mathrm{d}}{\mathrm{d}t} \Omega(t, V) = \frac{\mathrm{d}}{\mathrm{d}t} \int_V J^0_\omega(t, x, y, z)\, \mathrm{d}x\, \mathrm{d}y\, \mathrm{d}z$$

$$= \int_V \frac{\partial J^0_\omega}{\partial t} \mathrm{d}x\, \mathrm{d}y\, \mathrm{d}z. \tag{6.51}$$

Using $\frac{\partial J^0_\omega}{\partial t} = -\sum_{i=1}^{3} \frac{\partial J^i_\omega}{\partial x^i}$ we get

$$\begin{aligned}\frac{\mathrm{d}}{\mathrm{d}t}\Omega(t,V) &= -\int_V \sum_{i=1}^{3} \frac{\partial J^i_\omega}{\partial x^i} \mathrm{d}x\,\mathrm{d}y\,\mathrm{d}z \\ &= -\oint_S J^i_\omega n_i\, da, \qquad \text{Gauss' divergence theorem,}\end{aligned} \tag{6.52}$$

where S is the boundary of V, da is the surface element on S, $\hat{\boldsymbol{n}} = (n_1, n_2, n_3)$ is the unit outward normal vector to S, and we have dropped the summation symbol—Einstein's summation convention.

Whenever the last surface integral vanishes, $\Omega(t, V)$ is constant. A very important case of this vanishing of the surface integral is where the fields $J^i_\omega(t, \boldsymbol{r})$ go to zero as $r \to \infty$ faster than r^{-2}, where $r = \sqrt{(x^2+y^2+z^2)}$. This technically means that $\lim_{r\to\infty}\left(r^2 J^i_\omega(t,x,y,z)\right) = 0$. This behavior of J^i_ω is necessary, because the surface element da grows like r^2 as $r \to \infty$. (To be convinced, recall that the surface element of the sphere with radius r is $r^2 \sin\theta\, \mathrm{d}\theta\, \mathrm{d}\varphi$.)

If $\lim_{r\to\infty}\left(r^2 J^i_\omega\right) = 0$, then $\Omega(t, \mathbb{R}^3)$ is constant. Here by $V = \mathbb{R}^3$ we mean the whole space.

What is the transformation law of this Ω? That is, what is the value of this quantity in frame K'? The answer depends on what ω is. If ω is the electric charge, then, because charge is a Lorentz invariant, the quantity Ω—the total charge—is also a Lorentz invariant. Let us prove this:

$$\begin{aligned}\Omega(t,\mathbb{R}^3) &= \int_{\mathbb{R}^3} J^0_\omega(t,x,y,z)\,\mathrm{d}x\,\mathrm{d}y\,\mathrm{d}z \\ &= \int_{\mathbb{R}^3} \sum\nolimits_a \omega_a \delta^{(3)}\left(\boldsymbol{r} - \boldsymbol{r}_a(t)\right) \mathrm{d}x\,\mathrm{d}y\,\mathrm{d}z \\ &= \sum\nolimits_a \omega_a \int_{\mathbb{R}^3} \delta^{(3)}\left(\boldsymbol{r} - \boldsymbol{r}_a(t)\right) \mathrm{d}x\,\mathrm{d}y\,\mathrm{d}z \\ &= \sum\nolimits_a \omega_a =: \Omega(\mathbb{R}^3).\end{aligned} \tag{6.53}$$

That is, $\Omega(\mathbb{R}^3)$ is simply the sum of all values of the quantity ω of all the particles. Since the electric charge is a Lorentz invariant quantity, that is, it has the same value in K and K', the total charge is also a Lorentz invariant. (For another proof see Weinberg [1972, pp. 40–41].)

What if for ω we consider the *energy* of the particles? The same sequence of formulas above now shows that the total energy is now the sum of the total energies of particles. However, this sum is not a Lorentz invariant, because the energy of each particle transforms as the time component of a 4-vector. To find the transformation law, it is simpler if we think of ω_a to be the 4-momentum of particle a, that is, P^μ_a—the energy is now the zeroth or time component of this quantity. We know that $P_a^\mu = L^\mu{}_\nu P_a'^\nu$, so for the total quantity, which we denote by $\mathcal{P}^\mu$, we have the following expression:

$$\mathcal{P}^\mu = \sum_a P_a^\mu = \sum_a \left(L^\mu{}_\nu P_a'^\nu\right) = L^\mu{}_\nu \left(\sum_a P_a'^\nu\right) = L^\mu{}_\nu \mathcal{P}'^\nu. \tag{6.54}$$

Generally, let us think of the equation $\partial_\mu K_\alpha^{\beta\mu} = 0$, where $K_\alpha^{\beta\mu}$ is a tensor with some upper indices, including μ, and some lower indices. In exact analogy with J_ω^μ we can define

$$\mathcal{K}_\alpha^\beta(t, V) := \int_V K_\alpha^{\beta 0}(t, x, y, z)\, \mathrm{d}x\, \mathrm{d}y\, \mathrm{d}z. \tag{6.55}$$

And, in exact analogy, we can prove the following integral identity:

$$\frac{\mathrm{d}}{\mathrm{d}t}\mathcal{K}_\alpha^\beta(t, V) = -\oint_S K_\alpha^{\beta i}\, n_i\, da. \tag{6.56}$$

Again, if $\lim_{r\to\infty}\left(r^2 K_\alpha^{\beta i}(t, \boldsymbol{r})\right) = 0$, then we can prove that

$$\frac{\mathrm{d}}{\mathrm{d}t}\mathcal{K}_\alpha^\beta(t, \mathbb{R}^3) = 0. \tag{6.57}$$

That is, the quantities $\mathcal{K}_\alpha^\beta$ are constant, or conserved. These quantities are not Lorentz invariants, because now they have tensorial transformation law, which means that

$$\mathcal{K}_\alpha^\beta = L^\beta{}_\mu L_\alpha{}^\nu \mathcal{K}_\nu'^\mu. \tag{6.58}$$

This tensorial transformation law has been inherited from the tensorial transformation law of the density $K_\alpha^{\beta\mu}$. Note that while $K_\alpha^{\beta\mu}(t, \boldsymbol{r})$ is a tensor field—that is, a tensor depending on $(t, \boldsymbol{r})$—the quantity $\mathcal{K}_\alpha^\beta$ is not a *tensor field*, because it does not depend on $(t, \boldsymbol{r})$—there has been an integration on the space; $\mathcal{K}_\alpha^\beta$ does not depend on either $\boldsymbol{r}$ or t!

6.7 Conservation of electric charge

For the electric 4-current J_e^μ we have $\partial_\mu J_e^\mu = 0$, and besides, if we consider a system enclosed in a finite sphere, the requirement $\lim_{r\to\infty}\left(r^2 J_e^i(t, \boldsymbol{r})\right) = 0$ is fulfilled. Therefore the total charge of the system

$$Q(\mathbb{R}^3) := \int_{\mathbb{R}^3} J_e^0(t, \boldsymbol{r})\, \mathrm{d}^3x, \qquad J_e^0 = \rho_e, \tag{6.59}$$

is a constant. Since J_e^μ is a 4-vector, Q is a scalar, that is, invariant under Lorentz transformations.

6.8 Conservation of 4-momentum

We know that the total energy-momentum-stress tensor, $T^{\mu\nu} := T^{\mu\nu}_{\text{mat}} + T^{\mu\nu}_{\text{em}}$, satisfies $\partial_\nu T^{\mu\nu} = 0$. So let us define

$$P^\mu(t, V) := \int_V T^{\mu 0}(t, \boldsymbol{r})\,\mathrm{d}^3 x. \tag{6.60}$$

If V is a constant volume, $P^\mu(t, V)$ can only change if there is a nonvanishing flux of the field $T^{\mu i}(t, \boldsymbol{r})$ crossing the boundary of V (denoted by S). This quantity, $P^\mu(t, V)$, is called the 4-momentum of the system contained in V at time t. It is tempting to consider the 4-vector $P^\mu(t, \mathbb{R}^3)$, but we must be careful. To see this, consider a system composed of some finite number of charged particles, all of them enclosed in a finite sphere. The Sun is such a system. For such a system, surely we have $\lim_{r\to\infty}(r^2\, T^{\mu i}_{\text{mat}}(t, \boldsymbol{r})) = 0$, because $T^{\mu\nu}_{\text{mat}}$ has a nonvanishing value only at the places where matter exists—inside the Sun. However, the Sun is radiating, and for a radiating system $T^{\mu i}_{\text{em}}(t, \boldsymbol{r})$ behave as r^{-2}, so $\lim_{r\to\infty}(r^2\, T^{\mu i}_{\text{em}}(t, \boldsymbol{r}))$ does not exist. Therefore, $P^\mu(t, \mathbb{R}^3)$ is not well defined!

Now consider a point radiating system enclosed at the origin of coordinates. To calculate $P^\mu(t, \mathbb{R}^3)$ we consider a large sphere $x^2 + y^2 + z^2 = R^2$ enclosing the volume V_R, and by integration, we calculate $P^\mu(t, V_R)$. What we are searching for, $P^\mu(t, \mathbb{R}^3)$, is the limit of this $P^\mu(t, V_R)$ as $R \to \infty$. If we know that for some large enough R the density $T^{\mu 0}$ vanishes outside V_R, or is at least smaller than r^{-2}, then we can ignore the contribution of this so-called *neighborhood of infinity*. If the system has been radiating from the infinite past times, there is always some nonvanishing amount of $T^{\mu 0}$ in the neighborhood of infinity, causing $P^\mu(t, V)$ to diverge for $V \to \mathbb{R}^3$. But suppose the system was not radiating from $t = -\infty$ to some finite time t_i, but has been radiating only for times later than t_i. Since radiation moves with the speed of light, c, there is no radiation field outside V_R if $R = c\,(t - t_i)$. If the system is not pointlike, enclosed in a sphere $x^2 + y^2 + z^2 = a$ say, then R is a value greater than $c\,(t - t_i)$, but it is still finite. For such a system, $P^\mu(\mathbb{R}^3)$ is well defined and conserved.

Theorem. *For a system composed of charged particles and the electromagnetic field produced by them, $P^\mu(\mathbb{R}^3)$ is a conserved 4-vector if the radiation has been started at a finite time $t_i > -\infty$.*

6.9 Angular momentum and center of mass

Define

$$M^{\mu\nu\alpha} := x^\mu\, T^{\nu\alpha} - x^\nu\, T^{\mu\alpha}. \tag{6.61}$$

This is a tensor, and it is antisymmetric in its two first indices: $M^{\nu\mu\alpha} = -M^{\mu\nu\alpha}$. It can be shown that $\partial_\alpha M^{\mu\nu\alpha} = 0$. Let us prove this:

$$\begin{aligned}\partial_\alpha M^{\mu\nu\alpha} &= \partial_\alpha \left(x^\mu T^{\mu\alpha}\right) - \partial_\alpha \left(x^\nu T^{\mu\alpha}\right)\\ &= \left(\partial_\alpha x^\mu\right) T^{\mu\alpha} + x^\mu \partial_\alpha T^{\mu\alpha} - \left(\partial_\alpha x^\nu\right) T^{\mu\alpha} - x^\nu \partial_\alpha T^{\mu\alpha}\\ &= \delta^\mu_\alpha T^{\nu\alpha} - \delta^\mu_\alpha T^{\mu\alpha} = T^{\nu\mu} - T^{\mu\nu} = 0.\end{aligned} \tag{6.62}$$

Note that we have used both the symmetry of $T^{\mu\nu}$ and the continuity equation $\partial_\alpha T^{\beta\alpha} = 0$.

Before going to the interpretation of conservation laws associated with this "conserved tensor," let us think of the behavior of elements of $M^{\mu\nu\alpha}(t, \boldsymbol{r})$ as $r \to \infty$.

Since $\partial_\alpha M^{\mu\nu\alpha} = 0$, we can define

$$J^{\mu\nu}(t, V) = \int_V M^{\mu\nu 0}(t, \boldsymbol{r})\, \mathrm{d}^3x. \tag{6.63}$$

We now investigate the interpretation of $J^{\mu\nu}$. To be prepared for that, recall that if ρ is the density of mass of a body of volume V, $M = \int_V \rho\, \mathrm{d}^3x$ is its total mass and $M^{-1} \int_V x^i \rho\, \mathrm{d}^3x$ is the i-th component of the center of mass of that body. Now, $M^{0i0} = x^0 T^{i0} - x^0 T^{00}$, so that

$$J^{0i}(t, V) = t \int_V T^{i0}(t, \boldsymbol{r})\, \mathrm{d}^3x - \int_V x^i T^{00}(t, \boldsymbol{r})\, \mathrm{d}^3x, \tag{6.64}$$

where $\int_V T^{i0} \mathrm{d}^3x = P^i(t, V)$ is the i-th component of the momentum contained in V at time t and $\int_V x^i T^{00}\, \mathrm{d}^3x$ is $M X^i$, where M is the mass contained in volume V and X^i is the i-th component of the center of mass.

J^{ij}, which is an anti-symmetric 3×3 matrix, is the angular momentum of the system. To see this we note that since $T^{i0}(t, \boldsymbol{r})$ is the density of the i-th component of the momentum

$$M^{ij0}\, \mathrm{d}^3x = x^i T^{j0}(t, \boldsymbol{r})\, \mathrm{d}^3x - x^j T^{i0}(t, \boldsymbol{r})\, \mathrm{d}^3x \tag{6.65}$$

is the component of the angular momentum vector of the infinitesimal element d^3x at point $\boldsymbol{r}$ at time t.

Guide to Literature. The topics of this chapter are usually taught in texts on field theory or general relativity. See, for example, Weinberg (1972, pp. 39–47), Landau and Lifshitz (1975, pp. 69–83), and Jackson (1999, pp. 579–585 and 610–612).

References

Jackson, J.D., 1999. Classical Electrodynamics, 3rd edition. John Wiley & Sons, New York.

Landau, L.D., Lifshitz, E.M., 1975. The Classical Theory of Fields, fourth revised English edition. Landau and Lifshitz Course of Theoretical Physics, vol. 2. Pergamon Press, Oxford.

Weinberg, S., 1972. Gravitation and Cosmology. Principles and Applications of the General Theory of Relativity. John Wiley & Sons, New York.

CHAPTER

Noninertial frames

7

7.1 Curvilinear coordinates

The changes of coordinates we have used so far are all of the type $X = Lx + a$, where $X = (X^\mu)$, $L = [L^\mu_{\ \nu}]$, $x = (x^\mu)$, and $a = (a^\mu)$. As usual, L is the matrix of a Lorentz transformation. We know that such transformations have the property that

$$\mathrm{d}s^2 = \eta_{\mu\nu}\,\mathrm{d}X^\mu\,\mathrm{d}X^\nu = \eta_{\alpha\beta}\,\mathrm{d}x^\alpha\,\mathrm{d}x^\beta, \tag{7.1}$$

where

$$[\eta_{\mu\nu}] = \varsigma\,\mathrm{diag}(-c^2, 1, 1, 1) \tag{7.2}$$

is the Minkowski metric.

Now let us consider a general change of coordinates of the form

$$X^\mu = f^\mu(x^0, x^1, x^2, x^3), \tag{7.3}$$

where X^μs are the usual Cartesian coordinates and x^μs are general. Using the symbol X^μ for the function f^μ, a convention used mostly in the texts, by the usual rules of differentiation we get

$$\begin{aligned}\mathrm{d}s^2 &= \eta_{\alpha\beta}\,\mathrm{d}X^\alpha\,\mathrm{d}X^\beta \\ &= \eta_{\alpha\beta}\left(\frac{\partial X^\alpha}{\partial x^\mu}\,\mathrm{d}x^\mu\right)\left(\frac{\partial X^\beta}{\partial x^\nu}\,\mathrm{d}x^\nu\right) \\ &= \left(\eta_{\alpha\beta}\,\frac{\partial X^\alpha}{\partial x^\mu}\,\frac{\partial X^\beta}{\partial x^\nu}\right)\mathrm{d}x^\mu\,\mathrm{d}x^\nu,\end{aligned} \tag{7.4}$$

meaning that

$$\mathrm{d}s^2 = g_{\mu\nu}\,\mathrm{d}x^\mu\,\mathrm{d}x^\nu, \tag{7.5}$$

where

$$g_{\mu\nu} = \eta_{\alpha\beta}\,\frac{\partial X^\alpha}{\partial x^\mu}\,\frac{\partial X^\beta}{\partial x^\nu} \tag{7.6}$$

A Mathematical Approach to Special Relativity. https://doi.org/10.1016/B978-0-32-399708-9.00016-6

is called the metric of the space-time in coordinate system x^μ. Let us write this relation in matrix notation; defining matrices

$$S = \left[S^\alpha{}_\mu\right], \qquad S^\alpha{}_\mu = \frac{\partial X^\alpha}{\partial x^\mu}, \tag{7.7}$$

$$H = \left[\eta_{\alpha\beta}\right], \tag{7.8}$$

$$G = \left[g_{\mu\nu}\right] \tag{7.9}$$

and denoting S transposed by S^T we have

$$G = S^T\, H\, S. \tag{7.10}$$

Since both H and S are invertible, G is invertible.

Definition. The entries of the matrix G^{-1} are denoted by $g^{\mu\nu}$, that is,

$$g^{\mu\kappa}\, g_{\kappa\nu} = \delta^\mu_\nu. \tag{7.11}$$

It is instructive to see some examples.

Example 1. Let us examine the following:

$$T = t, \tag{7.12}$$

$$X = r\, \sin\theta\, \cos\phi, \tag{7.13}$$

$$Y = r\, \sin\theta\, \sin\phi, \tag{7.14}$$

$$Z = r\, \cos\theta. \tag{7.15}$$

It is easy to see that the line element $\mathrm{d}s^2$ now becomes

$$\mathrm{d}s^2 = \varsigma\left(-c^2\,\mathrm{d}t^2 + \mathrm{d}r^2 + r^2\,\mathrm{d}\theta^2 + r^2\,\sin^2\theta\,\mathrm{d}\phi^2\right), \tag{7.16}$$

which could be written as

$$\mathrm{d}s^2 = g_{\mu\nu}\,\mathrm{d}x^\mu\,\mathrm{d}x^\nu, \tag{7.17}$$

where

$$x^0 = t, \quad x^1 = r, \quad x^2 = \theta, \quad x^3 = \phi, \tag{7.18}$$

and

$$\left[g_{\mu\nu}\right] = \varsigma \begin{pmatrix} -c^2 & 0 & 0 & 0 \\ 0 & 1 & 0 & 0 \\ 0 & 0 & r^2 & 0 \\ 0 & 0 & 0 & r^2\,\sin^2\theta \end{pmatrix}, \tag{7.19}$$

$$[g^{\mu\nu}] = \varsigma \begin{pmatrix} -\frac{1}{c^2} & 0 & 0 & 0 \\ 0 & 1 & 0 & 0 \\ 0 & 0 & \frac{1}{r^2} & 0 \\ 0 & 0 & 0 & \frac{1}{r^2 \sin^2\theta} \end{pmatrix}. \tag{7.20}$$

Remark. When reading formulas, pay attention to the context: While x^2 means θ, by r^2 we mean $(r)^2$, which is the same as $(x^1)^2$.

Note that the set $(r, \theta, \phi) = (c_1, c_2, c_3)$ for constants c_1, c_2, c_3 is the world-line of a stationary particle, and the set $t = c_0$ for constant c_0 is the Euclidean space.

Example 2. Consider the following change of coordinates:

$$T = \frac{1}{\sqrt{2}\,c}(u + v), \tag{7.21}$$

$$X = r \cos\phi, \tag{7.22}$$

$$Y = r \sin\phi, \tag{7.23}$$

$$Z = \frac{1}{\sqrt{2}}(u - v). \tag{7.24}$$

It is easy to see that now we have

$$\varsigma\, \mathrm{d}s^2 = 2\,\mathrm{d}u\,\mathrm{d}v + \mathrm{d}r^2 + r^2\,\mathrm{d}\phi^2, \tag{7.25}$$

which means, defining $x^0 = u$, $x^1 = r$, $x^2 = \phi$, and $x^3 = v$, we have

$$g_{\mu\nu} = \varsigma \begin{pmatrix} 0 & 0 & 0 & 1 \\ 0 & 1 & 0 & 0 \\ 0 & 0 & r^2 & 0 \\ 1 & 0 & 0 & 0 \end{pmatrix}. \tag{7.26}$$

The set $x^3 = v = 0$ is the set $Z = cT$, so that the set $(x^0, x^1, x^2) = (c_0, c_1, c_2)$ is the world-line of a light signal moving along the Z-axis. Similarly, the set $(x^1, x^2, x^3) = (c_1, c_2, c_3)$ is the world-line $Z = -cT$. This could easily be seen from the form of the metric $g_{\mu\nu}$. We note that $g_{00} = 0$, so that if x^1, x^2, and x^3 are constants, which means $\mathrm{d}x^1 = \mathrm{d}x^2 = \mathrm{d}x^3 = 0$, we get $\mathrm{d}s^2 = g_{00}\left(\mathrm{d}x^0\right)^2 = 0$, for any values of $\mathrm{d}x^0$; and this means that the coordinate x^0 is light-like or null. The same is true for x^3, since $g_{33} = 0$.

Definition. If (x^0, x^1, x^2, x^3) is a set of general coordinates, the coordinate x^0 is:

1. time-like if $\operatorname{sgn}(g_{00}) = -\varsigma$,
2. space-like if $\operatorname{sgn}(g_{00}) = \varsigma$,
3. light-like (or null) if $g_{00} = 0$.

Problem 29. Let $\beta < 1$, and consider the coordinates $x^0 = u$, $x^1 = r$, $x^2 = \phi$, and $x^3 = v$, defined by

$$T = \frac{1}{\sqrt{2}\beta c}(u + v), \tag{7.27}$$

$$X = r \cos\phi, \tag{7.28}$$

$$Y = r \sin\phi, \tag{7.29}$$

$$Z = \frac{1}{\sqrt{2}}(u - v). \tag{7.30}$$

Write ds^2 and the metric $g_{\mu\nu}$. Show that x^0 and x^3 are both time-like, while x^1 and x^2 are both space-like. This example shows that one could easily have coordinate systems which have more than one time coordinate.

Problem 30. Show that it is possible to have coordinates with all coordinates time-like, space-like, or null.

Definition. By a general frame we mean a coordinate system x^μ for which x^0 is the only time-like coordinate and x^i, for $i = 1, 2, 3$, are all space-like.

7.2 Point observer vs. frame of reference

Point observer. Consider a spaceship moving in interstellar space. The astronaut has access to the following equipments:

1. the solid walls of the spacecraft defining the coordinate axes—a set of three unit vectors,
2. a precise clock registering the proper time,
3. a gyroscope to detect if the ship is rotating or not,
4. a theodolite to determine the direction of the incoming or outgoing particles (including photons),
5. a set of various kinds of telescopes and detectors, and
6. a set of radars.

In our mathematical model of the space-time, such a point observer is described by a world-line, three spatial axes attached to it, and of course, a clock to register the proper time.

Frame of reference. Now consider a squadron consisting of several spaceships, each equipped with the above equipment. This squadron is given by a set of world-lines and the following additional structure:

1. Labels for the different members of the squadron,
2. a convention for the synchronization of clocks and registering the time.

Upon a little reflection on this, we see that such a squadron is equivalent to what we call a frame—that is, a general coordinate system for which x^0 is the only time-like coordinate and x^is are space-like; $x^i = c_i$ (for constant c_i) specifies the spaceship, and x^0 is the time used by the squadron.

The size of our planet, Earth, and even the Solar System, compared to the scale of the universe is very small. In cosmology, which is the study of the structure of the universe on very large scales, our Solar System is just a point observer. However, when we are studying the atmosphere, or the oceans, or the seismic vibrations of the Earth, we are using a frame of reference—each lab is a member of this extended squadron.

7.3 Time, distance, and simultaneity in general coordinates

Consider a curvilinear coordinate system x^μ, such that x^0 is the only time-like coordinate. Let us write the line element:

$$ds^2 = g_{\mu\nu}\, dx^\mu\, dx^\nu \tag{7.31}$$

$$= g_{00}\left(dx^0\right)^2 + 2\, g_{0i}\, dx^0\, dx^i + g_{ij}\, dx^i\, dx^j, \tag{7.32}$$

where in the last terms, summation is understood to be on i and j in the range 1, 2, 3. Note also that $g_{\mu\nu}$ is, in general, a function of x^μ, that is, a function of time x^0 and spatial coordinates (x^1, x^2, x^3).

The curve $x^i =$ constant is the world-line of a particle located at x^i. If a clock is located at x^i, $dx^0 = dt$ is the coordinate time interval between ticks of the clock, and $d\tau$ is the proper time interval between the same ticks, then, since

$$d\tau^2 = -\frac{\varsigma}{c^2} ds^2, \tag{7.33}$$

we get

$$d\tau = \frac{1}{c}\sqrt{-\varsigma\, g_{00}}\, dt. \tag{7.34}$$

Coordinate simultaneity. Two events $E_1 = (x_1^0, x_1^i)$ and $E_2 = (x_1^0, x_2^i)$ are said to be x^μ-coordinate simultaneous if $x_1^0 = x_2^0$.

This definition simply means that all events with the same temporal coordinate x^0 are simultaneous. Let us denote the set of all events simultaneous with $x^0 = t$ by Σ_t. Clearly, Σ_t is a 3D surface in space-time. A point $p \in \Sigma_t$ is given by the event

$$\left(x^0 = t, x_p^1, x_p^2, x_p^3\right). \tag{7.35}$$

Now let q be a point in Σ_t, infinitesimally close to p, that is,

$$x_q^i = x_p^i + dx^i. \tag{7.36}$$

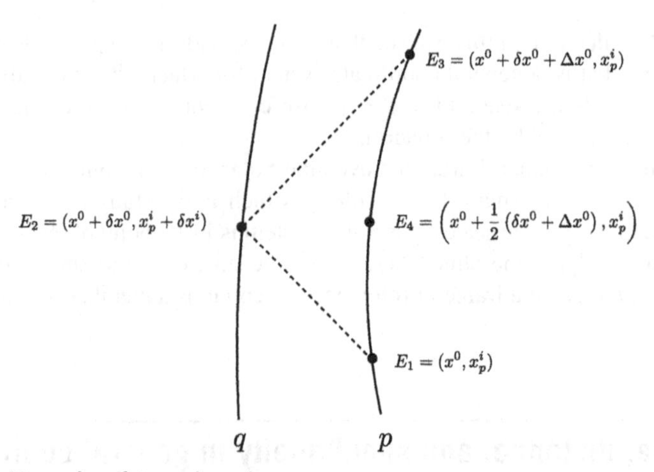

FIGURE 7.1 Measuring distance by radar.

The curved lines are world-lines of spaceships p and q. Spaceship p sends a light signal to spaceship q and receives the reflected light. Measuring the time interval, the distance is found.

Based on the coordinate simultaneity, we define the following.

Coordinate simultaneous distance. The coordinate simultaneous distance of two infinitesimally nearby points x_p^i and $x_p^i + \mathrm{d}x^i$ is defined by

$$\mathrm{d}\sigma^2 = g_{ij}(t, x_p)\,\mathrm{d}x^i\,\mathrm{d}x^j. \tag{7.37}$$

If p and q are two points on Σ_t, not necessarily very close, we can define the distance between them by the following procedure. Consider a curve C joining p to q, divide it by a large enough number N, and calculate the length of C by calculating the integral $\int \mathrm{d}\sigma$. Find the curve C which has the smallest length. The length of this curve, which is called the geodesic joining p to q, is the distance between p and q. This is the usual procedure done in differential geometry. The only point here is that the geometry of Σ_t could depend on t (if $g_{\mu\nu}$ depends explicitly on t); the result would be a distance depending on t. Therefore the procedure must be done at a single instant of time t.

Radar distance. Consider two spaceships p and q. The observer in p wants to measure the distance to q. At time x^0 (see Fig. 7.1), p sends a light signal to q. This is event E_1:

$$E_1 = (x^0, x_p^i), \tag{7.38}$$

where x_p^i are the spatial positions of p. The light signal reaches spaceship q at event E_2:

$$E_2 = \left(x_0 + \delta x^0, x_p^i + \delta x^i\right), \tag{7.39}$$

where δx^i are the components of the spatial vector joining p to q. The light is being reflected on q and arrives at p, at event E_3:

$$E_3 = \left(x^0 + \delta x^0 + \Delta x^0, x_p^i\right). \tag{7.40}$$

Now we calculate. Since $\overrightarrow{E_1 E_2} = \left(\delta x^0, \delta x^i\right)$ is a light-like vector, we have

$$g_{00} \left(\delta x^0\right)^2 + 2\,\delta x^0\, g_{0i}\, \delta x^i + g_{ij}\, \delta x^i\, \delta x^j = 0, \tag{7.41}$$

and δx^0 is the positive root of this quadratic equation:

$$\delta x^0 = \frac{1}{-\varsigma\, g_{00}} \left[\sqrt{\left(-g_{00}\, g_{ij} + g_{0i}\, g_{0j}\right) \delta x^i\, \delta x^j} + \varsigma\, g_{0i}\, \delta x^i\right]. \tag{7.42}$$

Remark. Note that for the time-like convention, $\varsigma = +1$ and $g_{00} < 0$, while for the space-like convention, $\varsigma = -1$ and $g_{00} > 0$, which means $\varsigma = -\operatorname{sgn}(g_{00})$, so that $-\varsigma\, g_{00} = |g_{00}|$.

Now the vector $\overrightarrow{E_2 E_3} = \left(\Delta x^0, -\delta x^i\right)$ is also light-like, which means

$$g_{00} \left(\Delta x^0\right)^2 - 2\,\Delta x^0\, g_{0i}\, \delta x^i + g_{ij}\, \delta x^i\, \delta x^j = 0, \tag{7.43}$$

where Δx^0 is the positive root of this quadratic equation:

$$\Delta x^0 = \frac{1}{-\varsigma\, g_{00}} \left[\sqrt{\left(-g_{00}\, g_{ij} + g_{0i}\, g_{0j}\right) \delta x^i\, \delta x^j} - \varsigma\, g_{0i}\, \delta x^i\right]. \tag{7.44}$$

The coordinate time interval between E_1 and E_3 is

$$\delta t = \delta x^0 + \Delta x^0 \tag{7.45}$$

$$= \frac{2}{-\varsigma\, g_{00}} \sqrt{\left(-g_{00}\, g_{ij} + g_{0i}\, g_{0j}\right) \delta x^i\, \delta x^j}. \tag{7.46}$$

By (7.34), the proper time between E_1 and E_3 is given by

$$c\, \mathrm{d}\tau = \frac{2}{\sqrt{-\varsigma\, g_{00}}} \sqrt{\left(-g_{00}\, g_{ij} + g_{0i}\, g_{0j}\right) \delta x^i\, \delta x^j} \tag{7.47}$$

$$= 2\sqrt{\varsigma \left(g_{ij} - \frac{g_{0i}\, g_{0j}}{g_{00}}\right) \delta x^i\, \delta x^j} \tag{7.48}$$

$$= 2\sqrt{\Upsilon_{ij}\, \delta x^i\, \delta x^j}, \tag{7.49}$$

where

$$\Upsilon_{ij}(t, x_p) = \varsigma \left(g_{ij} - \frac{g_{0i}\, g_{0j}}{g_{00}}\right). \tag{7.50}$$

Therefore using radar, an observer at p finds that the distance of q from p is

$$d\ell = c\, d\tau = \sqrt{\Upsilon_{ij}\, \delta x^i\, \delta x^j}, \tag{7.51}$$

which means that, using light rays, the geometry of the space is given by the metric

$$d\ell^2 = \Upsilon_{ij}(t, x_p)\, dx^i\, dx^j. \tag{7.52}$$

If $g_{0i} \neq 0$, this metric is not the same as g_{ij}.

7.4 Einstein synchronization in curvilinear coordinates

The event $E_2 = (x_0 + \delta x^0, x_q^i)$ on the world-line of q is *coordinate simultaneous* with an event on the world-line of p for which the coordinate time is $x_0 + \delta x^0$. Now, as shown in Fig. 7.1, consider the event

$$E_4 = \left(x^0 + \frac{1}{2}\left(\delta x^0 + \Delta x^0\right), x_p^i\right). \tag{7.53}$$

This event is, by Einstein's definition, simultaneous with E_2. The idea is simple: It takes $\delta x^0 + \Delta x^0$ coordinate-time for the light signal to go from p to q and return to p, so it is "natural" to define E_4 being simultaneous with E_2.[1] Now, since

$$E_2 = \left(x^0 + \delta x^0, x_p^i + \delta x^i\right), \tag{7.54}$$

we have

$$\overrightarrow{E_4\, E_2} = \left(\frac{1}{2}\left(\delta x^0 - \Delta x^0\right), \delta x^i\right). \tag{7.55}$$

It is easy to see from (7.42) and (7.44) that the time component of this 4-vector is

$$\frac{1}{2}\left(\delta x^0 - \Delta x^0\right) = \frac{1}{-g_{00}}\, g_{0i}\, \delta x^i. \tag{7.56}$$

From this we see that an event $(x^0 + dx^0, x_p^i + dx^i)$ happening at a nearby point $x_p^i + dx^i$ is Einstein simultaneous with the event (x^0, x_p^i) (on the world-line of the observer p) if

$$dx^0 = \frac{1}{2}\left(\delta x^0 - \Delta x^0\right) = \frac{1}{-g_{00}} g_{0i}\, dx^i, \tag{7.57}$$

[1] The following example might be helping: Suppose in the year CE 2020 we see an explosion on a star a distance $d = 20\,\text{ly}$ from us. It is natural to say that this explosion occurred in the year CE 2000.

which could be written as

$$g_{00}\,\mathrm{d}x^0 + g_{0i}\,\mathrm{d}x^i = 0. \tag{7.58}$$

According to the rules of raising and lowering of indices, the left-hand side of this equation is denoted by $\mathrm{d}x_0$, so we can define Einstein's simultaneity condition by the following differential relation.

Definition. Two events (x^0, x^i_p) and $(x^0 + \mathrm{d}x^0, x^i_p + \mathrm{d}x^i)$ are Einstein simultaneous if $\mathrm{d}x_0 = 0$, where $\mathrm{d}x_0 = g_{0\mu}\,\mathrm{d}x^\mu$.

Remark. $x^\mu(E)$ is a function assigning a real number to the event E. By $\mathrm{d}x^\mu$ we simply mean the differential of this function x^μ. However, by $\mathrm{d}x_\mu$ we mean $g_{\mu\nu}\,\mathrm{d}x^\nu$, which is a linear combination of differentials $\mathrm{d}x^\mu$, with coefficients $g_{\mu\nu}$ depending on x. The 1-form $\mathrm{d}x_\mu$ (as it is usually called in mathematical jargon) is not, in general, the differential of a function.

Remark. In the following we consider $g_{\mu\nu}$ to be independent of time x^0. The case when $g_{\mu\nu}$ depends on x^0 needs more care and is not needed in what follows.

Einstein synchronization of two stationary clocks. Consider two spaceships p_1 and p_2. Spaceship p_1, at position x^i_1, sends a light signal to p_2, which is at position $x^i_2 = x^i_1 + \mathrm{d}x^i_1$. The coordinate time for the light signal to reach p_2 is given by (7.57). From this, we see that for the clock at p_2 to be Einstein synchronized with the clock at p_1, the clock at p_2 must be advanced by the increment

$$\mathrm{d}x^0_2 = -\varsigma\frac{g_{0i}(p_1)}{g_{00}(p_1)}\,\mathrm{d}x^i_1. \tag{7.59}$$

Einstein synchronization of a chain of observers. Consider a chain of clocks, p_k, $k = 1, 2, \ldots N$, at positions

$$\begin{aligned} &x^i_1,\\ &x^i_2 = x^i_1 + \mathrm{d}x^i_1,\\ &x^i_3 = x^i_2 + \mathrm{d}x^i_2,\\ &\quad\ldots\\ &x^i_N = x^i_{N-1} + \mathrm{d}x^i_{N-1}. \end{aligned} \tag{7.60}$$

Let us see what happens if we try to synchronize all these clocks by Einstein's method. First, as we saw earlier, clock p_2 must be advanced by

$$\mathrm{d}x^0_2 = -\frac{g_{0i}(p_1)}{g_{00}(p_1)}\,\mathrm{d}x^i_1 \tag{7.61}$$

to be synchronized with clock p_1. Similarly, clock p_3 must be advanced by

$$\mathrm{d}x^0_3 = -\frac{g_{0i}(p_2)}{g_{00}(p_2)}\,\mathrm{d}x^i_2 \tag{7.62}$$

to be synchronized with clock p_2, which means that clock p_3 must be advanced by

$$\Delta t = \mathrm{d}x_1^0 + \mathrm{d}x_2^0 = -\frac{g_{0i}(p_1)}{g_{00}(p_1)}\mathrm{d}x_1^i - \frac{g_{0i}(p_2)}{g_{00}(p_2)}\mathrm{d}x_2^i \tag{7.63}$$

to be synchronized with clock p_1. In general, we have

$$\mathrm{d}x_{k+1}^0 = -\frac{g_{0i}(p_k)}{g_{00}(p_k)}\mathrm{d}x_k^i, \tag{7.64}$$

and clock p_N must be advanced by

$$\Delta t = -\sum_{k=1}^{N-1} \frac{g_{0i}(p_k)}{g_{00}(p_k)}\mathrm{d}x_k^i \tag{7.65}$$

to be synchronized with clock p_1. In the continuum limit of $N \to \infty$ we get

$$\Delta t = -\int \frac{g_{0i}(x)}{g_{00}(x)}\mathrm{d}x^i. \tag{7.66}$$

If the chain of clocks forms a closed path, that is, if clock N is the same as clock 1, then to Einstein-synchronize all clocks, starting from 1 and ending with N, we see that clock N, which is the same as clock 1, must by advanced relative to clock 1 by

$$\Delta t = -\oint \frac{g_{0i}(x)}{g_{00}(x)}\mathrm{d}x^i. \tag{7.67}$$

This integral may or may not be zero. It is zero if and only if

$$\frac{\partial}{\partial x^i}\left(\frac{g_{0j}}{g_{00}}\right) - \frac{\partial}{\partial x^j}\left(\frac{g_{0i}}{g_{00}}\right) = 0. \tag{7.68}$$

If this condition is not fulfilled, then we cannot implement Einstein synchronization on the closed chain of clocks.

7.5 A Galilean change of coordinates on the Minkowski space-time

Let (T, X, Y, Z) be Cartesian coordinates of an inertial frame and define coordinates (t, x, y, z) as given by the following Galilean change of coordinates[2]:

$$T = t, \tag{7.69}$$

[2] Compare with Fig. 1.1 on page 12 but note the different notations used in that figure.

$$X = x + vt, \tag{7.70}$$
$$Y = y, \tag{7.71}$$
$$Z = z, \tag{7.72}$$

where v is, for the moment, a parameter with the dimensions of velocity. Note that $t = T$ is the time, as measured by clocks in the inertial frame so that to realize this frame, we need to have a set of particles moving with velocity v along the X-axis, and each particle has a clock which shows T! The skeptical reader perhaps asks: How could the moving clocks show T? The answer is that a clock moving with velocity v shows the proper time $\mathrm{d}\tau = \sqrt{1 - v^2/c^2}\,\mathrm{d}T$; but it is possible to make a clock that shows T. Mathematically this means using the 1-1 function $T = f(\tau)$. For a digital clock, this is simply a computer code to multiply $\mathrm{d}\tau$ by the known factor $1/\sqrt{1 - v^2/c^2}$.

Calculating $\mathrm{d}s^2$ is easy and leads to

$$\varsigma\,\mathrm{d}s^2 = -c^2\left(1 - \frac{v^2}{c^2}\right)\mathrm{d}t^2 + \mathrm{d}x^2 + 2\,v\,\mathrm{d}x\,\mathrm{d}t + \mathrm{d}y^2 + \mathrm{d}z^2, \tag{7.73}$$

meaning that if we define $x^0 = t$, $x^1 = x$, $x^2 = y$, and $x^3 = z$, we have

$$\varsigma\left[g_{\mu\nu}\right] = \begin{pmatrix} -c^2\left(1 - \frac{v^2}{c^2}\right) & 0 & v & 0 \\ 0 & 1 & 0 & 0 \\ v & 0 & 1 & 0 \\ 0 & 0 & 0 & 1 \end{pmatrix}. \tag{7.74}$$

Observation 1. From the form of g_{00} it is obvious that x^0 is time-like if and only if $v < c$.

Observation 2. It is obvious that $\mathrm{d}\sigma^2 = \mathrm{d}x^2 + \mathrm{d}y^2 + \mathrm{d}z^2$, or $g_{ij} = \delta_{ij}$, which means that the geometry of Σ_t as given by g_{ij} is Euclidean.

Observation 3. From the definition of the proper time

$$\mathrm{d}\tau^2 = -\frac{\varsigma}{c^2}\mathrm{d}s^2, \tag{7.75}$$

we see that for a clock at the spatial position (x, y, z), we have

$$\mathrm{d}\tau = \sqrt{1 - \frac{v^2}{c^2}}\,\mathrm{d}t. \tag{7.76}$$

This is the relativistic time dilation. We see that even if we use Galilean change of coordinates properly, the metric of space-time dictates that a moving clock works slower than a stationary clock.

Observation 4. We have

$$\begin{aligned}\Upsilon_{11} &= \varsigma\left(g_{11} - \frac{(g_{01})^2}{g_{00}}\right) \\ &= 1 - \frac{v^2}{-c^2\left(1 - \frac{v^2}{c^2}\right)} = \frac{1}{1 - \frac{v^2}{c^2}},\end{aligned} \tag{7.77}$$

and the spatial metric Υ_{ij} is

$$\left[\Upsilon_{ij}\right] = \begin{pmatrix} \gamma^2 & 0 & 0 \\ 0 & 1 & 0 \\ 0 & 0 & 1 \end{pmatrix}, \tag{7.78}$$

where

$$\gamma = \frac{1}{\sqrt{1 - \frac{v^2}{c^2}}}. \tag{7.79}$$

This means that the line element, if measured by light ray signals (radar), is given by

$$\mathrm{d}\ell^2 = \gamma^2(v)\,\mathrm{d}x^2 + \mathrm{d}y^2 + \mathrm{d}z^2. \tag{7.80}$$

This is the relativistic length contraction. To see this, consider two points on the x-axis, with coordinates x and $x + \mathrm{d}x$. The distance between these two points, as measured by radar, is $\mathrm{d}\ell = \gamma\,\mathrm{d}x$, which is a factor of $\gamma(v)$ longer than $\mathrm{d}x$. This is because the lengths of moving rods (stationary in the (x, y, z) coordinates and along the x-axis) are contracted by a factor of $\gamma^{-1} = \sqrt{1 - \frac{v^2}{c^2}}$, so that more rods are required to cover $\mathrm{d}x$, and the number of the rods required is exactly the distance between the points.

7.6 An expanding universe

Consider the following change of coordinates:

$$T = t\cosh\frac{x}{a}, \tag{7.81}$$

$$X = ct\sinh\frac{x}{a}, \tag{7.82}$$

$$Y = y, \tag{7.83}$$

$$Z = z, \tag{7.84}$$

where (T, X, Y, Z) are the Cartesian coordinates of an inertial frame, c is the velocity of light, and a is a parameter with the dimensions of length. This change of coordinates is not valid for $t = 0$; only for $t > 0$ or for $t < 0$ can we use this change of

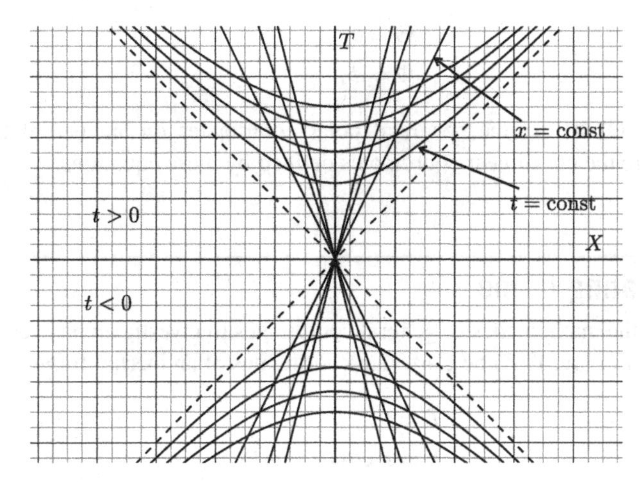

FIGURE 7.2 An expanding universe.

(T, X) are the Cartesian coordinates in an inertial frame, and (t, x) are defined by (7.81), (7.82).

coordinates. The point with constant coordinates (x, y, z) is moving with velocity

$$v = \frac{X}{T} = c \tanh \frac{x}{a} \tag{7.85}$$

parallel to the X-axis, so that the farther the point is from the $x = 0$ plane, the faster it moves; the points for which $x > 0$ go to the right, while the points for which $x < 0$ go to the left. The set $x = 0$, does not move.

Since $c^2 T^2 - X^2 = t^2$, in the (T, X)-plane the sets of constant t are hyperbolas (see Fig. 7.2).

It is easy to calculate the line element:

$$\varsigma \, ds^2 = -c^2 dt^2 + \frac{c^2 t^2}{a^2} dx^2 + dy^2 + dz^2. \tag{7.86}$$

The factor $(ct/a)^2$ indicates that the distance between points x and $x + dx$ is increasing with time, that is, the space is being stretched in the x-direction. We also note that

$$d\tau^2 = dt^2 - \left(\frac{t^2}{a^2} dx^2 + dy^2 + dz^2\right), \tag{7.87}$$

from which we see that the clock at (x, y, z) shows the proper time.

Problem 31. Investigate the following change of coordinates:

$$T = t \cosh \chi, \tag{7.88}$$

$$R = ct \sinh \chi, \tag{7.89}$$

$$\Theta = \theta, \tag{7.90}$$

$$\Phi = \phi, \tag{7.91}$$

where (T, R, Θ, Φ) are the spherical polar coordinates on an inertial frame and (t, χ, θ, ϕ) are the curvilinear coordinates defined by these equations.

7.7 A rotating frame

Now we use the mathematics of general coordinates to investigate physics in a rotating frame. Let (T, X, Y, Z) be Cartesian coordinates of an inertial frame and define

$$T = t, \tag{7.92}$$

$$X = r \cos(\theta + \omega t), \tag{7.93}$$

$$Y = r \sin(\theta + \omega t), \tag{7.94}$$

$$Z = z, \tag{7.95}$$

where ω is an angular velocity. If r, θ, and z are constants, we get the world-line of a particle moving on the circle $X^2 + Y^2 = r^2$ in the $Z = z$ plane, with velocity $v = r\,\omega$. This is the motion of a particle attached to a *rigid* disk, rotating with angular velocity ω.

Here $t = T$ is the time as measured in the inertial frame with coordinates (T, X, Y, Z). To realize this frame, we need a set of particles moving according to the above equations—rigid rotation. Each particle is accompanied by a clock showing T.

Calculating $\mathrm{d}T$, $\mathrm{d}X$, $\mathrm{d}Y$, and $\mathrm{d}Z$, we have

$$\mathrm{d}T = \mathrm{d}t, \tag{7.96}$$

$$\mathrm{d}X = \mathrm{d}r \cos(\theta + \omega t) - r(\mathrm{d}\theta + \omega\,\mathrm{d}t)\sin(\theta + \omega t), \tag{7.97}$$

$$\mathrm{d}Y = \mathrm{d}r \sin(\theta + \omega t) + r(\mathrm{d}\theta + \omega\,\mathrm{d}t)\cos(\theta + \omega t), \tag{7.98}$$

$$\mathrm{d}Z = \mathrm{d}z, \tag{7.99}$$

from which one easily gets

$$\varsigma\,\mathrm{d}s^2 = -c^2\left(1 - \frac{\omega^2 r^2}{c^2}\right)\mathrm{d}t^2 + \mathrm{d}r^2 + r^2\,\mathrm{d}\theta^2 + \mathrm{d}z^2 + 2r^2\omega\,\mathrm{d}\theta\,\mathrm{d}t, \tag{7.100}$$

which means that defining $x^0 = t$, $x^1 = r$, $x^2 = \theta$, and $x^3 = z$, we have

$$\varsigma\,[g_{\mu\nu}] = \begin{pmatrix} -c^2\left(1 - \frac{r^2\omega^2}{c^2}\right) & 0 & r^2\omega & 0 \\ 0 & 1 & 0 & 0 \\ r^2\omega & 0 & r^2 & 0 \\ 0 & 0 & 0 & 1 \end{pmatrix}. \tag{7.101}$$

We now investigate physics in a rotating frame.

Observation 1. From the form of g_{00}, we see that $x^0 = t$ is time-like if and only if $r\,\omega < c$.

Observation 2. From the form of $\mathrm{d}\tau$ we see that for a clock at point (r, θ, z) we have

$$\mathrm{d}\tau = \sqrt{1 - \frac{r^2\,\omega^2}{c^2}}\,\mathrm{d}t. \tag{7.102}$$

This is the relativistic time dilation: The clock at (r, θ, z) is moving with velocity $r\,\omega$ relative to the inertial frame (T, X, Y, Z), $\mathrm{d}\tau$ is the proper time, and $\mathrm{d}t = \mathrm{d}T$ is the time measured by clocks stationary in the inertial frame (T, X, Y, Z).

Observation 3. The coordinate simultaneous metric of the space is

$$\mathrm{d}\sigma^2 = \mathrm{d}r^2 + r^2\,\mathrm{d}\theta^2 + \mathrm{d}z^2, \tag{7.103}$$

which is the usual metric of the Euclidean space in the cylindrical coordinates.

Observation 4. Calculating Υ_{ij} is easy, and we get

$$\left[\Upsilon_{ij}\right] = \begin{pmatrix} 1 & 0 & 0 \\ 0 & \frac{r^2}{1-\frac{r^2\omega^2}{c^2}} & 0 \\ 0 & 0 & 1 \end{pmatrix}, \tag{7.104}$$

which means that the geometry of the space, as defined by the radar distance, is given by

$$\mathrm{d}\ell^2 = \mathrm{d}r^2 + \frac{r^2\,\mathrm{d}\theta^2}{1 - \frac{r^2\omega^2}{c^2}} + \mathrm{d}z^2. \tag{7.105}$$

From this we conclude that the rods tangent to the circle $r =$ constant are being contracted by the factor

$$\gamma^{-1} = \sqrt{1 - \frac{r^2\,\omega^2}{c^2}}. \tag{7.106}$$

As a consequence of this relativistic contraction of the lengths of rods, the circumference of the circle $r =$ constant is larger than $2\pi\,r$:

$$L = \int_0^{2\pi} \frac{r\,\mathrm{d}\theta}{\sqrt{1 - \frac{r^2\omega^2}{c^2}}} = \frac{2\pi\,r}{\sqrt{1 - \frac{r^2\omega^2}{c^2}}}. \tag{7.107}$$

This means that the set (r, θ, z) with the metric (7.104) is non-Euclidean—it is a curved surface! This non-Euclidean nature of the space in a rotating frame could be seen as follows.

Let C be a circle of radius r on the $Z = 0$ surface, with center on the Z-axis, and let R be a radius of this circle. Let us cover both R and C by rods of length ℓ. By properly choosing ℓ, we suppose that $r \ll \ell$ so that a large number of rods is needed to cover R and C: $N_1 = r/\ell$ rods cover R, and $N_2 = 2\pi r/\ell$ rods cover C, and we have $N_2/N_1 = 2\pi$, as expected from theorems of Euclidean geometry.

Now consider a rotating table. Let $z = Z$ be the rotation axis (normal to the table), and let ω denote the angular velocity. What happens if we try to measure a rotating radius R′ and the circumference of the circle C?

The length of a rod laying along the radius R does not change, because the velocity of such a rod is normal to it. So in this case also we need $N_1 = r/\ell$ rods to cover R. However, the rods on the circumference, C, are being contracted and we have

$$\ell' = \sqrt{1 - \frac{v^2}{c^2}}\,\ell = \sqrt{1 - \frac{r^2\omega^2}{c^2}}\,\ell. \tag{7.108}$$

Therefore using rotating rods, we need N_2' rods to cover C, where

$$N_2' = \frac{N_2}{\sqrt{1 - \frac{r^2\omega^2}{c^2}}}. \tag{7.109}$$

Thus the length of C, as measured by rotating rods, is

$$L' = N_2'\,\ell = \frac{N_2\,\ell}{\sqrt{1 - \frac{\omega^2 r^2}{c^2}}} = \frac{2\pi r}{\sqrt{1 - \frac{\omega^2 r^2}{c^2}}}. \tag{7.110}$$

From this, it follows that the ratio of the circumference of a circle of radius r to its radius, if measured by rotating rods, is $2\pi\,\gamma(r)$, meaning that the geometry of a rotating disk is not Euclidean—it is a surface with *negative* curvature (see §D.2).

Guide to Literature. It was Paul Ehrenfest who in 1909 pointed out that the ratio of circumference to the radius of a rotating cylinder is greater than 2π, but as a contradiction to Max Born's definition of rigidity, he did not conclude that this means that the space is curved. For a translation of Ehrenfest's note see Rizzi and Ruggiero (2004, pp. 3–4). It was Einstein who pointed out that the argument shows no contradiction, but that the geometry is not Euclidean. Einstein's argument could be found in Einstein (1980, pp. 57–59).

Observation 5. From $g_{02} = r^2\,\omega$ and $g_{01} = 0$ we see that

$$\frac{\partial}{\partial x^1}\left(\frac{g_{02}}{-g_{00}}\right) - \frac{\partial}{\partial x^2}\left(\frac{g_{01}}{-g_{00}}\right) = \frac{\partial}{\partial r}\left(\frac{r^2\,\omega}{1 - \frac{r^2\omega^2}{c^2}}\right) = \frac{2\,r\,\omega}{1 - \frac{r^2\omega^2}{c^2}}, \tag{7.111}$$

which means that it is impossible to Einstein synchronize all the clocks on the circumference of a rotating ring. So let us calculate the Einstein synchronization anomaly

$$\begin{aligned}
P_- &= [\mathrm{O} \to \mathrm{S} \to \mathrm{M}_1 \to \mathrm{M}_2 \to \mathrm{M}_3 \to \mathrm{S} \to \mathrm{D}]\\
P_+ &= [\mathrm{O} \to \mathrm{S} \to \mathrm{M}_3 \to \mathrm{M}_2 \to \mathrm{M}_1 \to \mathrm{S} \to \mathrm{D}]\\
C_- &= [\mathrm{S} \to \mathrm{M}_1 \to \mathrm{M}_2 \to \mathrm{M}_3 \to \mathrm{S}]\\
C_+ &= [\mathrm{S} \to \mathrm{M}_3 \to \mathrm{M}_2 \to \mathrm{M}_1 \to \mathrm{S}]
\end{aligned}$$

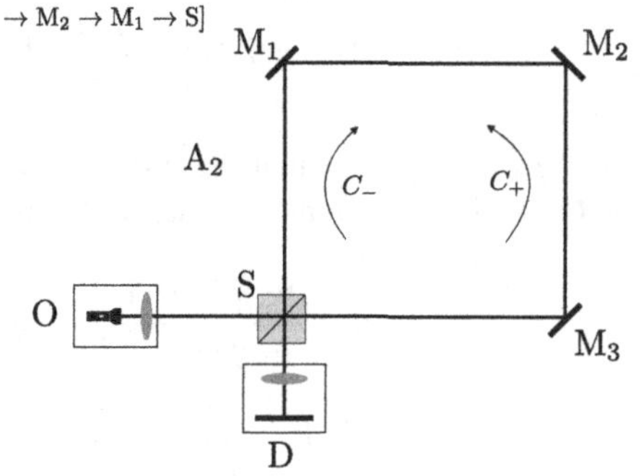

FIGURE 7.3 Sagnac interferometer.

Light from source O is split in two beams in the beam splitter S. Reflecting at mirrors M_1, M_2, and M_3 and in the beam splitter S, the beam travels on the path P_-. Reflecting at mirrors M_3, M_2, and M_1 and going through the beam splitter S, the other beam travels on the path P_+. These two paths differ on C_+ and C_-. The fringes observed at the detector D depend on the difference in time it takes for the light to complete C_+ or C_-.

(7.67) on a closed path $C_\pm$, where C_+ is counterclockwise oriented and C_- is clockwise oriented. We have

$$\begin{aligned}
\Delta t_\pm &= -\oint_{C_\pm} \frac{g_{0i}}{g_{00}} \mathrm{d}x^i\\
&= \frac{1}{c^2} \oint_{C_\pm} \frac{r^2 \omega}{1 - \frac{\omega^2 r^2}{c^2}} \mathrm{d}\theta\\
&\simeq \frac{\omega}{c^2} \oint_{C_\pm} r^2(\theta)\, \mathrm{d}\theta \qquad \text{if} \quad r\omega \ll c\\
&= \pm \frac{2\omega A}{c^2},
\end{aligned} \tag{7.112}$$

where A is the area enclosed by C and we have used the identity

$$\frac{1}{2} \oint_C r^2(\theta)\, \mathrm{d}\theta = A. \tag{7.113}$$

Sagnac effect. The Sagnac interferometer is an interferometer in which a light signal is split into two signals traveling a closed path in opposite directions as in Fig. 7.3. To find the fringe shift for a Sagnac interferometer placed on a rotating platform according to special relativity, we use the method of general curvilinear coordinates. We must first remember that when $g_{0i} \neq 0$, the coordinate time it takes a light signal to go from x^i to $x^i + \mathrm{d}x^i$ is δx^0 given by (7.42), while the time it takes for the light to go from $x^i + \mathrm{d}x^i$ to x^i is Δx^0 given by (7.44), and we have $\delta x^0 - \Delta x^0 = 2\,\mathrm{d}x^0$, where $\mathrm{d}x^0$ is given by (7.57), so that we have

$$\delta x^0 - \Delta x^0 = 2\,\mathrm{d}x^0 = \frac{2\,g_{0i}}{-g_{00}}\,\mathrm{d}x^i. \tag{7.114}$$

Using $\mathrm{d}\tau = \frac{1}{c}\sqrt{-\varsigma\, g_{00}}\,\mathrm{d}x^0$, we see that the *proper* time difference between a light signal moving on C_+ and a light signal moving on C_- is

$$\begin{aligned}
\Delta\tau &= \oint_{C_+} \frac{2\,\varsigma\, g_{0i}\,\mathrm{d}x^i}{-\varsigma\, g_{00}}\frac{1}{c}\sqrt{-\varsigma\, g_{00}} \\
&= \frac{\omega}{c^2}\oint_{C_+} \frac{r^2(\theta)}{\sqrt{1-\frac{\omega^2 r^2}{c^2}}}\,\mathrm{d}\theta \qquad (7.115)\\
&\simeq \frac{4\,A\,\omega}{c^2} \qquad \text{if} \quad r\,\omega \ll c. \qquad (7.116)
\end{aligned}$$

The difference between the optical paths is

$$2\,c\,\Delta\tau = \frac{4\,A\,\omega}{c}. \tag{7.117}$$

If λ is the wavelength of the light signal, the fringe shift is given by

$$N = \frac{4\,A\,\omega}{c\,\lambda}. \tag{7.118}$$

Remark. If $r\,\omega \ll c$, the leading terms of $2\,c\,\Delta t_+$ of (7.112) and $2\,c\,\Delta\tau$ are the same. However, we should note that the observer-independent (scalar) optical path is $c\,\Delta\tau$, and not $c\,\Delta t$. Remember that x^0 is a coordinate time, which could be changed with no effect on the interference pattern—the interference pattern does not depend on how the observer defines the coordinate time!

Guide to Literature. The approach we adopted in this chapter, which is now quite standard, could be found in Møller (1955, pp. 237–245) or Landau and Lifshitz (1975, pp. 234–237).

Special and general relativistic treatment of a rotating frame has a long history, with a lot of misunderstandings and apparent paradoxes discussed by great physicists. The interested reader is referred to Rizzi and Ruggiero (2004), which is a collection written by several authors.

For a criticism of the standard theory, which leads to non-Euclidean geometry of a rotating disk, see Strauss (1974); and for a criticism of his view, confirming the standard non-Euclidean geometry, see Grøn (1977).

A classical comprehensive review about the Sagnac effect is Post (1967). A more recent review which discusses both correct and incorrect explanations is Malykin (2000).

Sagnac did experiments in the first decade of the 20th century (1913). His experiments were considered by the skeptics of Einstein's special relativity as demonstrating the absolute space. It was Paul Langevin who, in 1921, explained the Sagnac effect according to special relativity. The interested reader is referred to (Pascoli, 2017).

Problem 32. Investigate the curvilinear coordinates $x^\mu = (t, r, \theta, z)$, where $X^\mu = (T, X, Y, Z)$ are Cartesian coordinates in an inertial frame K:

$$T = \frac{t}{\sqrt{1 - \frac{r^2\omega^2}{c^2}}}, \tag{7.119}$$

$$X = r\cos(\theta + \omega t), \tag{7.120}$$

$$Y = r\sin(\theta + \omega t), \tag{7.121}$$

$$Z = z. \tag{7.122}$$

1. What is the path of a particle with constant r, θ, and z?
2. What is the velocity of the particle with constant r, θ, and z, relative to K?
3. Show that $x^0 = t$ *is* a time-like coordinate for $0 \le r < c/\omega$.
4. Does this x^μ coordinate system describe a *uniformly* rotating frame?

Problem 33. Write coordinates for a *uniformly* rotating frame, in which t is the proper time measured by a clock at (r, θ, z) (usual cylindrical coordinates). Check your answer by computing $\varsigma\, g_{00}$—it should be $-c^2$.

7.8 Nonrigid hyperbolic motion

Consider a spaceship S_1 moving with constant positive proper acceleration α. Suppose at time $t = 0$, the spaceship is at point x_1 and momentarily at rest. As we saw in (3.120), the world-line of S_1 is

$$\left(x - x_1 + \frac{c^2}{\alpha}\right)^2 - c^2 t^2 = \frac{c^4}{\alpha^2}. \tag{7.123}$$

Evidently, if S_2 is another spaceship with the same positive proper acceleration being momentarily at rest at $(t = 0, x = x_2)$, the world-line of S_2 is

$$\left(x - x_2 + \frac{c^2}{\alpha}\right)^2 - c^2 t^2 = \frac{c^4}{\alpha^2}. \tag{7.124}$$

These two world-lines are drawn in Fig. 7.4. Note that the asymptotes are parallel, but not the same.

An important property of this situation is that the distance between S_1 and S_2, as measured in the stationary frame K, is always the same (see Fig. 7.4), in spite of the fact that the velocities of both spaceships are increasing so that the distance

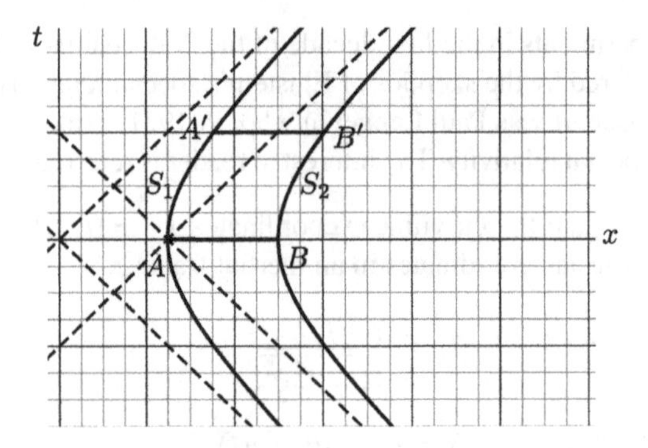

FIGURE 7.4 Nonrigid hyperbolic motion.

Particles S_1 and S_2 are moving with the same proper acceleration, both starting in the inertial frame K at time $t = 0$. The distance between them as measured in K is constant: At time $t = 0$ it is the distance between events A and B, and at later time t it is the distance between events A' and B'. Since at time t' the velocity of the rod $S_1 S_2$ is not the same at its velocity at $t = 0$, if measured by measuring rods comoving with them, the length of the rod $S_1 S_2$ at t' is longer than the length of it at t. Therefore such an accelerated frame is not rigid.

between S_1 and S_2 *in the rest frame of the spaceships* is increasing. To see this, note that because of the relativistic contraction of the lengths, the length of the moving rod is smaller than its proper length. In the case of S_1 and S_2, the length of the moving rod (joining S_1 to S_2) is constant, while its velocity is increasing, so that the proper length must be increasing as well.

Generalizing to a continuum of point observers is trivial. If X is a positive real number, the family of world-lines

$$\left(x - X + \frac{c^2}{\alpha}\right)^2 - c^2 t^2 = \frac{c^4}{\alpha^2}, \qquad 0 < X < \infty, \qquad (7.125)$$

describes a squadron moving with constant proper acceleration α. This squadron is *nonrigid* in the sense that the distance between nearby spaceships, as measured by the spaceships, is increasing with time. We are not going to write the change of coordinates leading to this squadron as a frame.

7.9 The Rindler frame

Note on notation. In this section (t, x) and (t', x') are Cartesian coordinates in inertial frames K and K'; the curvilinear coordinates are (θ, ξ). We do not write the (y, z)-coordinates for simplicity. The reader is asked to complete formulas for the $(3 + 1)$-dimensional space-time.

Let ξ be a fixed positive real number. The curve $x^2 - c^2 t^2 = \xi^2$ defines a hyperbola. Comparing with (3.120), we see that this is the world-line of a particle experiencing constant proper acceleration

$$\alpha = \frac{c^2}{\xi} \tag{7.126}$$

with initial conditions

$$x(0) = \xi, \quad \dot{x}(0) = 0. \tag{7.127}$$

Let us prove these assertions. First, if $t = 0$ we get $x^2 = \xi^2$ and therefore $x(0) = \xi$. Second, differentiating $x^2 - c^2 t^2 = \xi^2$ we get $x\,\mathrm{d}x - c^2 t\,\mathrm{d}t = 0$, from which we get

$$\frac{\mathrm{d}x}{\mathrm{d}t} = c\frac{ct}{x}. \tag{7.128}$$

It is evident that for $t = 0$ we get $\dot{x}(0) = 0$.

If the inertial frame K' is moving with velocity v along the x-axis relative to the inertial frame K, then we have

$$x^2 - c^2 t^2 = x'^2 - c^2 t'^2. \tag{7.129}$$

Therefore the world-line of this accelerating particle in K' is given by $x'^2 - c^2 t'^2 = \xi^2$. By the same reasoning given in the last paragraph, we know that at $t' = 0$ the particle has zero velocity with respect to K' and is a distance $x' = \xi$ from the origin.

Now consider the family of world-lines

$$x^2 - c^2 t^2 = \xi^2, \qquad 0 < \xi < \infty. \tag{7.130}$$

This family of world-lines covers the following subset of the space-time, which we denote by region I (see Fig. 7.5):

$$I = \left\{(t, x) \mid \ x > 0,\ x^2 - c^2 t^2 > 0\right\}. \tag{7.131}$$

This family of world-lines is called the Rindler frame and has the following features:

1. The particle with parameter ξ has constant proper acceleration $\alpha = c^2/\xi$, which means different values for different ξ.
2. All these hyperbolas have the same asymptotes:

$$\Delta^+, \qquad ct - x = 0, \tag{7.132}$$

$$\Delta^-, \qquad ct + x = 0. \tag{7.133}$$

3. All these accelerating particles have velocity 0 relative to K at time $t = 0$.
4. All these accelerating particles have velocity 0 relative to K' at time $t' = 0$.

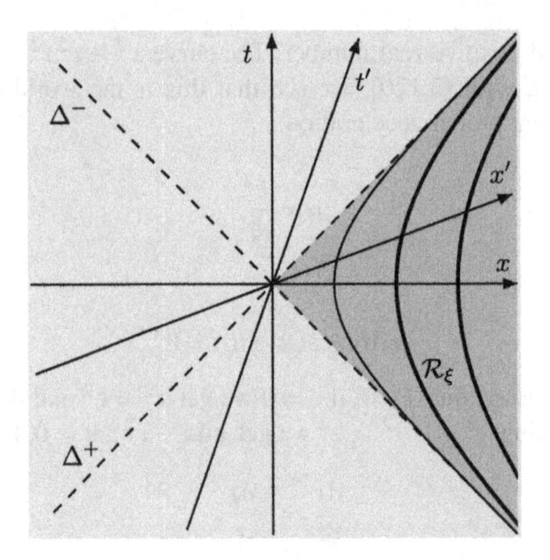

FIGURE 7.5 Rindler frame.

($c = 1$) The family of hyperbolas $x^2 - c^2 t^2 = \xi^2$ for positive ξ defines region I (shaded). All these hyperbolas have the same future and past event horizon $\Delta^{\pm}$. For the inertial observer K with coordinates (t, x), all these moving particles have velocity 0 at $t = 0$. For the inertial observe K' with coordinates (t', x'), which is moving with velocity v relative to K, all these accelerating particles have velocity 0 at $t' = 0$.

Rindler coordinates. If (t, x) are the usual coordinates on the Minkowski space-time, in an inertial frame, we define (θ, ξ) by the following pair of equations:

$$x = \xi \cosh\theta, \tag{7.134}$$

$$ct = \xi \sinh\theta. \tag{7.135}$$

It is easy to see that

$$x^2 - c^2 t^2 = \xi^2, \tag{7.136}$$

$$\frac{ct}{x} = \tanh\theta. \tag{7.137}$$

Using (3.125) and (7.126), if the world-line of a clock is $x^2 - c^2 t^2 = \xi^2$, the proper time, read by a clock comoving with the particle, is

$$\tau = \frac{\xi}{c} \sinh^{-1} \frac{ct}{\xi} = \frac{\xi}{c} \sinh^{-1} \left(\frac{ct}{x} \cdot \frac{x}{\xi} \right) \tag{7.138}$$

$$= \frac{\xi}{c} \sinh^{-1} (\tanh\theta \cdot \cosh\theta) \tag{7.139}$$

$$= \frac{\xi}{c} \sinh^{-1}(\sinh\theta) \tag{7.140}$$

$$= \frac{\xi\,\theta}{c}. \tag{7.141}$$

Here θ is called the Rindler time. Note that the astronaut in spaceship ξ has a clock to measure τ and knows ξ (which is constant for his/her position) and can calculate θ, which is called *the Rindler coordinate time.*

The same relation could be derived another way, which is worthy to mention. We have

$$\mathrm{d}\tau^2 = \frac{1}{c^2}\left[\xi^2\,\mathrm{d}\theta^2 - \mathrm{d}\xi^2\right]. \tag{7.142}$$

The interpretation of this equation is simple and important. A clock stationary in the Rindler frame has the constant spatial coordinate ξ. Two successive ticks of the clock are the events

$$E_1 = (\theta, \xi), \tag{7.143}$$

$$E_2 = (\theta + \mathrm{d}\theta, \xi), \tag{7.144}$$

and the proper time interval between these ticks is given by $\mathrm{d}\tau = \frac{\xi}{c}\,\mathrm{d}\theta$ from which we get $\tau = \xi\,\theta/c$ (since ξ is constant).

Rigidity of the Rindler frame. A very important property of the Rindler frame is that it is *rigid*, by which we mean that the distance between nearby particles remains the same in the comoving frame. This could be seen by observing this: Consider the continuum $x^2 - c^2 t^2 = \xi^2$. At a fixed time t differentiate this to get

$$x\,\Delta x = \xi\,\Delta\xi, \tag{7.145}$$

from which we get

$$\frac{\Delta x}{\Delta\xi} = \frac{\xi}{x} = \frac{\sqrt{x^2 - c^2 t^2}}{x} = \sqrt{1 - \frac{c^2 t^2}{x^2}} \tag{7.146}$$

$$= \sqrt{1 - u^2(\xi, t)}, \tag{7.147}$$

where $u(\xi, t)$ is the velocity of particle ξ at time t given in (7.128). Now this equation is nothing but the rigidity criterion. To see this, let us rewrite it as follows:

$$\Delta x = \Delta\xi\,\sqrt{1 - u^2}, \tag{7.148}$$

which is the relativistic length contraction relation, where Δx is the length of the moving rod, $\Delta\xi$ is the length of the rod in its rest frame, and $\sqrt{1 - u^2}$ is the inverse of the Lorentz γ factor.

Guide to Literature. For a short text on the proper acceleration and the rigidity of Rindler frame see Rindler (1969, pp. 49–51) or Stephani (2004, pp. 28–30).

Coordinate velocity of light. Consider a Rindler frame. Suppose a light ray is emitted at point ξ_0 at Rindler time θ_0, and this same light ray is absorbed at point ξ_1 at Rindler time θ_1. Suppose that

$$\xi_1 = \xi_0 + \Delta\xi, \quad \theta_1 = \theta_0 + \Delta\theta, \tag{7.149}$$

where $\Delta\xi$ and $\Delta\theta$ are infinitesimal. For these two events—emission and absorption of light—we have $\mathrm{d}s^2 = 0$, so that we get

$$\Delta s^2 = 0 = \Delta\xi^2 - \theta_0^2\,\Delta\theta^2, \tag{7.150}$$

from which it follows that

$$\beta_{\text{light}} = \frac{\Delta\xi}{\Delta\theta} = \theta_0. \tag{7.151}$$

The interpretation of this equation is that the *coordinate* velocity of light depends on the Rindler time θ_0. There is nothing odd here. Remember that it is just a consequence of defining time and spatial coordinates in a peculiar way.

Problem 34. Compare Rindler's coordinates (θ, ξ) of the space-time with polar coordinates (r, θ) of the Euclidean plane.

Problem 35. Consider the change of coordinates

$$x = \xi\,\cosh\theta, \quad ct = \xi\,\sinh\theta, \tag{7.152}$$

where (t, x) are the usual Minkowski coordinates. Draw a graph where ξ and θ are Cartesian orthogonal, ξ being the abscissa and θ being the ordinate. Draw the following world-lines, and interpret. Use $c = 1$, $x_0 = 1$, and $v = 0.6\,c$.

1. $x = x_0$,
2. $x = x_0 + ct$,
3. $x = x_0 - ct$,
4. $x = x_0 + vt$.

7.10 Motion of free particles

In the Minkowski space-time and in a Cartesian coordinate system X^μ, the equations of motion of a free particle are

$$\frac{\mathrm{d}^2 X^\alpha}{\mathrm{d}\tau^2} = 0, \tag{7.153}$$

where $\mathrm{d}\tau$ is the proper time. We now write this equation in terms of the curvilinear coordinates x^μ. We begin by

$$\frac{\mathrm{d}X^\alpha}{\mathrm{d}\tau} = \frac{\partial X^\alpha}{\partial x^\mu}\frac{\mathrm{d}x^\nu}{\mathrm{d}\tau}, \tag{7.154}$$

and then

$$\begin{aligned}\frac{\mathrm{d}^2X^\alpha}{\mathrm{d}\tau^2} &= \frac{\mathrm{d}}{\mathrm{d}\tau}\left(\frac{\partial X^\alpha}{\partial x^\mu}\frac{\mathrm{d}x^\mu}{\mathrm{d}\tau}\right)\\ &= \frac{\partial X^\alpha}{\partial x^\mu}\frac{\mathrm{d}^2x^\mu}{\mathrm{d}\tau^2} + \frac{\mathrm{d}x^\mu}{\mathrm{d}\tau}\frac{\mathrm{d}}{\mathrm{d}t}\frac{\partial X^\alpha}{\partial x^\mu}\\ &= \frac{\partial X^\alpha}{\partial x^\mu}\frac{\mathrm{d}^2x^\mu}{\mathrm{d}\tau^2} + \frac{\mathrm{d}x^\mu}{\mathrm{d}\tau}\frac{\partial^2X^\alpha}{\partial x^\nu\,\partial x^\mu}\frac{\partial x^\nu}{\partial\tau} = 0.\end{aligned} \tag{7.155}$$

Now we use the identity

$$\frac{\partial x^\kappa}{\partial X^\alpha}\frac{\partial X^\alpha}{\partial x^\mu} = \delta^\kappa_\mu \tag{7.156}$$

to write this equation as follows:

$$\frac{\mathrm{d}^2x^\kappa}{\mathrm{d}\tau^2} + \Gamma^\kappa_{\mu\nu}\frac{\mathrm{d}x^\mu}{\mathrm{d}\tau}\frac{\mathrm{d}x^\nu}{\mathrm{d}\tau} = 0, \tag{7.157}$$

where

$$\Gamma^\kappa_{\mu\nu} = \frac{\partial x^\kappa}{\partial X^\alpha}\frac{\partial^2X^\alpha}{\partial x^\mu\,\partial x^\nu}, \tag{7.158}$$

which are called the coefficients of the *connection*, or Christoffel symbols. Eq. (7.157) describes motion along a line in Minkowski space-time, in curvilinear coordinates x^μ. These equations are called equations of the geodesic motion.

Motions in a rotating frame. Earlier, in §7.7, we saw the cylindrical metric of a rotating frame. We now use the Cartesian coordinates. Consider the following change of coordinates:

$$\begin{cases} T = t,\\ X = x\cos\omega t - y\sin\omega t,\\ Y = x\sin\omega t + y\cos\omega t,\\ Z = z,\end{cases} \qquad \begin{cases} t = T,\\ x = X\cos\omega T + Y\sin\omega T,\\ y = -X\sin\omega T + Y\cos\omega T,\\ z = Z.\end{cases} \tag{7.159}$$

Problem 36. Define $x^\mu = (t, x, y, z)$. Find the metric $g_{\mu\nu}$, where $\mathrm{d}s^2 = g_{\mu\nu}\,\mathrm{d}x^\mu\,\mathrm{d}x^\nu$.

The calculation of the Christoffel symbols is straightforward, but lengthy. Here we do the calculation for Γ^1_{00}:

$$\begin{aligned}\Gamma^1_{00} &= \sum_\alpha \frac{\partial x^1}{\partial X^\alpha}\frac{\partial^2X^\alpha}{\partial x^0\,\partial x^0}\\ &= \frac{\partial x}{\partial X^\alpha}\frac{\partial^2X^\alpha}{\partial t^2}\\ &= \frac{\partial x}{\partial X^0}\frac{\partial^2X^0}{\partial t^2} + \frac{\partial x}{\partial X^1}\frac{\partial^2X^1}{\partial t^2} + \frac{\partial x}{\partial X^2}\frac{\partial^2X^2}{\partial t^2} + \frac{\partial x}{\partial X^3}\frac{\partial^2X^3}{\partial t^2}\end{aligned}$$

$$
\begin{aligned}
&= \frac{\partial x}{\partial T}\frac{\partial^2 T}{\partial t^2} + \frac{\partial x}{\partial X}\frac{\partial^2 X}{\partial t^2} + \frac{\partial x}{\partial Y}\frac{\partial^2 Y}{\partial t^2} + \frac{\partial x}{\partial Z}\frac{\partial^2 Z}{\partial t^2} \\
&= 0 + \cos\omega T\left(-\omega^2 X\right) + \sin\omega T\left(-\omega^2 Y\right) + 0 \\
&= -\omega^2\,(X\cos\omega T + Y\sin\omega T) \\
&= -\omega^2 x. \qquad (7.160)
\end{aligned}
$$

In this way one can show that the nonvanishing components of the connection are

$$\Gamma^1_{02} = \Gamma^1_{20} = -\omega, \qquad (7.161)$$

$$\Gamma^2_{01} = \Gamma^2_{10} = \omega, \qquad (7.162)$$

$$\Gamma^1_{00} = -\omega^2 x, \qquad (7.163)$$

$$\Gamma^2_{00} = -\omega^2 y. \qquad (7.164)$$

It is now easy to write the equations of geodesic motion:

$$\frac{d^2 t}{d\tau^2} = 0, \qquad (7.165)$$

$$\frac{d^2 x}{d\tau^2} = \omega^2 x\left(\frac{dt}{d\tau}\right)^2 + 2\omega\frac{dy}{d\tau}\frac{dt}{d\tau}, \qquad (7.166)$$

$$\frac{d^2 y}{d\tau^2} = \omega^2 y\left(\frac{dt}{d\tau}\right)^2 - 2\omega\frac{dx}{d\tau}\frac{dt}{d\tau}, \qquad (7.167)$$

$$\frac{d^2 z}{d\tau^2} = 0. \qquad (7.168)$$

For a nonrelativistic motion, we have $dt \simeq d\tau$, and these equations become the usual Newton equations in a rotating frame; $M(\omega^2 x, \omega^2 y, 0)$ are the components of the centrifugal force, and $M(2\omega\dot{y}, -2\omega\dot{x}, 0)$ are the components of the Coriolis force (M is the mass of the particle).

7.11 Covariant derivative

In this section we briefly mention how to write partial differential equations for vector and tensor fields in a curvilinear coordinate system. We do not go through the proof of some of the assertions. These topics are usually taught in courses on differential geometry and general relativity. Our aim here is to present how one can writes Maxwell's equations in curvilinear coordinates.

In earlier chapters, we introduced the notion of upper-index and lower-index 4-vectors and tensors. Let us recall the definition. First, a Lorentz transformation in the form of a change of coordinates is

$$X^\alpha = L^\alpha{}_\beta\, X'^\beta + a^\alpha, \qquad (7.169)$$

where a^αs are constant and $L^\alpha{}_\beta$ is the matrix of a Lorentz transformation. Noting that

$$\frac{\partial X^\alpha}{\partial X'^\beta} = L^\alpha{}_\beta \tag{7.170}$$

and using the usual chain rule of differentiation, we get the following two identities:

$$\mathrm{d}X^\alpha = L^\alpha{}_\beta\, \mathrm{d}X'^\beta, \tag{7.171}$$

$$\frac{\partial}{\partial X'^\beta} = L^\alpha{}_\beta \frac{\partial}{\partial X^\alpha}. \tag{7.172}$$

By definition, any set of four quantities transforming as $\mathrm{d}X^\alpha$ is said to be an upper-index 4-vector, and any set of four quantities transforming as $\partial_\alpha = \partial/\partial_{X^\alpha}$ is said to be a lower-index 4-vector. Now we generalize this to a general change of coordinates:

$$X^\alpha = f^\alpha(x), \qquad x = (x^0, x^1, x^2, x^3). \tag{7.173}$$

First, remember that

$$\mathrm{d}X^\alpha = \frac{\partial X^\alpha}{\partial x^\mu}\mathrm{d}x^\mu, \tag{7.174}$$

$$\frac{\partial}{\partial X^\alpha} = \frac{\partial x^\mu}{\partial X^\alpha}\frac{\partial}{\partial x^\mu}. \tag{7.175}$$

Note that we have written the last equation in a different equivalent form.

Definition. A general 4-vector is a set of four quantities changing under a general change of coordinates as $\mathrm{d}X^\alpha$ or $\frac{\partial}{\partial X^\alpha}$. This means that under a coordinate change $\tilde{x} = f(x)$,

$$\tilde{A}^\alpha = \frac{\partial \tilde{x}^\alpha}{\partial x^\mu} A^\mu, \tag{7.176}$$

$$\tilde{A}_\alpha = \frac{\partial x^\mu}{\partial \tilde{x}^\alpha} A_\mu. \tag{7.177}$$

Similarly, F is said to be a tensor if

$$\tilde{F}^{\alpha\beta} = \frac{\partial \tilde{x}^\alpha}{\partial x^\mu}\frac{\partial \tilde{x}^\beta}{\partial x^\nu} F^{\mu\nu}, \tag{7.178}$$

$$\tilde{F}^\alpha{}_\beta = \frac{\partial \tilde{x}^\alpha}{\partial x^\mu}\frac{\partial x^\beta}{\partial \tilde{x}^\nu} F^\mu{}_\nu, \tag{7.179}$$

$$\tilde{F}_{\alpha\beta} = \frac{\partial x^\mu}{\partial \tilde{x}^\alpha}\frac{\partial x^\nu}{\partial \tilde{x}^\beta} F_{\mu\nu}. \tag{7.180}$$

Now let us calculate $\tilde{A}^\alpha{}_{,\beta}$, where a comma "," means partial differentiation and $\tilde{x}$ is supposed to be Cartesian coordinates on the Minkowski space-time:

$$\tilde{A}^\alpha{}_{,\beta} = \frac{\partial \tilde{A}^\alpha}{\partial \tilde{x}^\beta} = \frac{\partial x^\mu}{\partial \tilde{x}^\beta}\frac{\partial}{\partial x^\mu}\left(\frac{\partial \tilde{x}^\alpha}{\partial x^\nu} A^\nu\right) \tag{7.181}$$

$$= \frac{\partial x^\mu}{\partial \tilde{x}^\beta}\frac{\partial \tilde{x}^\alpha}{\partial x^\nu}\frac{\partial A^\mu}{\partial x^\mu} + \frac{\partial x^\mu}{\partial \tilde{x}^\beta}\frac{\partial^2 \tilde{x}^\alpha}{\partial x^\mu \partial x^\nu} A^\alpha \tag{7.182}$$

$$= \frac{\partial x^\mu}{\partial \tilde{x}^\beta}\frac{\partial \tilde{x}^\alpha}{\partial x^\nu}\left\{\frac{\partial A^\nu}{\partial x^\mu} + \frac{\partial x^\nu}{\partial \tilde{x}^\sigma}\frac{\partial^2 \tilde{x}^\sigma}{\partial x^\mu \partial x^\kappa} A^\kappa\right\} \tag{7.183}$$

$$= \frac{\partial x^\mu}{\partial \tilde{x}^\beta}\frac{\partial \tilde{x}^\alpha}{\partial x^\nu} A^\nu{}_{;\mu}, \tag{7.184}$$

where

$$A^\nu{}_{;\mu} = A^\nu{}_{,\mu} + \Gamma^\nu_{\mu\kappa} A^\kappa, \tag{7.185}$$

$$\Gamma^\mu_{\nu\kappa} = \frac{\partial x^\nu}{\partial \tilde{x}^\sigma}\frac{\partial^2 \tilde{x}^\sigma}{\partial x^\mu \partial x^\kappa}. \tag{7.186}$$

Using the same procedure, we get

$$\tilde{A}_{\alpha,\beta} = \frac{\partial x^\mu}{\partial \tilde{x}^\alpha}\frac{\partial x^\nu}{\partial \tilde{x}^\beta} A_{\mu;\nu}, \tag{7.187}$$

$$A_{\mu;\nu} = A_{\mu,\nu} - \Gamma^\kappa_{\mu\nu} A_\kappa. \tag{7.188}$$

Definition. $A^\mu{}_{;\nu}$, $A_{\mu;\nu}$, $F^{\mu\sigma}{}_{;\nu}$, etc., are called covariant partial derivatives.

Guide to Literature. The material briefly presented in this section is most completely covered in texts on differential geometry such as Bishop and Goldberg (1980). But these texts are not usually useful for students who have not passed the required background material. The subject is also covered in textbooks on general relativity such as Weinberg (1972, pp. 91–129), which is an excellent text—both rigorous and concise. Battaglia and George (2013) is also recommended. It is addressed to undergraduate students, covers general tensors suitable for use in the theory of elasticity, and gives a list of textbooks on the subject.

7.12 Electrodynamics in curvilinear coordinates

In §4.6 we saw the Maxwell equations in covariant form. Using a tilde for quantities in the Cartesian coordinates $\tilde{x}^\mu$, we now rewrite the basic equations. First, we have the 4-potential $\tilde{A}^\mu$, from which we derived the electromagnetic field tensor

$$\tilde{F}_{\mu\nu} = \tilde{A}_{\nu,\mu} - \tilde{A}_{\mu,\nu}. \tag{7.189}$$

Denoting the 4-current by $\tilde{J}^\mu$, Maxwell's equations are

$$\tilde{F}^{\mu\nu}{}_{,\beta} = \varsigma\, \mu_0\, \tilde{J}^\nu, \tag{7.190}$$

$$\tilde{F}_{\mu\nu,\kappa} + \tilde{F}_{\kappa\mu,\nu} + \tilde{F}_{\nu\kappa,\mu} = 0. \tag{7.191}$$

Now, to write these equations in curvilinear coordinates x^μ, all we have to do is to replace partial differentiation with covariant partial differentiation. Let us see what happens. First, because $\Gamma^\alpha_{\mu\nu}$ is symmetric in μ and ν, we get

$$A_{\mu;\nu} - A_{\nu;\mu} = A_{\mu,\nu} - \Gamma^\alpha_{\mu\nu} A_\alpha - A_{\nu,\mu} + \Gamma^\alpha_{\nu\mu} A_\alpha \tag{7.192}$$

$$= A_{\mu,\nu} - A_{\nu,\mu}. \tag{7.193}$$

This means that in a general curvilinear coordinate we have

$$F_{\mu\nu} = \frac{\partial A_\nu}{\partial x^\mu} - \frac{\partial A_\mu}{\partial x^\nu}. \tag{7.194}$$

Second, using again the symmetry of $\Gamma^\alpha_{\mu\nu}$, one can show that

$$F_{\mu\nu;\kappa} + F_{\kappa\mu;\nu} + F_{\nu\kappa;\mu} = F_{\mu\nu,\kappa} + F_{\kappa\mu,\nu} + F_{\nu\kappa,\mu}. \tag{7.195}$$

Third, we write the inhomogeneous equations:

$$\varsigma \mu_0 J^\mu = F^{\mu\nu}{}_{;\nu} \tag{7.196}$$

$$= F^{\mu\nu}{}_{,\nu} + \overbrace{\Gamma^\mu_{\nu\beta} F^{\beta\nu}}^{=0} + \Gamma^\nu_{\nu\beta} F^{\mu\beta}. \tag{7.197}$$

It can be shown that

$$\Gamma^\nu_{\nu\beta} = \frac{1}{\sqrt{|g|}} \frac{\partial}{\partial x^\beta} \sqrt{|g|}, \tag{7.198}$$

where

$$g = \det\left[g_{\mu\nu}\right]. \tag{7.199}$$

Using this, Maxwell's inhomogeneous equation could be written in a more compact form:

$$\begin{aligned}
\varsigma \mu_0 J^\mu &= \partial_\nu F^{\mu\nu} + \frac{1}{\sqrt{|g|}} \partial_\beta \sqrt{|g|}\, F^{\mu\beta} \\
&= \partial_\beta F^{\mu\beta} + \frac{1}{\sqrt{|g|}} \partial_\beta \sqrt{|g|}\, F^{\mu\beta} \\
&= \frac{1}{\sqrt{|g|}} \partial_\beta \left(\sqrt{|g|}\, F^{\mu\beta}\right).
\end{aligned} \tag{7.200}$$

References

Battaglia, F., George, T.F., 2013. Tensors: a guide for undergraduate students. American Journal of Physics 81 (7), 498–511.

Bishop, R.L., Goldberg, S.I., 1980. Tensor Analysis on Manifolds. Dover, New York.

Einstein, A., 1980. The Meaning of Relativity. Chapman and Hall, London.
Grøn, Ø., 1977. Rotating frames in special relativity analyzed in light of a recent article by M. Strauss. International Journal of Theoretical Physics 16 (8), 603–614.
Landau, L.D., Lifshitz, E.M., 1975. The Classical Theory of Fields, fourth revised English edition. Landau and Lifshitz Course of Theoretical Physics, vol. 2. Pergamon Press, Oxford.
Malykin, G.B., 2000. The Sagnac effect: correct and incorrect explanations. Physics Uspekhi 43 (12), 1229–1252.
Møller, C., 1955. The Theory of Relativity. Clarendon Press, Oxford.
Pascoli, G., 2017. The Sagnac effect and its interpretation by Paul Langevin. Comptes Rendus. Physique 18 (9), 563–569.
Post, E.J., 1967. Sagnac effect. Reviews of Modern Physics 39 (2), 475–493.
Rindler, W., 1969. Essential Relativity. Special, General, and Cosmological. Springer-Verlag, New York.
Rizzi, G., Ruggiero, M.L., 2004. Relativity in Rotating Frames. Fundamental Theories of Physics, vol. 135. Springer, Dordrecht.
Stephani, H., 2004. Relativity. An Introduction to Special and General Relativity, 3rd edition. Cambridge University Press, Cambridge.
Strauss, M., 1974. Rotating frames in special relativity. International Journal of Theoretical Physics 11 (2), 107–123.
Weinberg, S., 1972. Gravitation and Cosmology. Principles and Applications of the General Theory of Relativity. John Wiley & Sons, New York.

CHAPTER

Gravity

8

Finding a theory for gravity that is consistent with special relativity was very challenging, and ultimately led Einstein to formulate the general theory of relativity. Today, general relativity is a cornerstone of our fundamental physical theories of nature—another one being quantum field theory. One should notice that both of these theories, general relativity and quantum field theory, agree on the special theory of relativity.

To understand the difficulties of finding a theory of gravitation consistent with special relativity and the solution, we begin by a review of the Newtonian gravitational theory, and then state the problem and Einstein's solution.

8.1 Newtonian gravity

Newtonian dynamics, very successful in describing planetary motions, is a combination of two important discoveries of Newton:

1. The rules of dynamics—Newton's first, second, and third laws. These laws enable us to write equations of motion in an inertial frame as $\boldsymbol{F} = m\,\boldsymbol{a}$, where m is the so-called inertial mass, or simply the inertia (see for example Kleppner and Kolenkow, 2014, pp. 48–57).
2. The universal gravitation—stating that any two bodies in the universe attract each other with a force proportional to the *inertial* masses of the bodies and inversely proportional to the square of the distance of the two bodies. The force is along the line joining the two bodies. Thus, the force which particle j exerts on particle i is

$$\boldsymbol{F}_{ij} = -\frac{G\,m_i\,m_j}{\left|\boldsymbol{r}_i - \boldsymbol{r}_j\right|^2} \cdot \frac{\boldsymbol{r}_i - \boldsymbol{r}_j}{\left|\boldsymbol{r}_i - \boldsymbol{r}_j\right|}. \tag{8.1}$$

The constant G in this equation is Newton's gravitational constant. Its value must be determined experimentally. The currently accepted value is

$$G = 6.674\,30(15) \times 10^{-11}\mathrm{m}^3\,\mathrm{kg}^{-1}\,\mathrm{s}^{-2} \qquad \text{(Mohr et al., 2016)}. \tag{8.2}$$

Here the two digits written in parentheses, 15, mean that there is a standard uncertainty of 0.00015 in this number, or in other words, a standard deviation of $0.000\,15\,\mathrm{m}^3\,\mathrm{kg}^{-1}\,\mathrm{s}^{-2}$.

A Mathematical Approach to Special Relativity. https://doi.org/10.1016/B978-0-32-399708-9.00017-8

Remark. That the mass of test particle i in (8.1), m_i, is the same mass defined in the dynamics is very important. We shall return to this in the subsequent sections.

Remark. Newtonian theory states that the force of gravity does not depend on the velocity of the particles. In contrast, the force two charged bodies exert on each other is the similar Coulomb force *only if* the charges are stationary.

8.2 Newtonian gravitational field

The superposition principle states that if there are N particles in the world, the force exerted on particle i by other particles is

$$\boldsymbol{F}_i = \sum_{j=1, j\neq i}^{N} \boldsymbol{F}_{ij} = -m_i \sum_{j=1, j\neq i}^{N} \frac{G m_j}{\left|\boldsymbol{r}_i - \boldsymbol{r}_j\right|^2} \cdot \frac{\boldsymbol{r}_i - \boldsymbol{r}_j}{\left|\boldsymbol{r}_i - \boldsymbol{r}_j\right|}, \tag{8.3}$$

where $\boldsymbol{F}_{ij}$ is the force exerted by particle j on particle i. We now define *the gravitational field* produced by all particles except particle i:

$$\boldsymbol{g}(\boldsymbol{r}, t) = -\sum_{j=1, j\neq i}^{N} \frac{G m_j}{\left|\boldsymbol{r} - \boldsymbol{r}_j(t)\right|^2} \cdot \frac{\boldsymbol{r} - \boldsymbol{r}_j(t)}{\left|\boldsymbol{r} - \boldsymbol{r}_j(t)\right|}. \tag{8.4}$$

The equation of motion of particle i is $m_i\, \boldsymbol{a}_i = m_i\, \boldsymbol{g}(\boldsymbol{r}_i, t)$, and since m_i is the same on both sides,

$$\boldsymbol{a}_i = \boldsymbol{g}(\boldsymbol{r}_i, t). \tag{8.5}$$

Generalizing to a continuous distribution of mass in the universe is straightforward, and the Newtonian gravity could be summarized as follows:

1. If $\rho(\boldsymbol{r}, t)$ is the density of mass in the universe, the gravitational field at point $\boldsymbol{r}$ at time t is

$$\boldsymbol{g}(\boldsymbol{r}, t) = -G \int \rho(\boldsymbol{s}, t) \frac{\boldsymbol{r} - \boldsymbol{s}}{|\boldsymbol{r} - \boldsymbol{s}|^3}\, \mathrm{d}V, \qquad \mathrm{d}V = \mathrm{d}^3 s. \tag{8.6}$$

2. If at time t a test particle is at point $\boldsymbol{r}$ and no other force is exerted on it, then its equation of motion is

$$\boldsymbol{a} = \boldsymbol{g}(\boldsymbol{r}, t). \tag{8.7}$$

By a *test* particle, we mean it is so small that it has almost no effect on $\rho(\boldsymbol{r}, t)$.

Definition. A test particle is said to be in free fall (or is freely falling) if there is no force on it except gravity.

From (8.6) it can be shown that $\boldsymbol{g}$ satisfies the following two partial differential equations:

$$\nabla \times \boldsymbol{g}(\boldsymbol{r},t) = 0, \tag{8.8}$$

$$\nabla \cdot \boldsymbol{g}(\boldsymbol{r},t) = -4\pi G \rho(\boldsymbol{r},t). \tag{8.9}$$

From the first equation we deduce that there is a function $\phi(\boldsymbol{r},t)$ such that

$$\boldsymbol{g}(\boldsymbol{r},t) = -\nabla\phi(\boldsymbol{r},t). \tag{8.10}$$

This function ϕ is called the Newtonian gravitational potential. It has the dimensions of c^2. Inserting ϕ in the second Eq. (8.9), we get the Poisson equation:

$$\nabla^2\phi(\boldsymbol{r},t) = 4\pi G \rho(\boldsymbol{r},t), \tag{8.11}$$

with solution

$$\phi(\boldsymbol{r},t) = -G\int \frac{\rho(\boldsymbol{s},t)}{|\boldsymbol{r}-\boldsymbol{s}|}\,\mathrm{d}^3 s. \tag{8.12}$$

Remark. If $\rho(\boldsymbol{r},t)$ is nonzero only in a finite region of space and we impose the boundary condition $\phi(\infty) = 0$, the Newtonian potential ϕ is unique.

Guide to Literature. Physics students usually learn the Poisson equation in electrodynamics. Jackson (1999) is one of the best texts on this subject. For another derivation of the solution (8.12), see Levine (1997, pp. 587–588). Newtonian gravity is usually covered in textbooks on classical or celestial mechanics; see for example Fitzpatrick (2012). Symon (1971, pp. 259–270) is recommended, but the reader should remember that in this text gravitational potential, denoted by $\mathcal{G}$, is defined by $\boldsymbol{g} = \nabla\mathcal{G}$.

Propagation. We know that a time-varying electric charge density or electric current density causes a disturbance in the electromagnetic fields, which propagates as a wave with velocity c (see (2.12) and (2.13) on page 33 or (4.117) on page 88). But in Newtonian theory of gravitation, according to (8.11), a disturbance in the mass density propagates instantly, or with infinite velocity. This means that Newtonian gravitational theory is not consistent with special relativity.

8.3 Gravitational energy

Consider a configuration of N point objects of masses $m_1,\ldots,m_N$ at positions $\boldsymbol{r}_1,\ldots,\boldsymbol{r}_N$. To this configuration, at any time instant t, we associate an energy, called the gravitational energy, which is the work needed to separate particles infinitely. Equivalently, this energy is the energy released when we bring objects from infinity to positions $\boldsymbol{r}_1,\ldots,\boldsymbol{r}_N$. It is easy to see that if the configuration consists of only two particles, the gravitational energy is

$$U_{\mathrm{G}} = -\frac{G m_1 m_2}{r_{12}}, \qquad r_{12} = |\boldsymbol{r}_1 - \boldsymbol{r}_2|. \tag{8.13}$$

The mathematics is exactly the one used in electrostatics to get the electrostatic energy,

$$U_{\mathrm{E}} = \frac{1}{4\pi\,\epsilon_0} \cdot \frac{q_1 q_2}{r_{12}}. \tag{8.14}$$

The difference in sign is because the gravitational force of two objects is always attractive, while the electrostatic force of two charges of the same sign is repulsive.

Constructing the general configuration step by step, that is, first bringing m_1 from infinity to $\boldsymbol{r}_1$, then m_2 to $\boldsymbol{r}_2$, etc., we find

$$U_{\mathrm{G}} = -\frac{1}{2} \sum_{i,j,i\neq j}^{N} \frac{G\, m_i\, m_j}{r_{ij}}. \tag{8.15}$$

Generalizing this to the continuum, we see that the gravitational field energy is

$$U = -\frac{G}{2} \iint \frac{\rho(\boldsymbol{r},t)\,\rho(\boldsymbol{s},t)}{|\boldsymbol{r}-\boldsymbol{s}|}\, \mathrm{d}^3 r\, \mathrm{d}^3 s \tag{8.16}$$

$$= \frac{1}{2} \int \rho(\boldsymbol{r},t)\,\phi(\boldsymbol{r},t)\, \mathrm{d}^3 r \tag{8.17}$$

$$= -\frac{1}{8\pi\, G} \int |\nabla\phi|^2\, \mathrm{d}^3 r. \tag{8.18}$$

So

$$u_{\mathrm{G}} = -\frac{g^2(\boldsymbol{r},t)}{8\pi\, G} = -\frac{|\nabla\phi|^2}{8\pi\, G} \tag{8.19}$$

has the interpretation of gravitational energy density.

Remark. The proof of all these relations is a word-by-word copy of the proof for similar expressions in the electrostatic field (see for example Jackson, 1999, pp. 40–41).

8.4 Spherically symmetric fields

Matter is said to be spherically symmetric if there is a point O in space such that $\rho = \rho(r,t)$, where r is the distance from O. For such a mass distribution, the gravitational field $\boldsymbol{g}$ must be spherically symmetric, that is, $\boldsymbol{g}(r,t) = g(r,t)\,\hat{r}$. If S is the surface $x^2 + y^2 + z^2 = r^2$,

$$\oint_S \boldsymbol{g} \cdot \hat{n}\, \mathrm{d}a = 4\pi\, g(r,t)\, r^2, \tag{8.20}$$

and using the divergence theorem for $\nabla \cdot \boldsymbol{g} = -4\pi\, G\, \rho$ we get

$$4\pi\, r^2\, g(r,t) = -4\pi\, G \int_0^r \rho(s,t)\, s^2\, \mathrm{d}s = -4\pi\, G\, M(r,t), \tag{8.21}$$

where $M(r)$ is the total mass in the region $x^2 + y^2 + z^2 \leq r^2$. Therefore the gravitational field at point $\boldsymbol{r}$ is

$$g(r,t) = -\frac{G\,M(r,t)}{r^2}. \tag{8.22}$$

Remark. Note that $g(r)$ depends on t only if $M(r)$ depends of t. Thus a pulsating star generates a static gravitational field outside the star if the pulsations are spherically symmetric.

An important consequence is that outside a spherically symmetric distribution of mass, *e.g.*, a star, the gravitational field is

$$\boldsymbol{g} = -\frac{G\,M}{r^2}\,\hat{r}. \tag{8.23}$$

One should note that this expression is valid only for $r > R$, where R is the radius of the star. Inside the star, if we know the density function, we can find the gravitational field.

Problem 37. Consider a spherical shell of radius a and mass M. Prove that outside the shell, the gravitational field is as if all the mass is located at the center, and inside the shell, the gravitational field is zero.

Problem 38. Consider a sphere of radius a with constant mass density ρ_0 and total mass M. Show that

$$\boldsymbol{g}(r) = \begin{cases} -\dfrac{G\,M}{a^3}\boldsymbol{r}, & r < a, \\[2ex] -\dfrac{G\,M}{r^3}\boldsymbol{r}, & r > a, \end{cases} \tag{8.24}$$

$$\phi(r) = \begin{cases} \dfrac{G\,M}{2a}\left(\dfrac{r^2}{a^2} - 3\right), & r < a, \\[2ex] -\dfrac{G\,M}{r}, & r > a. \end{cases} \tag{8.25}$$

Problem 39. Using (8.25), show that the gravitational energy of a sphere with uniform mass distribution is

$$U_{\mathrm{G}} = -\frac{3}{5}\cdot\frac{G\,M^2}{a}. \tag{8.26}$$

The dynamics of a test particle in the gravitational field of a spherically symmetric mass is quite well known; see for example Goldstein (1980) and Montenbruck and Gill (2000). For later use in our study, let us briefly state the following.

1. If two spherical objects of masses m_1 and m_2 form a binary with semimajor axis a, the period T of the orbit satisfies

$$\frac{G(m_1+m_2)}{a^3}=\frac{4\pi^2}{T^2}. \tag{8.27}$$

   This equation is the basis of determining the total mass of binaries.
2. If O is the center of a spherically symmetric mass distribution $\rho(r)$ and a test particle of mass m is moving on a circle with center O and radius r, the velocity v of the particle satisfies $\frac{GM(r)}{r^2}=\frac{v^2}{r}$, or

$$v(r)=\sqrt{\frac{GM(r)}{r}}. \tag{8.28}$$

   Using this equation, astronomers found that the mass of the spiral galaxies is more than what we can see.

If the spherically symmetric ρ is finite at $r=0$, then for small r we have $M(r)\propto\rho(0)\,r^3$, from which it follows that $g(0)=0$. If for small r and a positive n we have $\rho(r)\propto r^{-n}$, then $M(r)\propto r^{3-n}$ and $g(r)\propto r^{1-n}$. Here $n\geq 3$ leads to a divergent (infinite) mass. For $1<n<3$ the situation is interesting:

$$\lim_{r\to 0^+} M(r)\propto \lim_{r\to 0^+} r^{3-n}=0, \tag{8.29}$$

$$\lim_{r\to 0^+} g(r)\propto \lim_{r\to 0^+} r^{1-n}=\infty, \tag{8.30}$$

which says that although the mass content in a sphere of radius r is negligible for $r\to 0$, the gravitational field diverges at the center. For $n=1$, $M(r)\propto r^2$ and $g(r)\to$ constant! A steeply rising mass profile like this ($1\leq n<3$) is called a cusp. According to the current theories, an important component of the mass content of galaxies is dark matter. One important candidate for the density profile of dark matter in spiral galaxies is the Navarro–Frenk–White (NFW) profile, which has a cusp:

$$\rho_{\text{NFW}}(r)=\frac{\rho_0}{\frac{r}{a}\left(1+\frac{r}{a}\right)^2}\qquad \text{Navarro et al. (1996).} \tag{8.31}$$

Typical values suitable for the Milky Way are

$$\rho_0=1\times 10^{-21}\,\text{kg}\,\text{m}^{-3}, \tag{8.32}$$

$$a=5\times 10^{20}\,\text{m}. \tag{8.33}$$

8.5 Newton's equivalence principle

That the mass of the test particle i in (8.1), m_i, is the same mass defined in the dynamics—the inertial mass of the test particle or its inertia—is not obvious and must be based on experiment. Newton himself stated this and did experiments.

> *Furthermore, I mean this quantity whenever I use the term "body" or "mass" in the following pages. It can always be known from a body's weight, for—by making very accurate experiments with pendulums—I have found it to be proportional to the weight, as shall be shown below.*
>
> **Principia, Book 1, Definition 1 (Cohen et al., 1999, p. 404)**

> *All bodies gravitate toward each of the planets, and at any given distance from the center of any one planet the weight of any body whatever toward that planet is proportional to the quantity of matter which the body contains.*
>
> **Principia, Book 3, Proposition 6 (Cohen et al., 1999, p. 806)**

What Newton is saying is that if two *test* bodies with inertial masses m_1 and m_2 are at the same distance from a gravitating body (a planet), then their acceleration a_1 and a_2 would be the same. Let us define the so-called Eötvös[1] parameter for the two particles as follows:

$$\eta_{12} = 2\frac{a_1 - a_2}{a_1 + a_2}. \tag{8.34}$$

Temporarily, let us distinguish between the inertial mass m_{I}, appearing in Newton's second law, and the *passive* gravitational mass m_{G}, appearing in Newton's gravity, that is, suppose

$$m_{\mathrm{I}}\,\boldsymbol{a} = m_{\mathrm{G}}\,\boldsymbol{g}. \tag{8.35}$$

From this equation it follows that the Eötvös parameter is

$$\eta_{12} = 2\frac{a_1 - a_2}{a_1 + a_2} = 2\,\frac{(m_{\mathrm{G}}/m_{\mathrm{I}})_1 - (m_{\mathrm{G}}/m_{\mathrm{I}})_2}{(m_{\mathrm{G}}/m_{\mathrm{I}})_1 + (m_{\mathrm{G}}/m_{\mathrm{I}})_2}. \tag{8.36}$$

Note that $\eta_{12} = 0$ means that both particles have the same $m_{\mathrm{G}}/m_{\mathrm{I}}$ ratio. Experiments show that $|\eta| \ll 10^{-8}$ (Will, 2014). A recent measurement indicates that in the Earth's gravitational field

$$\eta_{\text{Be-Ti}} = (0.3 \pm 1.8) \times 10^{-13} \quad \text{(Schlamminger et al., 2008)}. \tag{8.37}$$

Newton's equivalence principle. Newton's equivalence principle states that $\eta_{ij} = 0$ for any pair of test particles in the gravitational field of any object. The consequence is that all test particles fall with the same acceleration in a gravitational field, no matter what the internal state and composition of the test particle is.

[1] After Loránd Eötvös de Vásárosnamény (1848–1919).

As we saw in §5.6, a consequence of special relativity is that the inertia of a body is not the sum of the rest masses of its constituents—the binding energy also contributes to the inertia of the system. Combining with Newton's equivalence principle, it follows that what determines the motion of a test particle in a gravitational field is not only the sum of the inertia of its constituents; the mass equivalent of the binding energy also contributes.

The origin of the binding energies is the interactions among the constituents of the system. For a hydrogen atom, the interaction is electromagnetic, and the binding energy (for the ground state) is $\sim 13.6\,\mathrm{eV}$. The mass equivalent of this is $\sim 10^{-8}\, M_{\mathrm{H}}$. Atomic nuclei are bound systems of several protons and neutrons, bound together by the nuclear forces. The fact that different atoms fall with the same acceleration in a gravitational field means that all the interaction energies have the same contribution to the gravitational mass.

What about gravitationally bound systems such as planets, stars, binaries, star clusters, or galaxies? Earth, for example, is a gravitationally bound system of mass $\sim 6 \times 10^{24}\,\mathrm{kg}$. Assuming a uniform mass density and using (8.26), the ratio of the gravitational binding energy to the rest energy of the Earth is

$$\frac{B}{M c^2} \sim \frac{G M^2}{R M c^2} = \frac{G M}{R c^2} = 10^{-9}. \tag{8.38}$$

For a binary

$$B \simeq \frac{G m_1 m_2}{2a}, \tag{8.39}$$

where a is the semimajor axis of the orbit. For the Earth–Moon system, $B \sim 4 \times 10^{28}\,\mathrm{J}$, equivalent to $B/c^2 \sim 4 \times 10^{11}\,\mathrm{kg}$ or $\sim 7 \times 10^{-14}(m_1 + m_2)$.

Remark. For objects such as the hydrogen atom or a binary, the binding energy is a sum of two energies: the interaction's potential energy and the kinetic energy of the particles (see §5.6 and page 105).

Remark. Whether or not the gravitational energy also contributes to the inertia as other energies is not obvious, and this is a very important question, experimentally as well as theoretically.

8.6 Tidal force

Consider a system S which consists of N objects with masses m_i at positions $\boldsymbol{r}_i(t)$. Let there be also another (external) distribution of mass $\rho_{\mathrm{e}}(\boldsymbol{r}, t)$ far enough from S. This external source has a gravitational field $\boldsymbol{g}_{\mathrm{e}} = -\nabla\phi_{\mathrm{e}}$. If $\boldsymbol{F}_{ij}$ is the force exerted by particle j on particle i—which could be gravitational or any other force—then the equation of motion of particle i is

$$m_i \ddot{\boldsymbol{r}}_i = \sum_{j=1 j \neq i}^{N} \boldsymbol{F}_{ij} + m_i \boldsymbol{g}_e(\boldsymbol{r}_i, t). \tag{8.40}$$

Remark. System S is said to be in free fall in the gravitational field generated by ρ_e if there is no external force on it except the gravity produced by ρ_e. Note that one should specify the system and the external matter. For example, consider the Moon. The external matter is the Sun and the Earth, and the Moon is in free fall in the gravitational field produced by the Sun and the Earth. However, the Moon is certainly not in free fall in the gravitational field of the Sun, because the Earth is also exerting a force on it.

As usual, let us define the total mass $M = \sum_{i=1}^{N} m_i$ and the center of mass $\boldsymbol{R} = \frac{1}{M}\sum_{i=1}^{N} m_i \boldsymbol{r}_i$. By Newton's third law, $\boldsymbol{F}_{ij} = -\boldsymbol{F}_{ji}$; if we sum all the equations we get

$$M\ddot{\boldsymbol{R}} = \sum_{i=1}^{N} m_i \ddot{\boldsymbol{r}}_i = \sum_{i=1}^{N} m_i \boldsymbol{g}_e(\boldsymbol{r}_i, t). \tag{8.41}$$

Defining $\boldsymbol{r}_i = \boldsymbol{R} + \boldsymbol{s}_i$ we Taylor expand $\boldsymbol{g}_e(\boldsymbol{R} + \boldsymbol{s}_i, t)$ to get

$$M\ddot{\boldsymbol{R}} = \sum_{i=1}^{N} m_i \boldsymbol{g}_e(\boldsymbol{R}, t) + \sum_{i=1}^{N} m_i \boldsymbol{s}_i \cdot \nabla \boldsymbol{g}_e(\boldsymbol{R}, t) = M\boldsymbol{g}_e(\boldsymbol{R}, t), \tag{8.42}$$

$$\ddot{\boldsymbol{R}} = \boldsymbol{g}_e(\boldsymbol{R}, t). \tag{8.43}$$

In getting this we have used $\sum_i m_i \boldsymbol{s}_i = 0$, which is true since $\boldsymbol{R}$ is the center of mass of S. Now, from the equation of motion of particle i, we get

$$m_i \ddot{\boldsymbol{s}}_i = m_i \ddot{\boldsymbol{r}}_i - m_i \ddot{\boldsymbol{R}} \tag{8.44}$$

$$= \sum_j \boldsymbol{F}_{ij} + \overbrace{m_i \boldsymbol{g}_e(\boldsymbol{R}, t) + m_i \boldsymbol{s}_i \cdot \nabla \boldsymbol{g}_e(\boldsymbol{R}, t)}^{=m_i \boldsymbol{g}_e(\boldsymbol{r}_i, t)} - m_i \boldsymbol{g}_e(\boldsymbol{R}, t) \tag{8.45}$$

$$= \boldsymbol{F}_i + m_i \boldsymbol{s}_i \cdot \nabla \boldsymbol{g}_e(\boldsymbol{R}, t). \tag{8.46}$$

This means that in the center of mass system, particle i senses two forces: $\boldsymbol{F}_i$, which is the force exerted by other particles of S on it, and a *tidal* force

$$\boldsymbol{F}_{Ti} = m_i \boldsymbol{h}(\boldsymbol{s}_i, t), \tag{8.47}$$

where

$$\boldsymbol{h}(\boldsymbol{s}, t) = \boldsymbol{s} \cdot \nabla \boldsymbol{g}_e(\boldsymbol{R}, t) \tag{8.48}$$

is the tidal acceleration field caused by the external gravitational field. The i-th component of this acceleration is

$$h_i = -\sum_{k=1}^{3} x_k \left. \frac{\partial^2 \phi_e}{\partial x^k \partial x^i} \right|_R . \tag{8.49}$$

Remark. If needed, one can Taylor expand $\boldsymbol{g}_e(\boldsymbol{R}+\boldsymbol{s},t)$ to higher orders. What we found here is the leading term.

Problem 40. Show that the tidal acceleration field $\boldsymbol{h}$ is "conservative,"[2] that is,

$$\nabla \times \boldsymbol{h}(\boldsymbol{s},t) = \nabla \times \boldsymbol{g}_e(\boldsymbol{R},t) = 0. \tag{8.50}$$

Behavior of the tidal field. The scale of the tidal acceleration (8.48) depends on two parameters:

1. The scale of the system, s (dimension L),
2. The scale of the tidal tensor $\partial_i \partial_j \phi_e$ (dimension T^{-2}).

Consider for example a spherically symmetric gravitating body of mass M. Differentiating (8.23) we see that

$$-\frac{\partial^2 \phi_e}{\partial x^i \partial x^j} = \frac{GM}{d^3}\left(3 n_i n_j - \delta_{ij}\right), \tag{8.51}$$

where $\boldsymbol{d} = d\,\hat{n}$ is the vector from the center of mass of S to the center of M. Therefore for $d \to 0$ the tidal tensor diverges as d^{-3}. It is easy to see that

$$\sum_j \left(3 n_i n_j - \delta_{ij}\right) n_j = 2 n_i, \tag{8.52}$$

which means that $\hat{n}$ is an eigenvector of the tidal tensor with eigenvalue $2GM/d^3$. If $\hat{p}$ is a vector normal to $\hat{n}$, then

$$\sum_j \left(3 n_i n_j - \delta_{ij}\right) p_j = -p_i, \tag{8.53}$$

which means that any vector normal to $\hat{n}$ is an eigenvector of the tidal tensor with eigenvalue $-GMs/d^3$. Therefore if $\boldsymbol{s}$ is parallel to $\boldsymbol{d}$, the tidal acceleration is $2GM/d^3$, away from the center of mass; and if $\boldsymbol{s}$ is perpendicular to $\boldsymbol{d}$, the tidal acceleration is GMs/d^3 toward the center of mass. In words, tidal force stretches the system S in the direction of $\boldsymbol{d}$ and compresses it perpendicular to $\boldsymbol{d}$ (see Fig. 8.1).

[2] In a strict sense it is conservative only if the field is independent of time.

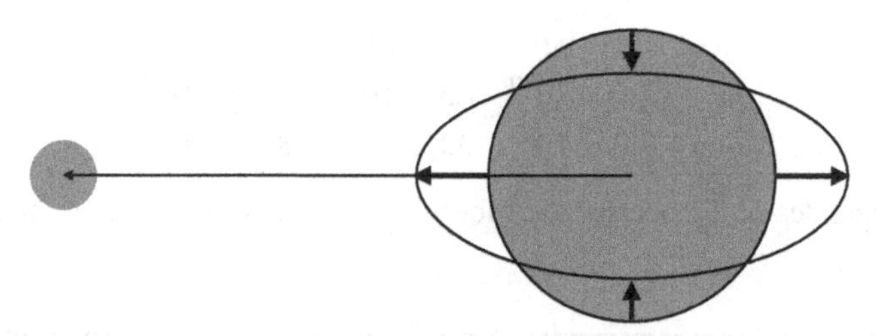

FIGURE 8.1 Tidal deformation of a spherical object by a point mass.

Tidal acceleration of the spherical mass M (left) on the system S (right) stretches S in the direction of M and compresses it normal to that.

Singularity near a point mass. Consider a point particle of mass M. For the moment we are not going through the question of the existence of such an object. Such a hypothetical object is called a black hole. Consider a hydrogen atom falling radially toward a black hole. Let us compare two forces: the electrostatic force the proton exerts on the electron and the tidal force of the black hole. If the radius of the atom is a and $M = M_\odot = 2.0 \times 10^{30}$ kg,

$$F_{\rm ep} = \frac{e^2}{4\pi\,\epsilon_0\,a^2}, \tag{8.54}$$

$$F_{\rm T} = m_{\rm e}\,\frac{2\,G\,M_\odot}{d^3}\,a, \tag{8.55}$$

$$\frac{F_{\rm T}}{F_{\rm ep}} = \frac{8\pi\,G\,M_\odot\,\epsilon_0\,m_{\rm e}}{e^2}\left(\frac{a}{d}\right)^3 \simeq 10^{18}\left(\frac{a}{d}\right)^3. \tag{8.56}$$

For $a/d \sim 10^{-6}$, this ratio is ~ 1. This means that if the distance of the hydrogen atom from a black hole of mass $M \sim M_\odot$ is $\sim 10^6\,a$, the tidal force of the black hole is as strong as the electrostatic force. Therefore when the atom is this close to the black hole, the tidal force would tear apart the atom! For $a \sim 0.1$ nm, we get $d \sim 0.1$ mm.

A proton is not a point particle; it is a bound state of three quarks. Its length scale is $a \sim 1$ fm. To estimate at what distance a proton would tear apart, we need to know the strength of the strong nuclear force, but the divergence of the tidal force indicates that at sufficiently close distance a proton would also tear apart.

Earth, Moon, and Sun. When we have a system S and an external gravitating body M, if they are in free fall, there is a tidal force due to M on S. It is not important whether or not M is more massive than S. As an example, consider the Earth as S. There are several external gravitating bodies: the Sun, the Moon, and the planets. The size of the Earth, s in our formulas, is of the order of $R_\oplus = 6.4 \times 10^6$ m. Let us compare the tidal acceleration coefficient ($G\,M/d^3$) due to the Sun and the Moon.

Object	M [kg]	d [m]	GM/d^3 [s^{-2}]
Sun	2.0×10^{30}	1.5×10^{11}	4.0×10^{-14}
Moon	7.3×10^{22}	3.8×10^{8}	8.9×10^{-14}

Thus, for the tides of the oceans, the Moon's influence is twice that of the Sun.

8.7 Search for a relativistic theory of gravity

Let us now think about a theory of gravity compatible with the special theory of relativity. Since classical electrodynamics is completely compatible with SR, our first approach is to write a theory analogous to electrodynamics. To this end, we introduce a field $\boldsymbol{h}$, analogous to the magnetic field, and write the following equations:

$$\boldsymbol{a} = \boldsymbol{g} + \frac{1}{c}\boldsymbol{u} \times \boldsymbol{h}, \tag{8.57}$$

$$\nabla \cdot \boldsymbol{h} = 0, \tag{8.58}$$

$$\nabla \times \boldsymbol{g} = -\frac{1}{c}\frac{\partial \boldsymbol{h}}{\partial t}, \tag{8.59}$$

$$\nabla \cdot \boldsymbol{g} = -4\pi G \rho, \tag{8.60}$$

$$\nabla \times \boldsymbol{h} = \frac{1}{c}\frac{\partial \boldsymbol{g}}{\partial t} - \frac{4\pi G}{c}\boldsymbol{J}. \tag{8.61}$$

In electrodynamics the 4-current density $J_{\mathrm{e}}^{\mu} = (\rho_{\mathrm{e}}, \boldsymbol{J}_{\mathrm{e}})$ is a 4-vector, and Maxwell's equations could be written in terms of the 4-potential A_{e}^{μ} satisfying (4.115)–(4.116):

$$\Box A_{\mathrm{e}}^{\mu} - \partial^{\mu}(\partial \cdot A_{\mathrm{e}}) = -\varsigma\, \mu_0\, J_{\mathrm{e}}^{\mu}. \tag{8.62}$$

So our Maxwell-type equations for the gravitation in the covariant form are

$$\Box A_{\mathrm{G}}^{\mu} - \partial^{\mu}(\partial \cdot A_{\mathrm{G}}) = 4\pi G \varsigma\, J_{\mathrm{m}}^{\mu}, \tag{8.63}$$

where J_{m}^{μ} is the mass-energy 4-current.

Though these equations seem to be exactly as the Maxwell equations, there is a big difference. While J_{e}^{μ} is a 4-vector, $J_{\mathrm{m}}^{\mu} = T_{\mathrm{m}}^{0\mu}$ is not a 4-vector—it is the 0-column of the energy-momentum-stress tensor. Therefore A_{G}^{μ} in this equation must also be the 0-column of a second rank tensor; otherwise, the equation is not Lorentz covariant. A truly Lorentz covariant set of equations for gravity would be of the form

$$\mathcal{O} g_{\mu\nu} = -4\pi G\, T_{\mu\nu}, \tag{8.64}$$

where $\mathcal{O}$ is a differential operator, perhaps of second order, $T_{\mu\nu}$ is the stress-energy-momentum tensor, and $g_{\mu\nu}$ is the gravitational potential. Perhaps the $\mu = \nu = 0$ component of this equation in the static limit reduces to the Poisson equation. The problem is to guess what $g_{\mu\nu}$ is and to find $\mathcal{O}$ and the correct equations of motion, which must reduce to $\boldsymbol{a} = \boldsymbol{g}$ for small velocities. Several physicists investigated this problem between 1905 and 1915, and finally, it was Einstein who found the solution.

8.8 Einstein's equivalence principle

By Newton's equivalence principle:

(1) In a uniform gravitational field $\boldsymbol{g}$ all bodies fall with the same acceleration $\boldsymbol{g}$. Also, because of $\boldsymbol{F} = m\,\boldsymbol{a}$, in a frame undergoing constant acceleration $-\boldsymbol{g}$, all test particles fall with the same acceleration $\boldsymbol{g}$.

(2) Since the tidal acceleration is proportional to the size scale of the system, in a system S of size scale a, freely falling in the external gravitational field $\boldsymbol{g}_\mathrm{e}$, no mechanical effect of gravity ($\boldsymbol{g}_\mathrm{e}$) is sensed if a is sufficiently small.

Einstein elevated this to a principle, which could be stated in two different, but equivalent forms.

Einstein's equivalence principle.

(1) The laws of physics are such that by no experiment—mechanical, optical, or anything—one can distinguish a uniform gravitational field from a uniformly accelerated frame in vacuum.

(2) In a small enough freely falling lab no effects of gravity—mechanical, optical, or anything—is sensed.

A geometric analogy. The following geometric analogy is useful to get a feeling of what we mean by saying that in a small enough freely falling frame there is *almost* no force of gravity.

Consider the sphere $x^2 + y^2 + z^2 = a^2$. Differentiating this we get

$$x\,\mathrm{d}x + y\,\mathrm{d}y + z\,\mathrm{d}z = 0, \tag{8.65}$$

from which we get

$$\mathrm{d}z = -\frac{x\,\mathrm{d}x + y\,\mathrm{d}y}{\sqrt{a^2 - x^2 - y^2}}. \tag{8.66}$$

Now inserting $\mathrm{d}z$ in $\mathrm{d}s^2 = \mathrm{d}x^2 + \mathrm{d}y^2 + \mathrm{d}z^2$, we get the metric of the sphere as follows:

$$\mathrm{d}s^2 = \frac{(a^2 - x^2)\,\mathrm{d}x^2 + (a^2 - y^2)\,\mathrm{d}y^2 + 2\,x\,y\,\mathrm{d}x\,\mathrm{d}y}{a^2 - (x^2 + y^2)} \tag{8.67}$$

$$= \mathrm{d}x^2 + \mathrm{d}y^2 + \frac{(y\,\mathrm{d}x + x\,\mathrm{d}y)^2}{a^2} + \frac{1}{a^4}(\ldots). \tag{8.68}$$

This is the expression of the line element (or metric) of a sphere of radius a, in terms of coordinates (x, y). Note that any point on the hemisphere $z > 0$ is uniquely determined by (x, y). The point $(x = 0, y = 0)$ is the point $N = (0, 0, 1)$ on the sphere, which we call the "north" pole, N. Expression (8.67) is exact, and we know that the sphere is a curved space with curvature $1/a^2$. Now, consider points near N, so close to N that $|x| \ll a$ and $|y| \ll a$. From the Taylor expansion of the line element (8.68), we see that to a very good approximation, in this *small* neighborhood of the north pole, the line element is $\mathrm{d}s^2 = \mathrm{d}x^2 + \mathrm{d}y^2$, which is the line element of the Euclidian plane. The coordinate lines become straight lines, and there is almost no distinction between this neighborhood of N and the Euclidean plane. We say that for neighborhoods with size $\ll a$, the curvature of the sphere is not important. For example, the surface of a pool on the Earth is considered flat, for any practical experiments of mechanics, but the surface of the Atlantic ocean with size comparable to Earth's radius is surely curved. It should be stressed that it is, in principle, possible to observe the curvature of the sphere in a very small neighborhood of any point if we have devices accurate enough to compare very small ratios of lengths. For example, one can measure the circumference of a circle of radius r and see if it is equal to $2\pi r$. It would not! Since the sphere is curved. It can be shown that (see §D.2) on a sphere of radius a, for a circle of radius r,

$$\frac{L - 2\pi r}{2\pi r} \simeq -\frac{r^2}{6a^2}, \qquad r \ll a. \tag{8.69}$$

For a circle of radius 1000 km on the surface of the Earth, this is $\sim 4 \times 10^{-3}$; for a circle of radius 10 m on the surface of the Earth, it is $\sim 4 \times 10^{-13}$.

Similarly, as we will see in the following, gravity is the curvature of space-time. Einstein's equivalence principle states that in a freely falling lab if the size of the lab is small enough (compared to the scale of the curvature of the space-time), the metric is almost the Minkowski metric $\eta_{\mu\nu}$, and no force of gravity (effect of curvature) is sensed.

8.9 Gravitational Doppler effect

Einstein used his equivalence principle in an article (Einstein, 1911) to deduce some important observable predictions about gravity. The first and perhaps the most important one is what we may call the gravitational Doppler effect. From this, Einstein deduced that gravity affects time, and he subsequently deduced that gravity is the curvature of space-time. We now present Einstein's reasoning.

By Einstein's equivalence principle, physics in a uniformly accelerated frame K' is the same as physics in a uniform gravitational field. So let us consider K' which has acceleration $g\hat{x}$ relative to the inertial frame K. At time $t = 0$ (as measured by clocks stationary in K), the velocity of K' relative to K is 0. In K' there are two clocks at positions $x' = h$ and $x' = 0$. (See Fig. 8.2.) A signal with frequency ν is sent from $x' = h$ toward $x' = 0$. The detector at $x' = 0$ will receive the signal at time $t = h/c$. During this time, the clock at $x' = 0$ has attained the velocity $v = (h/c)g$, so when the signal is received by this clock (at $x' = 0$), there is a Doppler shift of

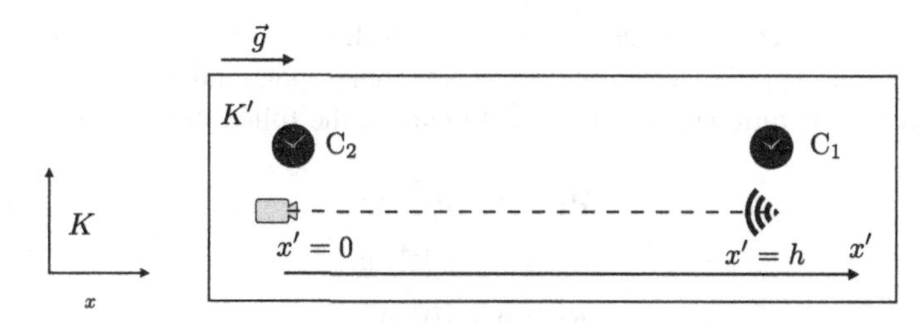

FIGURE 8.2

Frame K' is moving with constant acceleration $\boldsymbol{g}$ relative to the inertial frame K. Clocks C_1 and C_2 are at $x' = h$ and $x' = 0$, respectively. A light signal of frequency ν_1 is sent from $x' = h$ and a receiver at $x' = 0$ receives the signal.

$v/c = h\,g/c^2$, which means that the frequency detected by the clock at $x' = 0$ is

$$\nu' = \left(1 + \frac{h\,g}{c^2}\right)\nu. \tag{8.70}$$

By Einstein's equivalence principle the same must be true in a constant gravitational field $\boldsymbol{g} = -g\,\hat{x}$, for which $\phi(x) = \phi(0) + g\,x$. Since

$$\Delta\phi = \phi(x_2) - \phi(x_1) = g\,x_2 - g\,x_1, \tag{8.71}$$

if $x_1 = h$ and $x_2 = 0$, we have

$$\frac{h\,g}{c^2} = \frac{\phi(x_1) - \phi(x_2)}{c^2}. \tag{8.72}$$

So that, by Einstein's equivalence principle, if a light signal of frequency ν_1 is emitted at x_1 and is received at x_2, the frequency detected at x_2 is given by

$$\nu_2 = \left(1 + \frac{\phi(x_1) - \phi(x_2)}{c^2}\right)\nu_1. \tag{8.73}$$

Remark. Note that for $g < 10^4\,\mathrm{m\,s^{-2}}$ and $h < 10^4\,\mathrm{m}$, $g\,h/c^2 < 10^{-9}$. So for the situations we are considering (to use the equivalence principle) the ratio ϕ/c^2 is in fact very small.

This prediction of Einstein's equivalence principle has been confirmed experimentally. Pound and Rebka (1960) verified this experimentally for the following challenging situation:

$$h = 22.6\,\mathrm{m}, \qquad g = 9.80\,\mathrm{m\,s^{-2}}, \qquad \frac{\Delta\nu}{\nu} = \frac{\nu_2 - \nu_1}{\nu_1} = 2.46 \times 10^{-15}. \tag{8.74}$$

Problem 41. Consider two spherical objects with masses M_1 and M_2 and radii R_1 and R_2 at distance d from each other. Assume that $d \gg R_1$ and R_2. An atom at the surface of object 1 emits a photon, which then is detected at the surface of object 2. What is the ratio $|\phi_1/\phi_2|$, where ϕ_i is the Newtonian potential at the surface of object i? What is the ratio $(\nu_2 - \nu_1)/\nu_1$? Estimate for the following typical values:

$$M_1 = 2 \times 10^{30}\,\text{kg}, \tag{8.75}$$

$$M_2 = 6 \times 10^{24}\,\text{kg}, \tag{8.76}$$

$$R_1 = 6 \times 10^{8}\,\text{m}, \tag{8.77}$$

$$R_2 = 6 \times 10^{6}\,\text{m}, \tag{8.78}$$

$$d > 10^{11}\,\text{m}. \tag{8.79}$$

8.10 Gravity affects time

From (8.73) Einstein concluded that the clocks C_1 and C_2, at points x_1 and x_2, are also affected by gravity. The reasoning is as follows. Consider the sequence of pulses emitted at x_1 at the instants

$$\tau_1, \quad \tau_1 + \frac{1}{\nu_1}, \quad \tau_1 + \frac{2}{\nu_1}, \quad \tau_1 + \frac{3}{\nu_1}, \quad \ldots, \tag{8.80}$$

measured by clock C_1 at x_1. (Later we will justify the use of symbol τ instead of t.) These pulses arrive at x_2 at instances

$$\tau_2, \quad \tau_2 + \frac{1}{\nu_2}, \quad \tau_2 + \frac{2}{\nu_2}, \quad \tau_2 + \frac{3}{\nu_2}, \quad \ldots, \tag{8.81}$$

where $\nu_2 = \kappa\, \nu_1$ and

$$\kappa = 1 + \frac{\phi(x_1) - \phi(x_2)}{c^2} \tag{8.82}$$

is the factor in (8.73). The "time" interval between the successive ticks of clock C_1 is

$$d\tau_1 = \frac{1}{\nu_1}, \tag{8.83}$$

while the "time" interval between the successive pulses arriving at x_2 and measured by clock C_2 is

$$d\tau_2 = \frac{1}{\nu_2}. \tag{8.84}$$

We therefore have

$$\frac{d\tau_2}{d\tau_1} = \frac{\nu_1}{\nu_2} = \frac{1}{\kappa} = 1 + \frac{\phi(x_2) - \phi(x_1)}{c^2}. \tag{8.85}$$

Exaggerated numerical example. Suppose $\nu_1 = 1.00\,\text{Hz}$, corresponding to period $p_1 = 1.00\,\text{s}$. If $\kappa = 1.01$, $\nu_2 = 1.01\,\text{Hz}$, corresponding to period $p_2 = 0.99\,\text{s}$. Thus, for the observer at x_2 the ticks of the clock at x_1 are observed to be separated by 0.99 s, not 1.00 s. This period, for the observer at x_2, is the time between ticks of the clock at x_1, because, whatever the time of flight of a light signal from x_1 to x_2 is, it is the same for all pulses, and therefore the time interval between the pulses is the same 0.99 s. So for the observer at x_2, the time is passing faster in x_1. For the observer at x_1, the clock at x_2 is working more slowly. The relation $\mathrm{d}t_2 = 0.99\,\mathrm{d}t_1$ says the same. Note the interpretations: $\mathrm{d}\tau_1$ is the time interval between ticks of the clock at x_1, while $\mathrm{d}\tau_2$ is the time interval between the ticks of the clock at x_1 *observed at* x_2!

8.11 Proper and coordinate times

Based on our experience with special relativity and considerations with respect to general curvilinear coordinates, we know that we must distinguish between two concepts: coordinate time and proper time.

Consider first proper time—time as measured by a clock, which may or may not be moving. As usual, we denote the proper time by τ, with infinitesimal element $\mathrm{d}\tau$.

Let us first think about an idealized clock. An idealized clock is a very small oscillator and a counter. Standard clocks have standard oscillators oscillating with frequency ν_0. This frequency ν_0 is the frequency of the standard oscillation *in the absence of gravity*, that is, *if in free fall in a gravitational field*. By definition, the proper time is the time shown by the idealized clock. So if the clock's counter shows N ticks, the time shown by the clock is

$$\mathrm{d}\tau = \frac{N}{\nu_0}. \tag{8.86}$$

This is the definition of the proper time increment, whatever the state of motion of the clock is.

Coordinate time. The definition of coordinate time t is, however, more tricky. The following approach could perhaps clarify what coordinate time is (M. Korrami, personal communication, September 27, 2021).

Coordinate time t is a function assigning to any event E in some region of spacetime the real number $t(E)$, which enables us to order events in that region temporally. Assuming that t is a coordinate time in a static gravitational field, we want to study the relation between $\mathrm{d}\tau$ and $\mathrm{d}t$.

Consider two points x_1 and x_2 in a static gravitational field. A light pulse is emitted from x_1 and is absorbed at x_2. Suppose the coordinate time of emission is t_1 and the coordinate time of absorption is t_2. If similarly a light is emitted from x_1 at $t_1 + \mathrm{d}t_1$ and absorbed in x_2 at $t_2 + \mathrm{d}t_2$, we must have

$$(t_2 + \mathrm{d}t_2) - (t_1 + \mathrm{d}t_1) = t_2 - t_1, \tag{8.87}$$

because the gravitational field is static—nothing changes with time if both the transmitter and the receiver are not moving. Therefore we get

$$dt_2 = dt_1. \tag{8.88}$$

The width of the emitted pulse, that is, the duration of the pulse, as measured by an ideal clock at x_1 is $d\tau_1$. The width of the received pulse as measured by an ideal clock at x_2 is $d\tau_2$. From the arguments of the previous pages (see Eq. (8.85)), we know that

$$d\tau_2 = \left(1 + \frac{\phi(x_2) - \phi(x_1)}{c^2}\right) d\tau_1. \tag{8.89}$$

It should be noted that $d\tau_1$, $d\tau_2$, and $\Delta\phi = \phi(x_2) - \phi(x_1)$ are all observables. Using the fact that $\phi(x) \ll c^2$, this relation could be written as

$$\frac{d\tau_2}{1 + \frac{\phi(x_2)}{c^2}} = \frac{d\tau_1}{1 + \frac{\phi(x_1)}{c^2}}. \tag{8.90}$$

Now using $dt_2 = dt_1$ we get

$$\frac{d\tau_2}{\left[1 + \frac{\phi(x_2)}{c^2}\right] dt_2} = \frac{d\tau_1}{\left[1 + \frac{\phi(x_1)}{c^2}\right] dt_1}, \tag{8.91}$$

meaning that

$$\frac{d\tau}{\left[1 + \frac{\phi(x)}{c^2}\right] dt} = a = \text{constant}, \tag{8.92}$$

or

$$d\tau = a\left(1 + \frac{\phi(x)}{c^2}\right) dt. \tag{8.93}$$

The constant a could be absorbed in the definition of t (by redefining $a\,t \to t$). Thus we get

$$d\tau = \left(1 + \frac{\phi(x)}{c^2}\right) dt. \tag{8.94}$$

Gauge transformation. Neither the coordinate time t nor the gravitational potential $\phi(x)$ is observable—their values could be changed without affecting the physics. But $d\tau$ is observable. So a change in the zero point of the potential must be accompanied by a change in the definition of coordinate time, as follows:

$$\phi'(x) = \phi(x) + \phi_0, \tag{8.95}$$

$$t' = \left(1 - \frac{\phi_0}{c^2}\right) t, \tag{8.96}$$

from which we see

$$\begin{aligned}\left[1 + \frac{\phi'(x)}{c^2}\right] \mathrm{d}t' &= \left[1 + \frac{\phi(x) + \phi_0}{c^2}\right]\left(1 - \frac{\phi_0}{c^2}\right) \mathrm{d}t \\ &= \left[1 + \frac{\phi(x)}{c^2}\right] \mathrm{d}t.\end{aligned} \tag{8.97}$$

Therefore we can change the origin of the gravitational potential, but it must be accompanied by a change of the coordinate time.

Remark. A gravitational field which is constant throughout space is very unphysical. The gravitational field of a planet or star could be considered as constant only in a small region. The Newtonian gravitational potential of such a system must satisfy the boundary condition $\phi(\infty) = 0$ and is unique. For such a system the coordinate time is also fixed if we require that very far from the planet for a stationary coordinate clock, the coordinate time t be the same as the proper time τ.

Setting up the coordinate clocks. Now suppose in a static gravitational field we want to distribute a set of atomic clocks, such that they show the coordinate time t. This can be done by integrating (8.94) at a fixed point x. We have

$$t = t_0 + \frac{\tau - \tau_0}{1 + \frac{\phi(x)}{c^2}}. \tag{8.98}$$

It is easy to see that this is equivalent to the following procedure. Let ν_0 be the nominal frequency of the atomic clock, that is, the frequency of the standard oscillator if there is no gravity—in free fall. We assign a frequency $\nu(x)$ and set up the clock such that N oscillations of the clock correspond to increment $\mathrm{d}t$. Since $\mathrm{d}\tau = N/\nu_0$ and $\mathrm{d}t = N/\nu(x)$, we have

$$\frac{N}{\nu_0} = \mathrm{d}\tau = \left(1 + \frac{\phi(x)}{c^2}\right) \mathrm{d}t = \left(1 + \frac{\phi(x)}{c^2}\right) \frac{N}{\nu(x)}, \tag{8.99}$$

meaning that

$$\nu(x) = \left(1 + \frac{\phi(x)}{c^2}\right) \nu_0. \tag{8.100}$$

In practice. For the moment, let us assume that the Earth is a perfect nonrotating sphere, with $\phi(r) = -G M r^{-1}$. Assuming, for simplicity, that in (8.98) $t_0 = \tau_0 = 0$, let us calculate $\Delta t/t$. We have

$$\frac{\Delta t}{t} \simeq \frac{\Delta \tau}{\tau} + \frac{G M}{r c^2} \cdot \frac{\Delta r}{r}. \tag{8.101}$$

For points near the surface of the Earth, $GM/(rc^2) \sim 10^{-9}$, so that we have

$$\frac{\Delta t}{t} \simeq \frac{\Delta \tau}{\tau} + 10^{-9}\frac{\Delta r}{r}. \tag{8.102}$$

Suppose we have clocks with precision $\Delta\tau/\tau \lesssim 10^{-18}$. If we want to use these clocks to show coordinate time with the same precision, we have to know the location (r) so precisely that the second term in the expression above is also $\lesssim 10^{-18}$, which means knowing r with precision $\Delta r/r \lesssim 10^{-9}$. For the points near the surface of the Earth, this means $\Delta r \lesssim 1\,\mathrm{cm}$. Note that in practice, the situation is more complicated, because the Earth is a nonspherical rotating object, and the gravitational field of the Earth is not "very" weak. One should consider the considerations for a not very weak gravitational field.

Proper time for a moving clock. Eq. (8.94) shows the relation between $d\tau$ and dt if the clock showing $d\tau$ is stationary. Based on the equivalence principle and special relativity, the first guess for the general relation is

$$d\tau^2 = \left(1+\frac{\phi(x)}{c^2}\right)^2 dt^2 - \frac{1}{c^2}\left(dx^2+dy^2+dz^2\right). \tag{8.103}$$

The reason is that in the absence of gravity, that is, for $\phi = 0$, we should recover the usual special relativistic relation. Assuming that the relation between ds^2 and $d\tau^2$ is as usual, we get

$$ds^2 = -c^2\left(1+\frac{\phi(x)}{c^2}\right)^2 dt^2 + dx^2 + dy^2 + dz^2. \tag{8.104}$$

This relation has some observable consequences, one of which is the bending of light which we now discuss.

If x^μ is the event of emission of a photon and $x^\mu + dx^\nu$ is the event of absorption of the same photon, we have $ds^2 = 0$, from which we see that *the coordinate velocity of light* is

$$v_{\text{light}} = c\left(1+\frac{\phi(x)}{c^2}\right). \tag{8.105}$$

If we compare this with the usual relation $v_{\text{light}} = c/n$ for the light moving in a medium with refractive index n, we see that a weak gravitational field could be considered as a medium with refractive index

$$n(x) = \left(1+\frac{\phi}{c^2}\right)^{-1} \simeq 1 - \frac{\phi(x)}{c^2}. \tag{8.106}$$

From this equation, Einstein in his 1911 paper[3] concluded that the light rays are deflected if passing close to Sun. But the deflection angle derived from this is half the

[3] For an English translation, see Einstein et al. (1952, pp. 107–108).

correct value. The reason is that (8.104) is not correct. To find the correct expression, we must go beyond the simple statements of the equivalence principle.

Remark. By solving Einstein's field equations, Karl Schwarzschild found the metric outside a spherically symmetric body. If the spherical polar coordinates (r, θ, ϕ) are defined properly, this metric is as follows:

$$\varsigma\, ds^2 = -c^2\left(1 - \frac{2GM}{c^2 r}\right) dt^2 + \left(1 - \frac{2GM}{c^2 r}\right)^{-1} dr^2 + r^2\left(d\theta^2 + \sin^2\theta\, d\phi^2\right). \tag{8.107}$$

The factor multiplying dr^2 could not be explained by simple arguments based on the equivalence principle.

8.12 Strong gravitational field

Einstein later stated a stronger statement, which we now present. By the equivalence principle in a freely falling frame there is no force of gravity, and therefore the metric in a freely falling frame is the Minkowski metric $ds^2 = \eta_{\alpha\beta}\, dX^\alpha\, dX^\beta$. Here X^α are the Cartesian coordinates in the freely falling frame, related to the coordinates x^μ of a general frame (curvilinear, and not necessarily in free fall) by $X^\alpha = f(x)$. We know that in terms of x-coordinates, $ds^2 = g_{\mu\nu}\, dx^\mu\, dx^\nu$, where

$$g_{\mu\nu} = \eta_{\alpha\beta}\frac{\partial X^\alpha}{\partial x^\mu}\frac{\partial X^\beta}{\partial x^\nu}. \tag{8.108}$$

So Einstein stated that

$$d\tau^2 = -\frac{\varsigma}{c^2}\, g_{\mu\nu}\, dx^\mu\, dx^\nu, \tag{8.109}$$

which could be summarized as follows.

Postulate. *Gravity affects the metric of the space-time. In the presence of gravity*

$$ds^2 = g_{\mu\nu}\, dx^\mu\, dx^\nu, \tag{8.110}$$

where $g_{\mu\nu}$ is a symmetric matrix, such that at any point, it has one time-like and three space-like eigenvalues. The relation between the proper time increment $d\tau$ and ds is $d\tau^2 = -\varsigma\, c^{-2}\, ds^2$.

What this postulate states is stronger than the claim that in a small lab the line element is $ds^2 = g_{\mu\nu}\, dx^\mu\, dx^\nu$. It states that space-time is curved, and even in large frames, where tidal effects of gravity are important, the line element is of this form. The metric of space-time is defined for all events of space-time.

Einstein's theory of gravity. Newtonian gravity, fully characterized by the Newtonian gravitational potential ϕ, is the limit of a more accurate theory of gravity in which gravity is given by the metric $g_{\mu\nu}$ of space-time. In the Newtonian limit we have

$$g_{00} = -\varsigma c^2 \left(1 + \frac{\phi}{c^2}\right)^2, \quad g_{0i} = 0, \quad g_{ij} = \varsigma \delta_{ij}, \tag{8.111}$$

which means

$$\mathrm{d}\tau^2 = \left(1 + \frac{\phi}{c^2}\right)^2 \mathrm{d}t^2 - \left(\mathrm{d}x^2 + \mathrm{d}y^2 + \mathrm{d}z^2\right). \tag{8.112}$$

Now the next steps would be to find field equations governing the metric $g_{\mu\nu}$ in the general case, solve them to find $g_{\mu\nu}$, and extract information from $g_{\mu\nu}$.

Problem 42. For $\phi(x) = g\,x$, define

$$\xi = x + \frac{c^2}{g}, \qquad \theta = \frac{g\,t}{c}, \tag{8.113}$$

and write (8.112) in terms of coordinates (θ, ξ, y, z). Compare with the metric of the Rindler frame. Interpret this using the solution to Problem 35. A consequence of this relation is that the space-time with metric (8.112) is, in fact, not curved!

Geodesic motion. If we accept the postulate that gravity changes the metric and we generalize the Lagrangian (see §5.2 on page 97), the action (5.46) will be generalized to

$$S = -m\,c^2 \int_i^f \mathrm{d}\tau = -m\,c \int_i^f \sqrt{-\varsigma\, g_{\mu\nu}\, \mathrm{d}x^\mu\, \mathrm{d}x^\nu}. \tag{8.114}$$

It is well known in differential geometry that the Euler–Lagrange equations for this action are the geodesic equations (see Appendix D), so that by Einstein's equivalence principle, a freely falling particle follows the geodesics of the space-time (for another proof of this statement, not using the action, directly from the equivalence principle, see Weinberg (1972, pp. 70–73)).

Guide to Literature. For an English translation of Einstein (1911), see Einstein et al. (1952, pp. 97–108), which is quite readable and recommended strongly. Ryder (2014) is an excellent recent article. Weinberg (1972, pp. 67–70) is an excellent text about the equivalence principle. Note that different terminology is used in different texts. A modern terminology suitable for explaining experiments is that of Will (2014). Di Casola and Liberati (2015) clarified the differences between various equivalence principles and is recommended strongly.

Einstein's field equations. Now that we know gravity causes the metric to be changed from $\eta_{\mu\nu}$ to $g_{\mu\nu}$, the question is to find the appropriate field equations for $g_{\mu\nu}$, so that given the mass-energy distribution, one can find $g_{\mu\nu}$. Einstein found these field equations. Presenting Einstein's field equations is beyond the scope of this book. Here we would like to say a few words to just show the road to the reader.

Einstein had a friend named Marcel Grossmann (1878–1936), who was a distinguished mathematician and an expert in differential geometry. Einstein and Grossmann actually collaborated on developing a theory for gravity. In almost one decade Einstein learned a lot of differential geometry from Grossman, and thus he was able to formulate general relativity. We are stressing this historical note to conclude that if someone wants to learn general relativity, he/she must do the same and learn differential geometry very well. It is essential; otherwise, one cannot go further. Always remember that "there is no royal shortcut to geometry."

Guide to Literature. A concise and yet rigorous text about general relativity is Weinberg (2008, pp. 511–529). A modern introductory text on general relativity is (Ryder, 2009). Weinberg (1972) is a classical text. The popular, nontechnical text Will (1993) is strongly recommended to anyone who wants to learn general relativity. For the history of the role of Marcel Grossman in developing general theory of relativity see Sauer (2012).

References

Cohen, I.B., Whitman, A., Budenz, J., 1999. The Principia: Mathematical Principles of Natural Philosophy, 1st edition. University of California Press, Berkeley.

Di Casola, E., Liberati, S., 2015. Nonequivalence of equivalence principles. American Journal of Physics 83 (1), 39–46.

Einstein, A., 1911. Über den Einflußder Schwerkraft auf die Ausbreitung des Lichtes. Annalen der Physik 35, 898–908.

Einstein, A., Lorentz, H.A., Weyl, H., Minkowski, H., 1952. The Principle of Relativity. Dover, New York.

Fitzpatrick, R., 2012. An Introduction to Celestial Mechanics. Cambridge University Press, Cambridge.

Goldstein, H., 1980. Classical Mechanics, 2nd edition. Addison-Wesley, Reading, MA.

Jackson, J.D., 1999. Classical Electrodynamics, 3rd edition. John Wiley & Sons, New York.

Kleppner, D., Kolenkow, R., 2014. An Introduction to Mechanics, 2nd edition. Cambridge University Press, Cambridge.

Levine, H., 1997. Partial Differential Equations. Studies in Advanced Mathematics, vol. 6. American Mathematical Society, International Press, Providence.

Mohr, P.J., Newell, D.B., Taylor, B.N., 2016. Codata recommended values of the fundamental physical constants: 2014. Reviews of Modern Physics 88, 035009.

Montenbruck, O., Gill, E., 2000. Satellite Orbits. Models, Methods, and Applications. Springer-Verlag, Berlin.

Navarro, J.F., Frenk, C.S., White, S.D.M., 1996. The structure of cold dark matter halos. The Astrophysical Journal 462, 563–575.

Pound, R.V., Rebka, G.A., 1960. Apparent weight of photons. Physical Review Letters 4, 337–341.

Ryder, L., 2009. Introduction to General Relativity. Cambridge University Press, Cambridge.

Ryder, L., 2014. Acceleration and gravity: Einstein's principle of equivalence. In: Ashtekar, A., Petkov, V. (Eds.), Springer Handbook of Spacetime. Springer-Verlag, Berlin, pp. 61–70. Ch. 4.

Sauer, T., 2012. Marcel Grossmann and his contribution to the general theory of relativity. In: Jantzen, R.T., Rosquist, K., Ruffini, R. (Eds.), The Thirteenth Marcel Grossman Meeting on Recent Developments in Theoretical and Experimental General Relativity, Astrophysics, and Relativistic Field Theories. World Scientific, Singapore, pp. 456–503.

Schlamminger, S., Choi, K.-Y., Wagner, T.A., Gundlach, J.H., Adelberger, E.G., 2008. Test of the equivalence principle using a rotating torsion balance. Physical Review Letters 100, 041101.

Symon, K.R., 1971. Mechanics, 3rd edition. Addison-Wesley, Reading, MA.

Weinberg, S., 1972. Gravitation and Cosmology. Principles and Applications of the General Theory of Relativity. John Wiley & Sons, New York.

Weinberg, S., 2008. Cosmology. Oxford University Press, Oxford.

Will, C.M., 1993. Was Einstein Right? Putting General Relativity to the Test, 2nd edition. HarperCollins (Basic Books), New York.

Will, C.M., 2014. The confrontation between general relativity and experiment. Living Reviews in Relativity 17 (1), 4.

PART

Mathematics

2

CHAPTER

9 Mathematics of translations

9.1 Defining translations

By a translation in the Euclidean plane E_2 we mean the mapping $\boldsymbol{r} \mapsto \boldsymbol{r}+\boldsymbol{a}$, which we denote by the symbol $T_{\boldsymbol{a}}$. This transformation maps any figure to a figure equal or congruent to itself. We denote the set of all translations of the Euclidian plane by the symbol T(2). This set has some structures which we are now going to study.

First, one can combine translations, and the result is always a translation. In mathematics this is phrased as follows: *The set of translations is closed under the combination of mappings.* The combination, denoted by the symbol $\circ$, is simple (see Fig. 9.1). For two translations T_1 and T_2,

$$T_2 \circ T_1(\boldsymbol{r}) = T_2(T_1(\boldsymbol{r})) = T_2(\boldsymbol{r}+\boldsymbol{a}_1) = \boldsymbol{r}+\boldsymbol{a}_1+\boldsymbol{a}_2 = T_3(\boldsymbol{r}), \tag{9.1}$$

where $\boldsymbol{a}_3 = \boldsymbol{a}_1 + \boldsymbol{a}_2$.

Second, for any three translations T_1, T_2, and T_3, the following identity holds (see Fig. 9.2):

$$T_1 \circ (T_2 \circ T_3) = (T_1 \circ T_2) \circ T_3. \tag{9.2}$$

This property is called associativity.

Third, if $\boldsymbol{a}=0$, then obviously $\boldsymbol{a}+\boldsymbol{r}=\boldsymbol{r}$, which means that the translation corresponding to the zero vector is the identity map which we denote by $\mathbb{I}$.

Fourth, for any vector $\boldsymbol{a}$, the following identity holds:

$$T_{\boldsymbol{a}} \circ T_{-\boldsymbol{a}} = T_{-\boldsymbol{a}} \circ T_{\boldsymbol{a}} = \mathbb{I}. \tag{9.3}$$

This means that any translation has an inverse.

Fifth, the order of combination does not matter:

$$T_1 \circ T_2 = T_2 \circ T_1. \tag{9.4}$$

9.2 The algebraic structure

Let us now review the basic definitions of group theory.

A Mathematical Approach to Special Relativity. https://doi.org/10.1016/B978-0-32-399708-9.00019-1

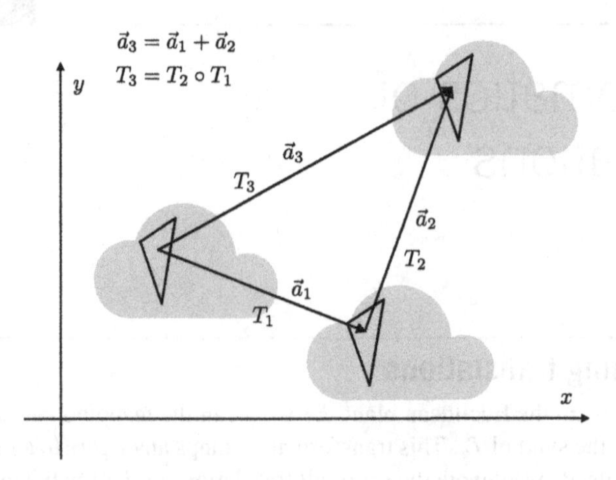

FIGURE 9.1 Combination of translations.

Under a translation any figure is mapped into a figure congruent (equal) to itself. The combination $T_2 \circ T_1$ means that first T_1 acts, and then T_2. The result of $T_2 \circ T_1$ is itself a translation.

Group. A set G together with an operation $\circ$ defined on G is said to be a group if $\circ$ has the following properties:

1. If g_1 and g_2 are in G, then $g_1 \circ g_2$ is defined and is an element of G.
2. The operation $\circ$ is associative, that is, for any three elements of G, we have

$$g_1 \circ (g_2 \circ g_3) = (g_1 \circ g_2) \circ g_3. \tag{9.5}$$

3. There is a unique element e in G, called the identity, such that for any g in G we have $e \circ g = g \circ e = g$.
4. For any g in G, there is a unique *inverse*, g^{-1}, such that

$$g \circ g^{-1} = g^{-1} \circ g = e. \tag{9.6}$$

Abelian groups. A group $(G, \circ)$ is said to be Abelian[1] if for any g_1 and g_2

$$g_1 \circ g_2 = g_2 \circ g_1. \tag{9.7}$$

Using this terminology we say that the set of translations of the Euclidean plane, T(2), is an Abelian group. The definition of the translations of the Euclidean space, T(3), or higher-dimensional spaces $\mathbb{R}^n$ is similar.

Now we are going to introduce some other groups and study them.

[1] After Niels Henrik Abel (1802–1829).

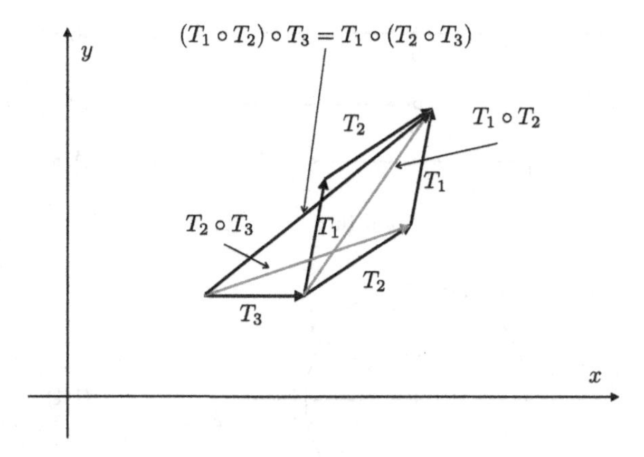

FIGURE 9.2 Associativity of the combination of translations.

Associativity of the combination of translations follows from the associativity of the addition of vectors.

The group Z_2. Consider the set $\{-1, 1\}$. This set with the usual multiplication of numbers is a group. It is easy to check that all the properties mentioned above are fulfilled. This is a finite group, meaning that it has a finite number of elements.

The group $(\mathbb{R}, +)$. Consider the set of real numbers $\mathbb{R}$. This set with the usual addition of numbers, "+," is an Abelian group. In fact, this group is T(1), that is, the translations of the 1D space. To think of this group, think of the real line. This is a continuous group (later, we will describe this).

The group $(\mathbb{R}^+, \cdot)$. Consider the set of positive real numbers. This set with the usual multiplication of numbers, "·," is an Abelian group.

Now consider the function $f(x) = e^x$, which has the property $e^{x_1} e^{x_2} = e^{x_1+x_2}$. Let us think about this relation. If x_1 and x_2 are two elements of $(\mathbb{R}, +)$, we can add them according to the rule of combination in $(\mathbb{R}, +)$, and the result is $x_3 = x_1 + x_2$. Now, let us first use f to send x_1 and x_2 to points e^{x_1} and e^{x_2} in $(\mathbb{R}^+, \cdot)$ and then combine them using the combination rule in $(\mathbb{R}^+, \cdot)$ to get $e^{x_1} \cdot e^{x_2}$. The result is the same as $e^{x_3} = e^{x_1+x_2}$.

Homomorphism. A function $f : G_1 \to G_2$ is said to be a homomorphism if it satisfies

$$f(g_1 \circ g_2) = f(g_1) \bullet f(g_2), \tag{9.8}$$

where $\circ$ is the combination in G_1 and $\bullet$ is the combination in G_2. If $f : G_1 \to G_2$ is a homomorphism, we say that G_2 is homomorphic to G_1.

If $f : G_1 \to G_2$ is a homomorphism and e is the identity element in G_1, then $f(e)$ is the identity element of G_2, because $f(e) \bullet f(g_2) = f(e \circ g_2) = f(g_2)$.

Isomorphism. An invertible homomorphism is called an isomorphism. If $f : G_1 \to G_2$ is an isomorphism, we say that G_1 and G_2 are isomorphic. In this case G_2 is homomorphic to G_1, and G_1 is homomorphic to G_2.

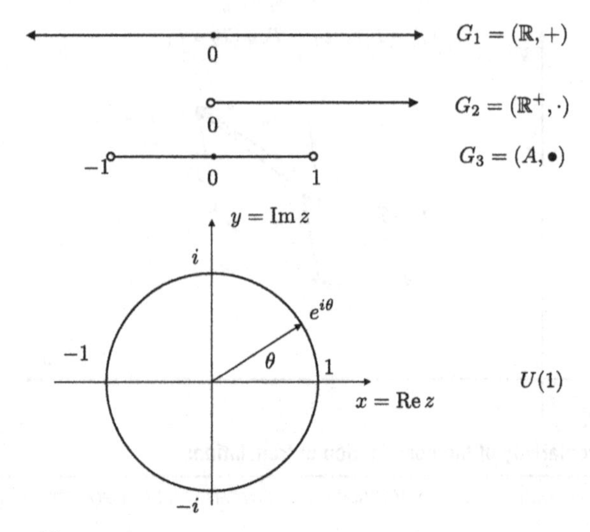

FIGURE 9.3 Some 1D groups.

G_1 is the real line $\mathbb{R}$ with the usual addition. G_2 is the half-real line $\mathbb{R}^+$ with the usual multiplication as the combination rule. G_3 is the open interval $(-1, 1)$ with the combination rule $x_1 \bullet x_2 = (x_1 + x_2)/(1 + x_1 x_2)$ (see Problem 43). Groups G_1, G_2, and G_3 are isomorphic! The group U(1) is the unit circle S and it is not isomorphic to G_1, G_2, or G_3. Also, S is *not* homeomorphic to $\mathbb{R}$.

Example. The function e^x from $\mathbb{R}$ to $\mathbb{R}^+$ is invertible, with the inverse given by $\log(x)$. It is therefore an isomorphism, and the groups $(\mathbb{R}, +)$ and $(\mathbb{R}, \cdot)$ are isomorphic. Note that $\log(x_1 \cdot x_2) = \log(x_1) + \log(x_2)$. Because of this property, $\log(x)$ defines a homomorphism from $(\mathbb{R}^+, \cdot)$ to $(\mathbb{R}, +)$.

Problem 43. Consider the set $A = (-1, 1)$, that is, the real numbers $-1 < x < 1$. If x_1 and x_2 are two points in A, we define

$$x_1 \bullet x_2 = \frac{x_1 + x_2}{1 + x_1 x_2}. \tag{9.9}$$

1. Show that A with this "addition" $\bullet$ is an Abelian group.
2. Show that $(A, \bullet)$ is isomorphic to $(\mathbb{R}, +)$. (See Fig. 9.3.)

The group **U(1).** Consider the set of complex numbers with absolute value equal to 1. That is, the set $x^2 + y^2 = 1$ in the complex z-plane, where $z = x + i\,y$. Any such number could be written as $e^{i\theta}$, where θ is a real number. We use the symbol U(1) for this set (the notation will be explained later). If ω_1 and ω_2 are in U(1), then $\omega_1 \omega_2$ is also in U(1), because $|\omega_1 \omega_2| = |\omega_1| \cdot |\omega_2| = 1$. The set U(1) with the usual multiplication of complex numbers is an Abelian group.

The set $|z| = 1$ is the circle $x^2 + y^2 = 1$. So to think of U(1), think of the unit circle, which sometimes is denoted by S^1.

Any complex number z could be considered as a 1×1 matrix $[z]$, and the matrix multiplication in this trivial case reads

$$[z_1][z_2] = [z_1 z_2]. \tag{9.10}$$

If u is a complex $N \times N$ matrix, by definition $u^\dagger$ is the complex conjugate of the transpose of u, that is,

$$u^\dagger = \bar{u}^T. \tag{9.11}$$

A matrix u is said to be unitary if

$$u^\dagger u = 1. \tag{9.12}$$

So the set U(1) could be considered as the set of 1×1 unitary matrices. For this reason it is denoted by U(1)—U is for unitary, and 1 indicates that the matrices are 1×1.

Homomorphism **$(\mathbb{R}, +) \to$ U(1).** Let θ be a real number and consider the function $f : \mathbb{R} \to \mathrm{U}(1)$ defined by

$$f(\theta) = e^{i\theta}. \tag{9.13}$$

This function has the property that if θ_1 and θ_2 are two real numbers, then

$$f(\theta_1 + \theta_2) = e^{i\theta_1} \cdot e^{i\theta_2}. \tag{9.14}$$

It is said that f is a homomorphism from $(\mathbb{R}, +)$ to U(1). Note that $f(0) = 1$, which means that the "identity" of $(\mathbb{R}, +)$ is sent to the identity of U(1). Also $f(-\theta) = (f(\theta))^{-1}$.

One may be tempted to consider the complex function $\log(z)$ to write a homomorphism in the reverse direction, that is, from U(1) to $(\mathbb{R}, +)$. But this does not work, for $\log(z) = |z| + i \arg(z) + 2n\pi i$ is not well defined on the complex plane $\mathbb{C}$, because n could be any integer. The groups $(\mathbb{R}, +)$ and U(1) *are not isomorphic*, which means that $(\mathbb{R}, +)$ *is not homomorphic to* U(1).

Guide to Literature. Group theory is very important in physics. What we mentioned above is just a very brief introduction to the basic definition. We encourage the reader to learn some group theory from texts such as Cornwell (1997), Joshi (2005), Meijer and Bauer (2004), and Ma (2007).

9.3 Topological structure

Topology is that part of mathematics that discusses properties that are invariant under continuous transformations. Of course, one should first define what a continuous transformation is. We assume the reader already knows that.

Homeomorphism. Two sets A and B are said to be *homeomorphic* if there is an invertible *continuous* map $f : A \to B$ and f^{-1} is also continuous.

Remark. Note the extra "e" in the spelling of homeomorphism. When reading texts one must pay attention to this. The concepts of homomorphism and homeomorphism are quite different. Homomorphism is an algebraic similarity, homeomorphism is a topological similarity.

The following examples are to clarify the meaning of homeomorphism. Before that, let us fix a useful notation.

Notation. Closed and open intervals of $\mathbb{R}$. If a and b are real numbers and $a < b$, we define

$$[a, b] = \{x \in \mathbb{R} \mid a \le x \le b\}, \tag{9.15}$$

$$(a, b) = \{x \in \mathbb{R} \mid a < x < b\}, \tag{9.16}$$

$$[a, b) = \{x \in \mathbb{R} \mid a \le x < b\}, \tag{9.17}$$

$$(a, b] = \{x \in \mathbb{R} \mid a < x \le b\}. \tag{9.18}$$

Examples of homeomorphic and nonhomeomorphic sets.

1. $f : \mathbb{R} \to (-1, 1)$, defined by $f(x) = \tanh x$, is a homeomorphism, because it is both invertible and continuous and $f^{-1}(x) = \tanh^{-1} x$ is also continuous so that $(0, 1)$ is homeomorphic to $\mathbb{R}$.
2. $f : (0, 1) \to (0, 2)$, defined by $f(x) = 2x$, is a homeomorphism—it is both invertible and continuous, and the inverse function $f^{-1}(x) = \frac{1}{2}x$ is also continuous. This example shows that any two intervals of the type (a, b) are homeomorphic, no matter how big or small they are.
3. $f : (0, 1) \to [0, 1]$, defined by $f(x) = x$, *is not* a homeomorphism, for it is not onto, that is, it does not cover all of B. The reason is that the points 0 and 1 are in B, but there are no points in A that are being mapped to these points. Although the function f is one to one and continuous, it is not invertible, and therefore it is not a homeomorphism. One may ask: Could we find some other function that is a homeomorphism from $(0, 1)$ to $[0, 1]$? The answer is negative. These two sets are not homeomorphic. The set $[0, 1]$ is *compact*, while the set $(0, 1)$ is not compact. A noncompact set is not homeomorphic to a compact set. The definition of compactness is technical and we are not going through it. We encourage the reader to learn it. In the theory of groups it is important whether or not a group is compact.
4. Let S be the unit circle in the (x, y)-plane. The function $f : [0, 2\pi] \to S$, defined by $f(\theta) = (\cos\theta, \sin\theta)$, is continuous and onto, but it is not one to one, because $f(0) = f(2\pi)$. Therefore f *is not* a homeomorphism. The set S *is not* homeomorphic to $[0, 1]$ (or $[0, 2\pi]$), because $[0, 1]$ has two boundary points, 0 and 1, but S has no boundary points. All points of S are interior points! Both S and $[0, 1]$ are compact.

5. Let $A = (0, \frac{1}{2}) \cup (\frac{1}{2}, 1)$. The function $f : A \to (0, 1)$, defined by $f(x) = x$, *is not* a homeomorphism—f is not continuous at $x = \frac{1}{2}$. Remember that a function $g(x)$ is continuous at a point x_0 if the following three values are the same:

$$\lim_{x \to x_0^-} g(x) = g(x_0) = \lim_{x \to x_0^+} g(x). \tag{9.19}$$

In the case of f, the left and right limits are equal, but $f(x_0)$ does not exist. Again one may ask: Could we find some other function that is a homeomorphism from A to $(0, 1)$? The answer is negative; A and $(0, 1)$ are not homeomorphic.

Guide to Literature. The concept of continuity is introduced in any textbook on calculus, and it is covered in almost any textbook on advanced calculus, such as Duistermaat and Kolk (2004, pp. 1–35), and textbooks on real and complex analysis, such as Apostol (1957, pp. 40–86) or Rudin (1967, pp. 24–43 and 83–102). Textbooks on topology such as Armstrong (1983) are also recommended.

9.4 Continuous and Lie groups

Let G be a group. Denote the elements of G by symbols like x and y. On G two important operations are defined. First the combination: If x and y are two elements of G, then $x \circ y$ is another element of G. This means that there is a function denoted by mul, defined on $G \times G$, which sends the pair (x, y) into $z = x \circ y = \text{mul}(x, y)$. In concise mathematical notation,

$$\begin{aligned} &\text{mul} : G \times G \to G, \\ &\text{mul} : (x, y) \mapsto \text{mul}(x, y) = x \circ y. \end{aligned} \tag{9.20}$$

If x is in G, there is an element called x^{-1} which is the inverse of x, with the property $x \circ x^{-1} = x^{-1} \circ x = e$, where e is the identity of G. Again we can think of a function

$$\begin{aligned} &\text{inv} : G \to G, \\ &\text{inv} : x \mapsto \text{inv}(x) = x^{-1}. \end{aligned} \tag{9.21}$$

Continuous groups. The group G is said to be continuous if the functions $\text{mul}(x, y)$ (multiplication) and $\text{inv}(x)$ (inversion) are continuous.

Example. The group $\text{T}(1) = (\mathbb{R}, +)$ is a continuous group, because the functions $\text{mul}(x, y) = x + y$ and $\text{inv}(x) = -x$ are continuous functions. From Problem 43 we know that this group is isomorphic to the group $(A, \bullet)$, which is also a continuous group. The sets $\mathbb{R}$ and $A = (-1, 1)$ are homeomorphic. The group U(1) is also continuous, and it is homomorphic to T(1), but U(1) is the unit circle, while T(1) is the real line, and they are not homeomorphic.

Smoothness. A function of several real variables $\boldsymbol{x}$, defined on an open set U, is said to be *smooth* if it is differentiable any number of times on U.

Analyticity. A function is said to be *analytic* if it is smooth, it has a convergent Taylor series, *and* the function itself is the limit of its Taylor series.

Remark. We are talking about real vector functions of several real variables. Do not confuse this usage of the term "analytic" with the one usually used for complex functions of one complex variable.

A function of a single real variable is analytic if it is given by an absolutely convergent series of the form

$$f(x) = \sum_{n=0}^{\infty} c_n (x - a)^n. \tag{9.22}$$

For the functions of several variables $\boldsymbol{x} = (x_1, \ldots, x_N)$, the Taylor series is of the form

$$f(\boldsymbol{x}) = \sum_{n_1, n_2, \ldots n_N} C_{n_1 n_2 \ldots n_N} (x_1 - a_1)^{n_1} (x_2 - a_2)^{n_2} \ldots (x_N - a_N)^{n_N}. \tag{9.23}$$

Since the series is absolutely convergent, the order of summations is not important.

The most important example of an analytic function is the function e^x. Remember that this function has the Taylor series

$$1 + x + \frac{x^2}{2!} + \frac{x^3}{3!} + \ldots, \tag{9.24}$$

which is convergent for any x, and the limit of the series is e^x.

Smoothness does not imply analyticity. A famous example of a function which is smooth, but not analytic is the function (Fig. 9.4)

$$f(x) = \begin{cases} 0 & x = 0, \\ e^{-1/x^2} & x \neq 0. \end{cases} \tag{9.25}$$

The reader who is not 100% sure that this *is* a function, because "it is given by two relations," is strongly recommended to review the meaning of function (see for example Rudin (1967, p. 24) or Apostol (1957, p. 27)). The reader is also asked to prove that all the derivatives of f at the point 0 vanish, that is,

$$f'(0) = f''(0) = \cdots = f^{(n)}(0) = 0. \tag{9.26}$$

Therefore this function f has the convergent Taylor series at 0

$$0 + 0x + 0x^2 + \cdots = 0. \tag{9.27}$$

However, the Taylor series is not convergent to the value of $f(x)$ for any non-vanishing x. The interested reader is referred to the readable, but mathematical article by Boas (1989).

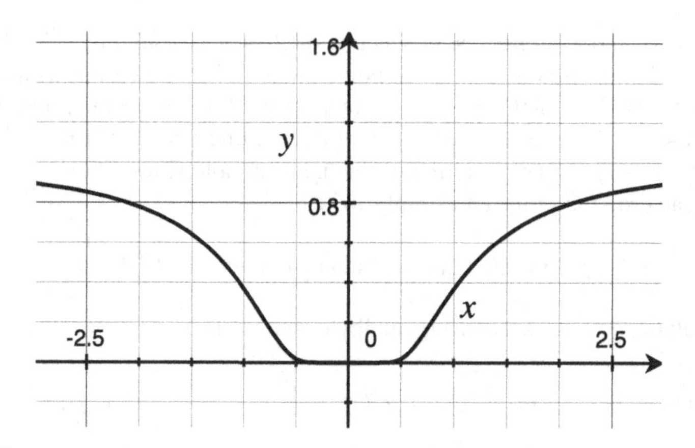

FIGURE 9.4

The function defined in Eq. (9.25). Although all its derivatives are zero at the point $x = 0$, it is not the function 0 (mapping every x to 0).

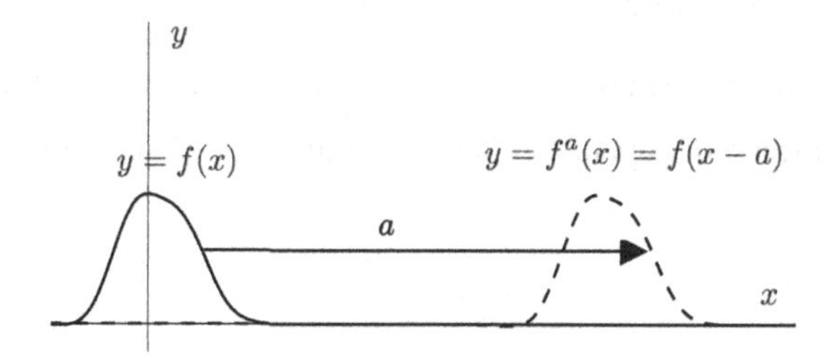

FIGURE 9.5

The value of the translated function f^a at point x is the value of the original function f at point $x - a$, where a is the parameter specifying the translation.

Lie groups. The continuous group G is said to be a Lie group[2] if the multiplication $\mathrm{mul}(x, y)$ and inversion $\mathrm{inv}(x)$ are *analytic* functions.

9.5 Generators of translations

Consider a function f defined on the real axis x. The graph of this function is the set of points $\{(x, f(x))\}$. Let us translate this function by a, which is a real number. We denote the resulting function by $U_a f$ or f^a (see Fig. 9.5). Therefore

$$U_a f(x) = f^a(x) = f(x - a). \tag{9.28}$$

[2] After Marius Sophus Lie (1842–1899).

We investigate an operator U_a, which, acting on f, would result in f^a. We know that the value of the function $U_a f = f^a$ at point $x + a = T_a x$ is the same as the value of f at x, so that we have $f^a(T_a x) = f(x)$, or simply $f^a(x+a) = f(x)$. Let us temporality define $x' = T_a x = x + a$ and solve it for x, getting $x = (T_a)^{-1} x' = x' - a$. The relation $f^a(x+a) = f(x)$ in terms of this new variable reads $f^a(x') = f(x' - a)$. Now, we can drop the prime and simply write

$$(U_a f)(x) = U_a f(x) = f^a(x) = f(x-a) = f\left(T_a^{-1} x\right). \tag{9.29}$$

If f is analytic, *i.e.*, it has a convergent Taylor series, then

$$U_a f(x) = f(x-a) = f(x) - a\, f'(x) + \frac{a^2}{2!} f''(x) - \frac{a^3}{3!} f'''(x) + \ldots \tag{9.30}$$

$$= \left(1 - a\frac{\mathrm{d}}{\mathrm{d}x} + \frac{a^2}{2!}\frac{\mathrm{d}^2}{\mathrm{d}x^2} - \frac{a^3}{3!}\frac{\mathrm{d}^3}{\mathrm{d}x^3} + \ldots\right) f(x) \tag{9.31}$$

$$= \exp\left(-a\frac{\mathrm{d}}{\mathrm{d}x}\right) f(x) = \hat{T}_a\, f(x). \tag{9.32}$$

The operator $\hat{T}_a$ has the form of $\exp(aX)$, where a is the parameter of translation and $X = -\dfrac{\mathrm{d}}{\mathrm{d}x}$ is a differential operator. Following the conventions of quantum mechanics, we define

$$P_x := \frac{1}{i}\frac{\mathrm{d}}{\mathrm{d}x}, \tag{9.33}$$

so that

$$T_a = \exp(-i\, a\, P_x). \tag{9.34}$$

It is straightforward to show that the operator corresponding to translation in the 3D space, $\mathbb{R}^3$, has the form

$$\boldsymbol{T_a} = -\boldsymbol{a} \cdot \nabla = -i\, \boldsymbol{a} \cdot \boldsymbol{P}, \tag{9.35}$$

where

$$\boldsymbol{P} = \frac{1}{i}\nabla. \tag{9.36}$$

Remark. In quantum mechanics a factor of $\hbar$ (the reduced Planck constant) is also multiplied, which causes the dimensions of P to be that of the momentum.

The generators of the translations, P_x, P_y, and P_z, obviously satisfy the following commutation relations:

$$\left[P_x, P_y\right] = \left[P_x, P_z\right] = \left[P_y, P_z\right] = 0, \tag{9.37}$$

where

$$[A, B] = A\,B - B\,A. \tag{9.38}$$

Let us see what this *commutativity* means. Consider a function $f(x, y, z)$, and compute $P_x P_y f$ and $P_y P_x f$ and compare the two:

$$P_x P_y f = \left(-i \frac{\partial}{\partial x}\right)\left(-i \frac{\partial}{\partial y}\right) f(x, y, z) = -\frac{\partial^2 f}{\partial x \, \partial y}, \tag{9.39}$$

$$P_y P_x f = \left(-i \frac{\partial}{\partial y}\right)\left(-i \frac{\partial}{\partial x}\right) f(x, y, z) = -\frac{\partial^2 f}{\partial y \, \partial x}. \tag{9.40}$$

Commutativity of P_x and P_y simply means that we can exchange the order of partial differentiations with respect to x and y.

Remark. That the order of partial derivatives $\partial_x \partial_y f$ and $\partial_y \partial_x f$ can be exchanged is not guaranteed by merely the existence of them; $\partial_x \partial_y f$ must be continuous too. The statement is true for smooth functions, but not for all functions. For a counterexample and the proof of the statement for smooth functions see Apostol (1957, pp. 120–121).

References

Apostol, T.M., 1957. Mathematical Analysis. A Modern Approach to Advanced Calculus. Addison-Wesley, Reading, MA.

Armstrong, M.A., 1983. Basic Topology. Springer, New York.

Boas, R.P., 1989. When is a C^∞ function analytic? The Mathematical Inteligencer 11 (4), 34–37.

Cornwell, J.F., 1997. Group Theory in Physics, an Introduction. Academic Press, San Diego.

Duistermaat, J.J., Kolk, J.A.C., 2004. Multidimensional Real Analysis I. Differentiation. Cambridge University Press, Cambridge.

Joshi, A.W., 2005. Elements of Group Theory for Physicists, 4th edition. New Age International (P) Limited, Publishers, New Delhi.

Ma, Z.-Q., 2007. Group Theory for Physicists. World Scientific, Singapore.

Meijer, P.H.E., Bauer, E., 2004. Group Theory, the Application to Quantum Mechanics. Dover, Mineola, New York.

Rudin, W., 1967. Principles of Mathematical Analysis, 3rd edition. McGraw-Hill, London.

CHAPTER

The rotation group

10

10.1 Rotations and inversions of the plane

Let E^2 be the Euclidean plane. Under a translation T_a any shape in E^2 maps into a shape identical, or more technically *congruent* to itself. There are two other transformations that transform any shape into one congruent to the original—rotations and inversion in lines. By a Euclidean transformation we mean a rotation, an inversion in a line, a translation, or a combination of them. In this section we investigate rotations and inversions of the plane. In the next section we deal with the rotations and inversions of the space.

Remark. In this chapter we use subscript indices for the components of vectors.

10.1.1 Rotations

Let $\theta \in \mathbb{R}$ be a fixed angle and consider the following action on the (x_1, x_2)-plane:

$$\begin{cases} x_1' = x_1 \cos\theta - x_2 \sin\theta, \\ x_2' = x_1 \sin\theta + x_2 \cos\theta. \end{cases} \tag{10.1}$$

We define the following matrices:

$$X = \begin{bmatrix} x_1 \\ x_2 \end{bmatrix}, \quad X' = \begin{bmatrix} x_1' \\ x_2' \end{bmatrix}, \quad R(\theta) = \begin{bmatrix} \cos\theta & -\sin\theta \\ \sin\theta & \cos\theta \end{bmatrix}. \tag{10.2}$$

The transformation above is thus written as $X' = RX$, or $R : X \mapsto X'$.

Before proceeding, note some properties of this map. First, it is linear, and it maps the origin into the origin. Second, it is periodic in θ with period 2π, that is, $R(\theta + 2N\pi) = R(\theta)$. Third, $R(0) = R(2N\pi) = \mathbb{I}$ is the identity transformation—sending any point to itself.

The geometrical picture of this action is very simple: the point $X = (x_1, x_2)$ is *rotated* around the origin by the angle θ (see Fig. 10.1). Obviously, under this action the circle $(x_1)^2 + (x_2)^2 = r^2$, where r is a constant, is being mapped to itself.

Transformation (10.1) has a very important property: If under the rotation $X = (x_1, x_2) \mapsto X' = (x_1', x_2')$ and $Y = (y_1, y_2) \mapsto Y' = (y_1', y_2')$, then

$$x_1' y_1' + x_2' y_2' = x_1 y_1 + x_2 y_2. \tag{10.3}$$

A Mathematical Approach to Special Relativity. https://doi.org/10.1016/B978-0-32-399708-9.00020-8

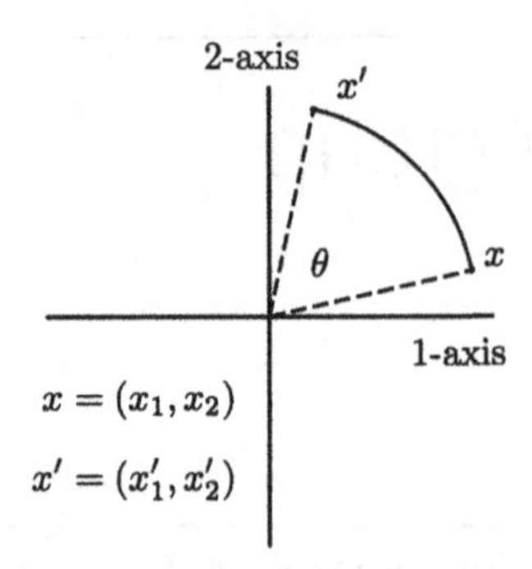

FIGURE 10.1

Under the action of $R(x_3, \theta)$, the point $x = (x_1, x_2)$ is being mapped to point $x' = (x_1', x_2')$. Some textbooks call this "an active rotation."

The proof is a very straightforward calculation, but let us prove it in a slightly different way. We note that the right-hand side of the above equation is

$$x_1\, y_1 + x_2\, y_2 = \begin{bmatrix} x_1 & x_2 \end{bmatrix} \begin{bmatrix} y_1 \\ y_2 \end{bmatrix} = X^T\, Y, \tag{10.4}$$

and we have a similar formula for the left-hand side. Therefore the equality of the above equation is the assertion that $X'^T\, Y' = X^T\, Y$. Now, since $X' = R(\theta)X$ and $Y' = R(\theta)\, Y$, we have

$$X'^T\, Y' = (R(\theta)\, X)^T\, (R(\theta)\, Y) = X^T\, R(\theta)^T\, R(\theta)\, Y. \tag{10.5}$$

And we have

$$R(\theta)^T\, R(\theta) = \begin{bmatrix} \cos\theta & \sin\theta \\ -\sin\theta & \cos\theta \end{bmatrix} \begin{bmatrix} \cos\theta & -\sin\theta \\ \sin\theta & \cos\theta \end{bmatrix} = \mathbb{I}. \tag{10.6}$$

Note that $x_1\, y_1 + x_2\, y_2$ is simply the inner product of two vectors $\boldsymbol{a} = \overrightarrow{OA}$ and $\boldsymbol{b} = \overrightarrow{OB}$. Besides, as a consequence of this, $(x_1')^2 + (x_2')^2 = (x_1)^2 + (x_2)^2$.

We note that the origin is invariant under $R(\theta)$. Is there any point $A \neq 0$ which is left invariant under $R(\theta)$? Such a point exists if the equation $RA = A$ has nontrivial solution for the components of A, and such a vector A is called an eigenvector of R with eigenvalue 1. We know, from the theory of linear equations, that $RA = \lambda A$ has nontrivial solutions only for some values of λ—those that are roots of the characteristic polynomial $P(\lambda) = \det(R - \lambda\mathbb{I})$. Let us calculate $P(\lambda)$:

$$P(\lambda) = \det(R - \lambda\mathbb{I}) = \begin{vmatrix} \cos\theta - \lambda & -\sin\theta \\ \sin\theta & \cos\theta - \lambda \end{vmatrix} \tag{10.7}$$

$$= \lambda^2 - 2\cos\theta\, \lambda + 1. \tag{10.8}$$

The roots of $P(\lambda)$ are $e^{\pm i\theta} = \cos\theta \pm i \sin\theta$. Therefore if $\theta \neq 0$, there is no invariant point. For $\theta = 0$ any point is invariant, simply because $R(0) = \mathbb{I}$.

Let us now look at the combination formula. By matrix multiplication and using trigonometric identities, we see

$$R(\alpha)R(\beta) = \begin{bmatrix} \cos\alpha & -\sin\alpha \\ \sin\alpha & \cos\alpha \end{bmatrix} \begin{bmatrix} \cos\beta & -\sin\beta \\ \sin\beta & \cos\beta \end{bmatrix} \tag{10.9}$$

$$= \begin{bmatrix} \cos\alpha\,\cos\beta - \sin\alpha\,\sin\beta & -\cos\alpha\,\sin\beta - \sin\alpha\,\sin\beta \\ \sin\alpha\,\cos\beta + \cos\alpha\,\sin\beta & \cos\alpha\,\cos\beta - \sin\alpha\,\sin\beta \end{bmatrix}$$

$$= \begin{bmatrix} \cos(\alpha+\beta) & -\sin(\alpha+\beta) \\ \sin(\alpha+\beta) & \cos(\alpha+\beta) \end{bmatrix} = R(\alpha+\beta). \tag{10.10}$$

From this it follows that

$$R^{-1}(\alpha) = R(-\alpha) = R^T(\alpha). \tag{10.11}$$

Note that $R(\theta)$ described above is a rotation of $\mathbb{R}^2$. By a rotation of E^2, we mean any rotation around any point of E^2. Let us fix a Cartesian orthonormal coordinate system on E^2. Then any point is described by a vector. Let $B = (y_1, y_2)$ be a point in E^2. The rotation around B by angle θ is the mapping

$$A \mapsto A' = B + R(\theta)(A - B), \tag{10.12}$$

where by "+" we mean the usual addition of vectors and $R(\theta)(A - B)$ is the action of the matrix $R(\theta)$ on the vector $A - B$.

10.1.2 Inversions

Let Δ be a line in E^2. We define the *inversion in* Δ as follows: If A is a point in E^2, there is one and only one point A' in E^2 such that the line Δ is perpendicular to the segment AA' and A and A' are the same distance from Δ, or in usual terms, Δ is the normal bisector of the segment AA'.

Let $\hat{\boldsymbol{n}} = (n_1, n_2,)$ be a unit vector *normal* to the line Δ passing through the origin (see Fig. 10.2). Any vector $\boldsymbol{a}$ can be decomposed into two perpendicular vectors $\boldsymbol{a}_{||}$ and $\boldsymbol{a}_{\perp}$, where $\boldsymbol{a}_{||} = (\boldsymbol{a} \cdot \hat{\boldsymbol{n}})\,\hat{\boldsymbol{n}}$ is parallel to $\hat{\boldsymbol{n}}$ and $\boldsymbol{a}_{\perp} := \boldsymbol{a} - \boldsymbol{a}_{||}$ is normal to $\hat{\boldsymbol{n}}$.

Remark. We define normal and parallel with respect to the vector $\hat{n}$, which is normal to the line Δ. The reason is that later we want to generalize to inversions in planes, and a plane is identified by its normal vector. Note that in the following, α is the angle between $\hat{n}$ and the x_1-axis.

Now, by definition, the result of inverting $\boldsymbol{a}$ in Δ is the vector

$$\boldsymbol{a}' = \boldsymbol{a}_{\perp} - \boldsymbol{a}_{||} = \boldsymbol{a}_{\perp} + \boldsymbol{a}_{||} - 2\,\boldsymbol{a}_{||} = \boldsymbol{a} - 2(\boldsymbol{a} \cdot \hat{\boldsymbol{n}})\,\hat{\boldsymbol{n}}, \tag{10.13}$$

$$a_i' = a_i - 2\,n_i \sum a_j\,n_j = \sum_j \left(\delta_{ij} - 2\,n_i\,n_j\right) a_j. \tag{10.14}$$

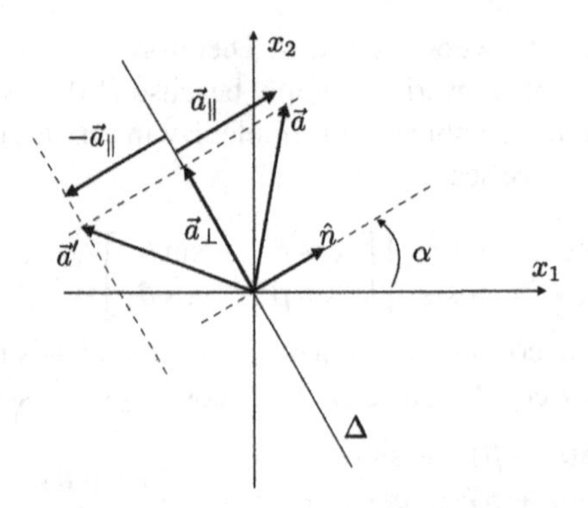

FIGURE 10.2 Inversion in a line.

Unit vector $\hat{n}$, which makes the angle α with the x_1-axis, is normal to the line Δ. The vector $\boldsymbol{a}'$ is the result of inversion of $\boldsymbol{a}$ in Δ. Note that the symbols $\parallel$ and $\perp$ refer to $\hat{n}$. One can visualize this as follows: The line Δ is a mirror, and $\boldsymbol{a}'$ is the image of $\boldsymbol{a}$ in the mirror.

From this, it is seen that the matrix representing inversion in line Δ, which is normal to the unit vector $\hat{\boldsymbol{n}}$, is

$$\left[S(\hat{\boldsymbol{n}})\right]_{ij} = \delta_{ij} - 2\, n_i\, n_j. \tag{10.15}$$

By changing $\hat{\boldsymbol{n}} \to -\hat{\boldsymbol{n}}$, this matrix does not change, as was expected since both $\hat{\boldsymbol{n}}$ and $-\hat{\boldsymbol{n}}$ describe the same line Δ. For $\hat{\boldsymbol{n}} = (\cos\alpha,\ \sin\alpha)$ we get

$$S = \begin{bmatrix} 1 - 2\cos^2\alpha & -2\cos\alpha\,\sin\alpha \\ -2\sin\alpha\,\cos\alpha & 1 - 2\sin^2\alpha \end{bmatrix} = \begin{bmatrix} -\cos 2\alpha & -\sin 2\alpha \\ -\sin 2\alpha & \cos 2\alpha \end{bmatrix}. \tag{10.16}$$

It is easy to see that the inversion also leaves $x_1\, y_1 + x_2\, y_2$ invariant, that is, $x_1\, y_1 + x_2\, y_2 = x_1'\, y_1' + x_2'\, y_2'$.

The combination of two inversions is a rotation! This can be proved either by elementary geometry or by matrix multiplication. The reader is asked to perform the matrix multiplication:

$$S(\alpha)\, S(\beta) = R(2\alpha - 2\beta). \tag{10.17}$$

From this identity it follows that $S^2(\alpha) = R(0) = \mathbb{I}$, that is,

$$S^{-1}(\alpha) = S(\alpha) = S^T(\alpha). \tag{10.18}$$

It also follows that

$$S(\alpha) = R(2\alpha)\, S(0). \tag{10.19}$$

We can now prove that the combination of a rotation and an inversion is an inversion. To see this, multiply the above equation from the left by $R(\theta)$:

$$R(\theta)\, S(\alpha) = R(\theta)\, R(2\alpha)\, S(0)$$

$$= R(2\alpha + \theta)\, S(0)$$
$$= S(\alpha + \tfrac{\theta}{2}). \tag{10.20}$$

10.1.3 O(2) and SO(2)

Basic definitions. By O(2) we mean the set of 2×2 matrices R satisfying $R^T R = \mathbb{I}$. This means that R^T is the left inverse of R, and since the left inverse of a square matrix must be also its right inverse, we have $R R^T = \mathbb{I}$. Let us write these two conditions:

$$R = \begin{bmatrix} a & b \\ c & d \end{bmatrix} \quad \Rightarrow \quad R^T = \begin{bmatrix} a & c \\ b & d \end{bmatrix}, \tag{10.21}$$

$$R^T R = \begin{bmatrix} a^2 + c^2 & a\,b + c\,d \\ b\,a + d\,c & b^2 + d^2 \end{bmatrix} = \begin{bmatrix} 1 & 0 \\ 0 & 1 \end{bmatrix}, \tag{10.22}$$

$$R\,R^T = \begin{bmatrix} a^2 + b^2 & a\,c + b\,d \\ c\,a + d\,b & c^2 + d^2 \end{bmatrix} = \begin{bmatrix} 1 & 0 \\ 0 & 1 \end{bmatrix}. \tag{10.23}$$

Condition $R^T R = \mathbb{I}$ is that the columns of R are two orthonormal vectors, that is, they are of unit length ($a^2 + c^2 = 1$ and $b^2 + d^2 = 1$) and they are perpendicular ($a\,b + c\,d = 0$). Condition $R\,R^T = \mathbb{I}$ is that the rows of R are two orthonormal vectors, that is, they are of unit length ($a^2 + b^2 = 1$ and $c^2 + d^2 = 1$) and they are perpendicular ($a\,c + b\,d = 0$). Because of this property, these matrices are called orthogonal. The letter O in O(2) is for orthogonal.

The same conditions, $R^T R = R\,R^T = \mathbb{I}$, in index notation are as follows:

$$\left[R^T R\right]_{ij} = \sum_{n=1}^{2} \left[R^T\right]_{in} [R]_{nj} = \sum_{n=1}^{2} R_{ni}\, R_{nj} = \delta_{ij}, \tag{10.24}$$

$$\left[R\,R^T\right]_{ij} = \sum_{n=1}^{2} [R]_{in} \left[R^T\right]_{nj} = \sum_{n=1}^{2} R_{in}\, R_{jn} = \delta_{ij}. \tag{10.25}$$

Remember that $\det(AB) = \det A \cdot \det B$ and that $\det A^T = \det A$. It is now easy to show that if $R^T R = \mathbb{I}$, we have $(\det R)^2 = 1$, that is, $\det R$ is either $+1$ or -1. The orthogonal matrices with $\det R = 1$ form a subgroup of O(2) named the special orthogonal group and denoted by SO(2). That SO(2) is a subgroup easily follows from $\det(RR') = \det R \cdot \det R'$. Note that the complement of SO(2)—the elements with $\det R = -1$—is not closed under multiplication, because if $\det R = \det R' = -1$, then $\det(R\,R') = +1$.

The geometric action. Let R be a 2×2 orthogonal matrix. To find the geometrical interpretation of the action $R : A \mapsto A' = RA$, we study the eigenvalues and eigenvectors of R. This requires a study of the characteristic equation. We have

$$R - \lambda \mathbb{I} = R - \lambda\, R\,R^T = R(\mathbb{I} - \lambda\, R^T) = -\lambda\, R(R^T - \lambda^{-1}\,\mathbb{I}), \tag{10.26}$$

$$P(\lambda) = \det(R - \lambda \mathbb{I}) = \det\left\{-\lambda R \left(R^T - \lambda^{-1}\right)\right\} \tag{10.27}$$

$$= (-1)^2 \lambda^2 \det R \cdot \det\left(R - \lambda^{-1}\right), \tag{10.28}$$

$$P(\lambda) = (-1)^2 \lambda^2 \det R \cdot P(\lambda^{-1}). \tag{10.29}$$

If $\det R = -1$, then from the above equation it follows that $P(1) = -P(1)$ and $P(-1) = -P(-1)$. Therefore $P(1) = P(-1) = 0$, that is, R has one eigenvalue $+1$ and one eigenvalue -1. The geometric interpretation of R is then obvious—it is inversion in the line corresponding to $\lambda = +1$—vectors in this direction are left invariant, while vectors normal to this direction are multiplied by -1.

If $\det R = +1$, nothing follows from the above identity for $P(\lambda)$, and we return to the form of R and calculate

$$P(\lambda) = \begin{vmatrix} a-\lambda & b \\ c & d-\lambda \end{vmatrix} = \lambda^2 - (a+d)\lambda + 1. \tag{10.30}$$

The roots of $P(\lambda)$ are $\lambda = a + d \pm \sqrt{(a+d)^2 - 4}$. The term under the square root is $(a+d)^2 - 4 = (a+d-2)(a+d+2)$. It is easy to see that $a+d-2 \leq 0$ and $a+d+2 \geq 0$; therefore $P(\lambda)$ has no real roots. To see this, recall that a and d are both real components of a vector of unit length, so $|a| \leq 1$ and $|d| \leq 1$. Therefore $(a+d)^2 - 4 \leq 0$. Note that $(a+d)^2 - 4 = 0$ only if $a = d = \pm 1$, which corresponds to $R = \pm\mathbb{I}$.

Now let us parametrize O(2). If $R = \left(\begin{smallmatrix} a & b \\ c & d \end{smallmatrix}\right)$ is in O(2), we know that $a^2 + c^2 = 1$. So we can find a real number θ such that $a = \cos\theta$ and $c = \sin\theta$. Now we must find a vector (b, d) normal to (a, c). It can be either $(b, d) = (-\sin\theta, \cos\theta)$ or $(b, d) = (\sin\theta, -\cos\theta)$. So R is of one of two forms:

$$R = \begin{bmatrix} \cos\theta & -\sin\theta \\ \sin\theta & \cos\theta \end{bmatrix} \qquad \det R = 1, \qquad \text{rotation}, \tag{10.31}$$

$$R = \begin{bmatrix} \cos\theta & \sin\theta \\ \sin\theta & -\cos\theta \end{bmatrix} \qquad \det R = -1, \qquad \text{inversion}. \tag{10.32}$$

10.1.4 The generator of the rotations

We now explore a formula for the *rotation* R, which is very suitable for generalization. Recall the identity $R(\alpha)R(\beta) = R(\alpha + \beta)$, from which it follows that $[R(\alpha)]^N = R(N\alpha)$. Now let θ be a finite angle. Divide θ by N and let $\alpha = \theta/N$. We now have $R(\theta) = [R(\alpha)]^N$, so we get $R(\theta) = \left[R\left(\frac{\theta}{N}\right)\right]^N$, which is also true as $N \to \infty$. We therefore have

$$R(\theta) = \lim_{N\to\infty} \left[R\left(\frac{\theta}{N}\right)\right]^N. \tag{10.33}$$

If we Taylor-expand $R(\theta)$ to the first order, we get

$$R(\theta)=\begin{bmatrix} 1 & -\theta \\ \theta & 1 \end{bmatrix}=\mathbb{I}+\theta\begin{bmatrix} 0 & -1 \\ 1 & 0 \end{bmatrix}, \qquad |\theta|\ll 1, \tag{10.34}$$

so the matrix $R(\theta)$, for small θ, is of the form $\mathbb{I}+\theta Z$, where Z is the antisymmetric matrix

$$Z=\begin{bmatrix} 0 & -1 \\ 1 & 0 \end{bmatrix}. \tag{10.35}$$

Now we can write

$$R(\theta)=\lim_{N\to\infty} R\left(\frac{\theta}{N}\right)=\lim_{N\to\infty}\left(\mathbb{I}+\frac{\theta}{N}Z\right)^N=\exp(\theta Z)=e^{\theta Z}. \tag{10.36}$$

Computing $e^{\theta Z}$ is simple, because

$$Z^2=\begin{bmatrix} 0 & -1 \\ 1 & 0 \end{bmatrix}\begin{bmatrix} 0 & -1 \\ 1 & 0 \end{bmatrix}=\begin{bmatrix} -1 & 0 \\ 0 & -1 \end{bmatrix}=-\mathbb{I}, \tag{10.37}$$

from which it follows that

$$Z^3=-Z, \quad Z^4=\mathbb{I}, \quad Z^5=Z, \quad Z^6=-Z, \quad \ldots, \tag{10.38}$$

$$\exp(\theta Z)=\mathbb{I}+\theta Z+\frac{\theta^2}{2!}Z^2+\frac{\theta^3}{3!}Z^3+\ldots \tag{10.39}$$

$$=\mathbb{I}+\theta Z-\frac{\theta^2}{2!}\mathbb{I}-\frac{\theta^3}{3!}Z+\frac{\theta^4}{4!}\mathbb{I}+\frac{\theta^5}{5!}Z+\ldots \tag{10.40}$$

$$=\mathbb{I}\left(1-\frac{\theta^2}{2!}+\frac{\theta^4}{4!}+\ldots\right)+Z\left(\theta-\frac{\theta^3}{3!}+\frac{\theta^5}{5!}+\ldots\right) \tag{10.41}$$

$$=\cos\theta\,\mathbb{I}+\sin\theta\,Z \tag{10.42}$$

$$=\begin{bmatrix} \cos\theta & -\sin\theta \\ \sin\theta & \cos\theta \end{bmatrix}=R(\theta). \tag{10.43}$$

The antisymmetric matrix Z is called the generator of SO(2). The set of all antisymmetric real matrices θZ is called "the Lie algebra $\mathfrak{so}(2)$."

Representation by complex numbers. Let us define a complex plane $\mathbb{C}$ with real and imaginary axes x and y, that is, $z=x+iy$. The action of rotation $R(\theta)$ on the real (x, y)-plane could be easily applied to the complex z-plane. It is easy to show that under $R(\theta)$ we would get

$$R:\begin{cases} z\mapsto z'=e^{i\theta}\,z, \\ \overline{z}\mapsto \overline{z'}=e^{-\theta}\,\overline{z}. \end{cases} \tag{10.44}$$

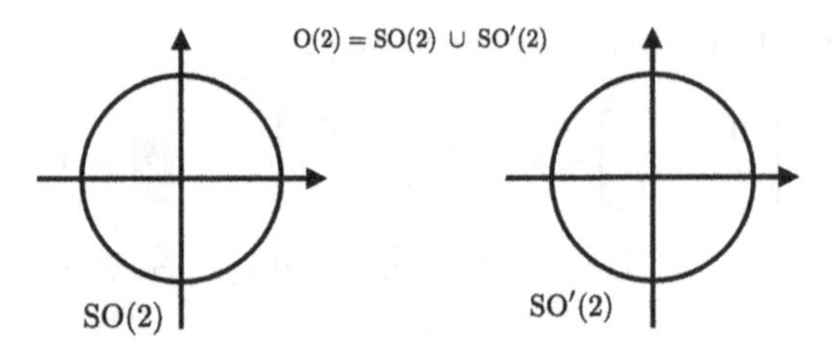

FIGURE 10.3 Topology of the group O(2).

The group O(2) is the union of two circles; one is SO(2), the other corresponds to the complement set SO′(2) consisting of matrices $A = RP$, where R is in SO(2) and P is the matrix (10.47).

It follows that the complex function z^n is being mapped to $e^{in\theta}\, z^n$ and the function $\overline{z}^n$ into $e^{-in\theta}\, \overline{z}^n$. In mathematical jargon one says that the complex function z^n is an eigenfunction of rotation $R(\theta)$ with eigenvalue $e^{in\theta}$.

10.1.5 Topology of SO(2) and O(2)

Any element of SO(2) is uniquely determined by an angle θ. The set of all angles could well be imagined as the points of the unit circle. As a subset of the (x, y)-plane, this circle is the set

$$S^1 = \left\{(x, y) \in \mathbb{R}^2 \mid x^2 + y^2 = 1\right\}. \tag{10.45}$$

As a subset of the complex $\mathbb{C}$-plane, it is the set of points $e^{i\theta}$ with $\theta \in \mathbb{R}$. The correspondence is one to one: To any matrix in SO(2) there corresponds one and only one point of the circle S^1; and to any point on S^1 corresponds one and only one matrix in SO(2). So if we want to imagine SO(2) as a set of points, we can think of the circle S^1:

$$\mathrm{SO}(2) \leftrightarrow S^1. \tag{10.46}$$

What about O(2)? To see this, let A be in O(2). If $\det A = 1$, then A is in SO(2). If $\det A = -1$, then $\det(AP) = +1$, where P is the matrix representing inversion in the x-axis, that is,

$$P = \begin{bmatrix} 1 & 0 \\ 0 & -1 \end{bmatrix}, \tag{10.47}$$

so AP is in SO(2). As a subset of O(2), the set of all inversions is the complement to SO(2); let us denote it by SO′(2). By the preceding argument SO′(2) is also a unit circle (see Fig. 10.3). We therefore have the following picture:

O(2) is the union of two disjoint circles, one representing SO(2), the other representing those elements of the form AP where A is in SO(2) and P is the inversion in the x-axis. O(2) has two connected components, one that contains the identity $\mathbb{I}$, that is, SO(2), and one that does not contain the identity, that is, the complement of SO(2)—it is not a subgroup of O(2).

10.1.6 O(2) and SO(2) are Lie groups

Recall that for any group G, we define two maps, mul and inv. The function mul is the multiplication map $\text{mul} : G \times G \to G$ with definition $\text{mul}(g_1, g_2) = g_1 \circ g_2$, while inv is the "inversion" $\text{inv} : G \to G$ sending $g \mapsto g^{-1}$.

For SO(2), any element is fully specified by an angle, α, and we know that $R(\alpha)\,R(\beta) = R(\alpha + \beta)$ and that $R^{-1}(\alpha) = R(-\alpha) = R(2\pi - \alpha)$. Therefore for SO(2), the functions mul and inv have the following simple form:

$$\text{mul}(\alpha, \beta) = \alpha + \beta \pmod{2\pi}, \tag{10.48}$$

$$\text{inv}(\alpha) = 2\pi - \alpha, \quad \pmod{2\pi}. \tag{10.49}$$

These two functions are indeed analytic and therefore smooth, though because of the $\text{mod}\,2\pi$, it is not trivial. If we write the matrices explicitly, we can see this analyticity more clearly. Note that the entries of $R(\alpha)$ are $\cos\alpha$ and $\pm\sin\alpha$. The entries of $R(\alpha)R(\beta)$ are $\cos(\alpha+\beta)$ and $\pm\sin(\alpha+\beta)$. Now these functions are analytic functions of (α, β).

O(2) is also a Lie group. To see this, we note that both rotations $R(\alpha)$ and inversions $S(\beta)$ are matrices with entries $\pm\cos$ or $\pm\sin$ of the angles, and these functions are analytic. When two elements of O(2) are multiplied, we get matrices with entries a combination of sin and cos functions. Inverting O(2) matrices is even more simple—it is just transposing the matrices. So both functions mul and inv are analytic.

10.2 Rotations and inversions of the space

Let E^3 be the Euclidean space. By a Euclidean transformation, we mean a translation, a rotation, an inversion *in a plane*, or a combination of them. These are the only transformations that map any set into one congruent with itself. We already know what a translation is. In this section we will study rotations and inversions in planes. To be prepared, it is better if we begin by two simple cases: rotation around the x_3-axis and inversion in the plane normal to the unit vector $\hat{\boldsymbol{n}}$.

10.2.1 Rotation around $\hat{n}$

For $\hat{\boldsymbol{n}} = \hat{z}$ this is the following transformation:

$$\begin{cases} x_1' = x_1 \cos\theta - x_2 \sin\theta, \\ x_2' = x_2 \cos\theta + x_1 \sin\theta, \\ x_3' = x_3, \end{cases} \tag{10.50}$$

which is nothing but the rotation we studied earlier in the (x_1, x_2)-plane. This rotation of the space rotates all surfaces $x_3 = \text{constant}$ by the same angle.

By defining $\boldsymbol{a} = (x_1, x_2, x_3)$, $\boldsymbol{a}' = (x_1', x_2', x_3')$, and $\hat{\boldsymbol{n}} = (0, 0, 1)$, the above formulas could be written as

$$\begin{cases} a_i' = a_i \cos\theta + (\hat{\boldsymbol{n}} \times \boldsymbol{a})_i \sin\theta, \quad i = 1, 2 \\ a_3' = a_3. \end{cases} \tag{10.51}$$

We denote this rotation by $R(x_3, \theta)$. Here x_3 reminds us of the axis of rotation. Note that vectors of the form $(0, 0, x_3)$ are eigenvectors of this transformation with eigenvalue $+1$. It can be easily shown that for $\theta \neq 2n\pi$ this transformation has no other real eigenvalues—there are no other invariant lines (for $\theta \neq 2n\pi$).

10.2.2 Inversion in the plane normal to $\hat{\boldsymbol{n}}$

In E^3 a plane passing through the origin is uniquely determined by a line or vector normal to it. Now if Σ is the plane normal to $\hat{\boldsymbol{n}} = (n_1, n_2, n_3)$ (and passing through the origin), we define inversion in Σ by the formula

$$\boldsymbol{a} \mapsto \boldsymbol{a}' = \boldsymbol{a} - 2\hat{\boldsymbol{n}}\,\hat{\boldsymbol{n}} \cdot \boldsymbol{a}. \tag{10.52}$$

The geometry is exactly as in the case of inversion in line in E^2.

Therefore

$$\left[S(\hat{\boldsymbol{n}})\right] = \left[\delta_{ij} - 2n_i n_j\right] = \begin{bmatrix} 1 - 2n_1^2 & -n_1 n_2 & -n_1 n_3 \\ -n_2 n_2 & 1 - 2n_2^2 & -n_2 n_3 \\ -n_3 n_1 & -n_3 n_2 & 1 - 2n_3^2 \end{bmatrix}. \tag{10.53}$$

To check this, let us think of $\hat{\boldsymbol{n}} = (0, 0, 1) = \hat{\boldsymbol{x}}_3$. Now it is easy to see that

$$S(\hat{\boldsymbol{x}}_3) = \begin{bmatrix} 1 & 0 & 0 \\ 0 & 1 & 0 \\ 0 & 0 & -1 \end{bmatrix}. \tag{10.54}$$

We saw in §10.1.2 that the combination of a rotation and an inversion in a line is always an inversion. This is not the case in 3D Euclidean space. To see this let us see what $R(x_3, \theta)S(x_3)$ is:

$$R(x_3, \theta)S(x_3) = \begin{bmatrix} \cos\theta & -\sin\theta & 0 \\ \sin\theta & \cos\theta & 0 \\ 0 & 0 & 1 \end{bmatrix} \begin{bmatrix} 1 & 0 & 0 \\ 0 & 1 & 0 \\ 0 & 0 & -1 \end{bmatrix} \tag{10.55}$$

$$= \begin{bmatrix} \cos\theta & -\sin\theta & 0 \\ \sin\theta & \cos\theta & 0 \\ 0 & 0 & -1 \end{bmatrix}. \tag{10.56}$$

This matrix has only one real eigenvalue, and it is -1. There are no invariant vectors (other than 0), so it is not the inversion in any plane.

10.2.3 The parity operator

The matrix

$$P = R(x_3, \pi)S(x_3) = -\mathbb{I} \tag{10.57}$$

is called the parity matrix. We have $\det P = -1$, so it is not a rotation. It is not an inversion in any plane. Take any line; parity is inversion in the plane normal to that line, followed by a rotation by π around that line. Parity can also be interpreted as "inversion in the origin," which means the mapping $\boldsymbol{r} \to -\boldsymbol{r}$.

Note that in 2D, that is, for 2×2 matrices of O(2), $\det(-\mathbb{I}) = 1$, so in 2D the matrix $-\mathbb{I}$ is not suitable to generate the complement of SO(2).

10.3 Lie algebras

This section is a brief introduction to Lie algebras, which are the mathematical structure to describe continuous symmetries in physics. The basic ideas are simple. Familiarity with the terminology of Lie algebras is a great help in understanding the literature.

Vector space. A vector space is a set of objects, called vectors, which we can add and we can multiply by numbers. The most familiar vector space is perhaps $\mathbb{R}^3$. Vectors can be added, $\boldsymbol{a} + \boldsymbol{b}$, and vectors can be multiplied by numbers, $\lambda \boldsymbol{a}$. These two operations satisfy certain rules that we suppose the reader is familiar with.

Algebra. An *algebra* is a vector space with one additional structure—a multiplication of vectors by vectors. For example, the usual $\mathbb{R}^3$ with the usual cross-product $\boldsymbol{a} \times \boldsymbol{b}$ is an algebra.

Remark. The result of multiplication of two vectors must be a vector! Therefore $\mathbb{R}^3$ with the usual inner product $\boldsymbol{a} \cdot \boldsymbol{b}$ *is not* an algebra!

The multiplication of vectors is a vector function of two arguments. Let us denote the multiplications of A and B (which are vectors in our algebra) by $[A, B]$. As in the cross-product, the function $[A, B]$ must be linear in each of its two arguments, that is,

$$[A + B, C] = [A, C] + [B, C], \tag{10.58}$$

$$[A, B + C] = [A, B] + [A, C], \tag{10.59}$$

$$[\lambda A, B] = \lambda [A, B] = [A, \lambda B]. \tag{10.60}$$

Lie algebra. A Lie algebra is an algebra whose multiplication, which is customarily named *the commutator* or the Lie bracket, satisfies the following two additional properties: anticommutativity,

$$[A,B]=-[B,A], \tag{10.61}$$

and the Jacobi identity,[1]

$$[A,[B,C]]+[B,[C,A]]+[C,[A,B]]=0. \tag{10.62}$$

As our first example, $\mathbb{R}^3$ with the cross-product is a Lie algebra, because we have $\boldsymbol{a}\times\boldsymbol{b}=-\boldsymbol{b}\times\boldsymbol{a}$ and $\boldsymbol{a}\times(\boldsymbol{b}\times\boldsymbol{c})+\boldsymbol{b}\times(\boldsymbol{c}\times\boldsymbol{a})+\boldsymbol{c}\times(\boldsymbol{a}\times\boldsymbol{b})=0$. It should be noted that the cross-product *is not* associative, that is, $\boldsymbol{a}\times(\boldsymbol{b}\times\boldsymbol{c})\neq(\boldsymbol{a}\times\boldsymbol{b})\times\boldsymbol{c}$.

Structure constants of a Lie algebra. Since a Lie algebra is a vector space, it is natural to pick a basis $\{T_i\}$ and write any vector A as a linear combination $A=\sum_i A^i\,T_i$. In expressions like this A^i is a number and T_i is a vector—an element of the *algebra*. Now that we have chosen a basis, to find the commutator of any two vectors, we need to know the commutators of basis elements, that is, $\left[T_i,T_j\right]$, but $\left[T_i,T_j\right]$ is a vector in the algebra, and we can write it as a linear combination of the basis elements. Therefore we must have

$$\left[T_i,T_j\right]=\sum_k C_{ij}^k\,T_k. \tag{10.63}$$

The constants C_{ij}^k are called the structure constants. The anticommutativity and the Jacobi identity impose the following two conditions on the structure constants:

$$C_{ij}^k=-C_{ji}^k, \tag{10.64}$$

$$\sum_n\left(C_{in}^m\,C_{jk}^n+C_{jn}^m\,C_{ki}^n+C_{kn}^m\,C_{ij}^n\right)=0. \tag{10.65}$$

The Lie algebra $\mathbb{R}^3$. Returning to our example, $\mathbb{R}^3$ with the cross-product $[\boldsymbol{a},\boldsymbol{b}]:=\boldsymbol{a}\times\boldsymbol{b}$, let $(\hat{\boldsymbol{e}}_1,\hat{\boldsymbol{e}}_2,\hat{\boldsymbol{e}}_3)$ be a right-handed orthonormal basis. We have

$$\left[\hat{\boldsymbol{e}}_i,\hat{\boldsymbol{e}}_j\right]=\hat{\boldsymbol{e}}_i\times\hat{\boldsymbol{e}}_j=\sum_k\epsilon_{ijk}\,\hat{\boldsymbol{e}}_k. \tag{10.66}$$

Therefore the structure constants of this Lie algebra are the familiar ϵ_{ijk}. The antisymmetry in $i\leftrightarrow j$ is obvious. Let us check the Jacobi identity. We have

$$\sum_n\left(C_{in}^m\,C_{jk}^n+C_{jn}^m\,C_{ki}^n+C_{kn}^m\,C_{ij}^n\right) \tag{10.67}$$

$$=\sum_n\left(\epsilon_{inm}\,\epsilon_{jkn}+\epsilon_{jnm}\,\epsilon_{kin}+\epsilon_{knm}\,\epsilon_{ijn}\right) \tag{10.68}$$

[1] After Karl Gustav Jacob Jacobi (1804–1851).

$$= \sum_n \left(\epsilon_{min}\,\epsilon_{jkn} + \epsilon_{mjn}\,\epsilon_{kin} + \epsilon_{mkn}\,\epsilon_{ijn}\right) \tag{10.69}$$

$$= \delta_{mj}\,\delta_{ik} - \delta_{mk}\,\delta_{ij} + \delta_{mk}\,\delta_{ji} - \delta_{mi}\,\delta_{jk} + \delta_{mi}\,\delta_{kj} - \delta_{mj}\,\delta_{ki} \tag{10.70}$$

$$= 0. \tag{10.71}$$

Adjoint action. If A is an element of the Lie algebra $\mathfrak{g}$, then we can send any B in $\mathfrak{g}$ into $[A, B]$ which is in $\mathfrak{g}$. In other words, the element A defines a function from $\mathfrak{g}$ to itself, which is linear—because $[A, B]$ is linear in B. The function generated by A is called *the adjoint action of* A and is denoted by $\mathrm{ad}(A)$. In concise mathematical notation,

$$\begin{aligned} &\mathrm{ad}(A) : \mathfrak{g} \to \mathfrak{g}, \\ &\mathrm{ad}(A) : B \mapsto [A, B]. \end{aligned} \tag{10.72}$$

To clarify notation, when we write $\mathrm{ad}(A)B$ or $\mathrm{ad}(A)(B)$, we mean the action of $\mathrm{ad}(A)$ on the vector B, that is, $[A, B]$.

For our $\mathbb{R}^3$, as a Lie algebra, with a right-handed orthonormal basis, we have $\mathrm{ad}(\hat{\boldsymbol{e}}_i)\,\hat{\boldsymbol{e}}_j = \sum_k \epsilon_{ijk}\,\hat{\boldsymbol{e}}_k$. From this it is obvious that the matrix representing $\mathrm{ad}(\hat{\boldsymbol{e}}_i)$ is the following[2]:

$$\left[\mathrm{ad}(\hat{\boldsymbol{e}}_i)\right]_{kj} = \epsilon_{ijk} = -\epsilon_{ikj}. \tag{10.73}$$

Explicitly,

$$\mathrm{ad}(\hat{\boldsymbol{e}}_1) = \begin{bmatrix} 0 & 0 & 0 \\ 0 & 0 & -1 \\ 0 & 1 & 0 \end{bmatrix}, \tag{10.74}$$

$$\mathrm{ad}(\hat{\boldsymbol{e}}_2) = \begin{bmatrix} 0 & 0 & 1 \\ 0 & 0 & 0 \\ -1 & 0 & 0 \end{bmatrix}, \tag{10.75}$$

$$\mathrm{ad}(\hat{\boldsymbol{e}}_3) = \begin{bmatrix} 0 & -1 & 0 \\ 1 & 0 & 0 \\ 0 & 0 & 0 \end{bmatrix}. \tag{10.76}$$

The Lie algebra of real antisymmetric matrices. The set of real $N \times N$ antisymmetric matrices forms a vector space of dimension $\frac{1}{2}N(N-1)$. The proof is simple: The linear combination of any number of antisymmetric matrices is again antisymmetric and real (we consider real numbers) and there are $\frac{1}{2}N(N-1)$ independent antisymmetric matrices. Now let us define *the multiplication of this algebra.* We define the commutator of two matrices as

$$[A, B] = AB - BA. \tag{10.77}$$

---

[2] Recall that if S is a linear transformation, the matrix S_{ij} representing it in the basis $\{\hat{\boldsymbol{e}}_i\}$ is defined by $S(\hat{\boldsymbol{e}}_j) = \sum_k S_{kj}\,\hat{\boldsymbol{e}}_k$. Note that the summation is on the first index of S.

Note that this multiplication, like the cross-product, *is not* associative, that is,

$$[A, [B, C]] \neq [[A, B], C]. \tag{10.78}$$

It is easy to see that both anticommutativity and Jacobi identity are satisfied. The anticommutativity follows directly from the definition of the commutator:

$$[B, A] = BA - AB = -(AB - BA) = -[A, B]. \tag{10.79}$$

The Jacobi identity follows from

$$[A, [B, C]] = [A, BC - CB] = A(BC - CB) - (BC - CB)A \tag{10.80}$$

$$= ABC - ACB - BCA + CBA. \tag{10.81}$$

Changing $A \to B \to C \to A$ twice, we get

$$[A, [B, C]] = ABC - ACB - BCA + CBA, \tag{10.82}$$

$$[B, [C, A]] = BCA - BAC - CAB + ACB, \tag{10.83}$$

$$[C, [A, B]] = CAB - CBA - ABC + BAC. \tag{10.84}$$

Adding these three lines we get the Jacobi identity.

It is easy to see that if A and B are antisymmetric, then $[A, B]$ is antisymmetric. The proof is simple: If $A = -A^T$ and $B = -B^T$, then we have

$$[A, B]^T = (A\,B)^T - (B\,A)^T = B^T\,A^T - A^T\,B^T \tag{10.85}$$

$$= (-B)(-A) - (-A)(-B) = B\,A - A\,B = -[A, B]. \tag{10.86}$$

In words, the commutator of two antisymmetric matrices is again an antisymmetric matrix. We therefore have the following theorem.

Theorem. *The set of $N \times N$ real antisymmetric matrices, with commutator as the multiplication, is a real Lie algebra of dimension $\frac{1}{2}N(N-1)$.*

The symbol to denote this Lie algebra is $\mathfrak{so}(N)$. This algebra is related to the group of $N \times N$ orthogonal matrices with determinant 1, the so-called SO(N). In the next section we elaborate on this relation for $N = 3$.

10.4 The Lie algebra $\mathfrak{so}(3)$

10.4.1 Generators of SO(3)

For the rotation around the x_3-axis, it is straightforward to see that

$$R(x_3, \theta) = \begin{bmatrix} \cos\theta & -\sin\theta & 0 \\ \sin\theta & \cos\theta & 0 \\ 0 & 0 & 1 \end{bmatrix} = \exp(\theta Z), \tag{10.87}$$

where

$$Z = \begin{bmatrix} 0 & -1 & 0 \\ 1 & 0 & 0 \\ 0 & 0 & 0 \end{bmatrix}. \tag{10.88}$$

The proof of this is simple, once we check that $Z^3 = -Z$ (the details of the proof are presented in §10.4.2).

It is easy to show that for the rotations around the x- and y-axes, we would get similar formulas with the following matrices:

$$R(x_1, \alpha) = e^{\alpha X} = \begin{bmatrix} 1 & 0 & 0 \\ 0 & \cos\alpha & -\sin\alpha \\ 0 & \sin\alpha & \cos\alpha \end{bmatrix}, \qquad X = \begin{bmatrix} 0 & 0 & 0 \\ 0 & 0 & -1 \\ 0 & 1 & 0 \end{bmatrix}, \tag{10.89}$$

$$R(x_2, \beta) = e^{\beta Y} = \begin{bmatrix} \cos\beta & 0 & \sin\beta \\ 0 & 1 & 0 \\ -\sin\beta & 0 & \cos\beta \end{bmatrix}, \qquad Y = \begin{bmatrix} 0 & 0 & 1 \\ 0 & 0 & 0 \\ -1 & 0 & 0 \end{bmatrix}. \tag{10.90}$$

It is easy to check that the following commutation relations hold:

$$[X, Y] = Z, \quad [Y, Z] = X, \quad [Z, X] = Y. \tag{10.91}$$

The matrices X, Y, and Z are called *the generators of the rotations*. This terminology will be clear at the end of this section. Instead of X, Y, and Z, the customary notation for the generators of the rotations is $\mathcal{S}_1 := X$, $\mathcal{S}_2 := Y$, and $\mathcal{S}_3 := Z$. These three matrices are exactly $\mathrm{ad}(\hat{\boldsymbol{e}}_i)$s of the $\mathbb{R}^3$ with the usual cross-product. That is, the Lie algebra $\mathfrak{so}(3)$ is the same algebra $\mathbb{R}^3$ with the cross-product as the commutator.

10.4.2 Exponentiating an antisymmetric matrix

We define a vector *with matrix components*, $\boldsymbol{\mathcal{S}} = (\mathcal{S}_1, \mathcal{S}_2, \mathcal{S}_3)$, where $\mathcal{S}_1 = X$, $\mathcal{S}_2 = Y$, and $\mathcal{S}_3 = Z$ are the antisymmetric matrices defined above. Using this vector, the most general 3×3 real antisymmetric matrix Ω could be written as $\Omega = \boldsymbol{\alpha} \cdot \boldsymbol{\mathcal{S}}$, where $\boldsymbol{\alpha} = (\alpha_1, \alpha_2, \alpha_3)$ is a real vector:

$$\Omega = \boldsymbol{\alpha} \cdot \boldsymbol{\mathcal{S}} = \alpha_1 \mathcal{S}_1 + \alpha_2 \mathcal{S}_2 + \alpha_3 \mathcal{S}_3 = \begin{bmatrix} 0 & -\alpha_3 & \alpha_2 \\ \alpha_3 & 0 & -\alpha_1 \\ -\alpha_2 & \alpha_1 & 0 \end{bmatrix}. \tag{10.92}$$

The vector $\boldsymbol{\alpha}$ can be written as $\boldsymbol{\alpha} = \alpha \hat{\boldsymbol{n}}$, where $\alpha = \sqrt{\alpha_1^2 + \alpha_2^2 + \alpha_3^2}$ and $\hat{\boldsymbol{n}} = (n_1, n_2, n_3)$ is the corresponding unit vector. Now let us define

$$A = \hat{\boldsymbol{n}} \cdot \boldsymbol{\mathcal{S}} = \begin{bmatrix} 0 & -n_3 & n_2 \\ n_3 & 0 & -n_1 \\ -n_2 & n_1 & 0 \end{bmatrix}, \quad \text{that is,} \quad A_{ij} = -\sum_k \epsilon_{ijk} n_k. \tag{10.93}$$

With this definition, $\Omega = \alpha A$, and we have

$$e^{\Omega} = \mathbb{I} + \alpha A + \frac{\alpha^2}{2!} A^2 + \frac{\alpha^3}{3!} A^3 + \dots . \qquad (10.94)$$

To calculate this exponential, we must know the powers of A. From the theorem of Hamilton and Cayley (see, for example, Duistermaat and Kolk, 2004, p. 239) we know that any matrix satisfies its own characteristic equation $P(\lambda) = \det(A - \lambda \mathbb{I})$. Calculating this polynomial for A is simple. The result is $P(\lambda) = -\lambda^3 - \lambda$, from which it follows that $A^3 = -A$. However, since straightforward calculation is both easy and informative, let us calculate the powers of A. We have

$$A^2 = \begin{bmatrix} n_1^2 - 1 & n_1 n_2 & n_1 n_3 \\ n_2 n_1 & n_2^2 - 1 & n_2 n_3 \\ n_3 n_1 & n_3 n_2 & n_3^2 - 1 \end{bmatrix} =: B, \quad \text{that is,} \quad B_{ij} = n_i n_j - \delta_{ij}. \qquad (10.95)$$

In obtaining the above matrix we used $n_1^2 + n_2^2 + n_3^2 = 1$, from which relations like $-n_2^2 - n_3^2 = n_1^2 - 1$ follow. One more matrix multiplication reveals that $A^3 = -A$, and this is the key of our computing e^{Ω}. From $A^2 = B$ and $A^3 = B A = A B = -A$, it follows that $A^4 = -B$, $A^5 = A$, etc.:

$$A^2 = B, \quad A^3 = -A, \quad A^4 = -B, \quad A^5 = A, \quad \dots . \qquad (10.96)$$

We can now find e^{Ω}. We have

$$\begin{aligned} R := e^{\Omega} &= \mathbb{I} + \alpha A + \frac{\alpha^2}{2!} A^2 + \frac{\alpha^3}{3!} A^3 + \frac{\alpha^4}{41} A^4 + \frac{\alpha^5}{5!} A^5 + \dots \\ &= \mathbb{I} + \alpha A + \frac{\alpha^2}{2!} B - \frac{\alpha^3}{3!} A - \frac{\alpha^4}{4!} B + \frac{\alpha^5}{5!} A + \frac{\alpha^6}{6!} B + \dots \\ &= \mathbb{I} + (1 - \cos\alpha) B + \sin\alpha \, A, \\ R_{ij} &= \delta_{ij} + (1 - \cos\alpha)\left(n_i n_j - \delta_{ij}\right) - \sin\alpha \sum_k \epsilon_{ijk} n_k. \end{aligned} \qquad (10.97)$$

This matrix is the matrix of rotation around $\hat{\boldsymbol{n}}$ with angle α. To prove this, first note that

$$\begin{aligned} \sum_j R_{ij} n_j &= \sum_j \{\delta_{ij} n_j + (1 - \cos\alpha)(n_i n_j n_j - \delta_{ij} n_j) - \sin\alpha \sum_k \epsilon_{ijk} n_k n_j\} \\ &= n_i + (1 - \cos\alpha)(n_i - n_i) - 0 = n_i. \end{aligned} \qquad (10.98)$$

That is, the vector $\hat{\boldsymbol{n}}$ is invariant under the action of R. This is also true for any vector $a \hat{\boldsymbol{n}}$. Now consider a vector $\boldsymbol{a}$, normal to $\hat{\boldsymbol{n}}$, which means $\sum_j n_j a_j = 0$. For such a vector we have

$$\sum_j R_{ij} a_j = \sum_j \left\{ \delta_{ij} a_j + (1 - \cos\alpha)(n_i n_j - \delta_{ij}) a_j - \sin\alpha \sum_k \epsilon_{ijk} n_k a_j \right\}$$

$$
\begin{aligned}
&= \cos\alpha\, a_i + \sin\alpha \sum_k \epsilon_{ikj} n_k\, a_j \\
&= \cos\alpha\, a_i + \sin\alpha\, (\hat{\boldsymbol{n}} \times \boldsymbol{a})_i, \qquad \text{if} \quad \hat{\boldsymbol{n}} \cdot \boldsymbol{a} = 0. \qquad (10.99)
\end{aligned}
$$

Therefore R represents rotation by α around $\hat{\boldsymbol{n}}$ (see (10.51) on page 208).
Hermitian generators of **SO(3).** If $[A, B] = C$, then $[iA, iB] = i(iC)$. Let us define the following complex matrices:

$$S_1 = i\mathcal{S}_1, \quad S_2 = i\mathcal{S}_2, \quad S_3 = i\mathcal{S}_3, \qquad (10.100)$$

$$S_1 = \begin{bmatrix} 0 & 0 & 0 \\ 0 & 0 & -i \\ 0 & i & 0 \end{bmatrix}, \quad S_2 = \begin{bmatrix} 0 & 0 & i \\ 0 & 0 & 0 \\ -i & 0 & 0 \end{bmatrix}, \quad S_3 = \begin{bmatrix} 0 & -i & 0 \\ i & 0 & 0 \\ 0 & 0 & 0 \end{bmatrix}. \qquad (10.101)$$

For these complex Hermitian matrices, the commutation relations read

$$[S_1, S_2] = iS_3, \quad [S_2, S_3] = iS_1, \quad [S_3, S_2] = iS_1, \qquad (10.102)$$

which we write in the compact form

$$[S_n, S_m] = i\epsilon_{nmk}\, S_k. \qquad (10.103)$$

The rotation matrices are not changed by this convention, but the relations to the "generators" read

$$R(x_1, \alpha) = e^{-i\alpha S_1}, \quad R(x_2, \beta) = e^{-i\beta S_2}, \quad R(x_3, \theta) = e^{-i\theta S_3}. \qquad (10.104)$$

Remark. In physics it is customary and convenient to work with these Hermitian matrices. The reason is that now their eigenvalues are real numbers, and they could be interpreted directly as observables of quantum systems. In some texts the Hermitian S_ns are called generators, while in some texts, like this book, $-iS_n$s are called the generators.

Remark. In quantum mechanics it is customary to define $S_n = i\hbar\mathcal{S}_n$. The extra $\hbar$ causes the dimension of these generators to be angular momentum. In this book we are going to drop the $\hbar$. We must remember this whenever we want to use these generators as angular momentum. Thus for us the angular momentum is $\hbar\boldsymbol{S}$.

10.5 Topology of SO(3) and O(3)

Let us first remember some definitions.

3D balls and spheres. The 3D ball with radius a is the set

$$B_a^3 = \left\{x^2 + y^2 + z^2 \le a^2\right\}. \tag{10.105}$$

The sphere

$$S_a^2 = \left\{x^2 + y^2 + z^2 = a^2\right\} \tag{10.106}$$

is the boundary of B_a^3. If the point (x, y, z) is on S_a^2, the point $(-x, -y, -z)$ is also on S_a^2, and the line segment joining these two points passes through the origin. We say that the points (x, y, z) and $(-x, -y, -z)$ are *antipodal*. The unit ball and the unit sphere are denoted by B^3 and S^2, respectively.

3D real projective space. $\mathbb{RP}^3$ is the unit ball B^3 on which the antipodal points of its surface, the S^2, are considered to be equivalent, that is, the point $(x, y, z) \in S^2$ *is the same point* $(-x, -y, -z)$.

The set of all rotations. A rotation of $\mathbb{R}^3$ is completely specified by giving a unit vector $\hat{\boldsymbol{n}}$ and an angle α. The unit vector $\hat{\boldsymbol{n}}$ specifies the axis of rotation, and the angle α specifies the angle of rotation. It is important that $(\hat{\boldsymbol{n}}, \alpha)$ and $(-\hat{\boldsymbol{n}}, 2\pi - \alpha)$ specify the same rotation. In words, rotation around the vector $\hat{\boldsymbol{n}}$ by angle α is the same as rotation around the vector $-\hat{\boldsymbol{n}}$ by angle $2\pi - \alpha$. To be convinced, think of this: The rotation around $\hat{z}$ by 10° is the same as the rotation around $-\hat{z}$ by 350°. Therefore to specify all rotations we must consider all unit vectors (directions) and for any direction rotations by angle $0 \le \alpha \le 180°$.

Define the vector

$$\boldsymbol{s} = \alpha\,\hat{\boldsymbol{n}}. \tag{10.107}$$

The length of this vector is the angle of rotation, and its direction is the axis of rotation. The identity map, that is, rotation by $\alpha = 0$, corresponds to the origin of the $\boldsymbol{s}$ space. Now if we want to obtain a set, each point of which specifies one and only one rotation, we must consider all vectors $\boldsymbol{s}$ such that $|\boldsymbol{s}| = \alpha \le \pi$. This is a 3D ball B_π^3, the surface of which is S_π^2.

Let $\boldsymbol{a}$ be a point on S_π^2, that is, a vector of length π. Let $\boldsymbol{a}' = -\boldsymbol{a}$ be the antipodal point. Both $\boldsymbol{a}$ and $\boldsymbol{a}'$ specify the same rotation, because the rotation by π around $\hat{\boldsymbol{n}}$ is the same as rotation by π around $-\hat{\boldsymbol{n}}$. In words, every antipodal pair of points on the surface of B_π^3 specifies the same rotation. Therefore we must consider any pair $(\boldsymbol{a}, -\boldsymbol{a})$ as a single rotation.

All this could be summarized by the following statement.

The topology of the rotation group. The group SO(3) as a topological set is the 3D real projective space $\mathbb{RP}^3$.

Let A be in O(3). If $\det A = 1$, then A is in SO(3). If $\det A = -1$, then $\det(AP) = 1$, where $P = -\mathbb{I}$ is the parity operator. Now AP is in SO(3). We therefore conclude that O(3) is the union of two disjoint copies of $\mathbb{RP}^3$.

10.6 Rotating the fields

10.6.1 Scalar fields

Consider a function $f(x) = f(x_1, x_2, x_3)$. Such a function is usually called a *scalar field*. To have an idea, imagine f as the temperature or the pressure at different points of the lab. Now if we rotate the lab, another function would describe the temperature. By rotating f we mean constructing this other function, which we denote by f^R. It should be obvious that

$$f^R(Rx) = f(x). \tag{10.108}$$

That is, the value of f^R at point Rx is the same as the value of the original function f at the original point x. This is a straightforward generalization of the definition of translation of a function. We want to obtain f^R from f. To do this we do exactly the same as in the case of translation. We define $x' = Rx$, solve it for x, which yields $x = R^{-1}x'$, and write $f^R(Rx) = f(x)$ using this new variable. It follows that $f^R(x') = f(R^{-1}x')$. Dropping the prime we now have

$$f^R(x) = f\left(R^{-1}x\right). \tag{10.109}$$

Let us consider analytic functions and search for the operator U_R such that

$$f^R(x) = (U_R f)(x). \tag{10.110}$$

A point about notation: f is a function, the value of which at point x is $f(x)$. Under rotation R, this function f maps to another function f^R, which is also written as $U_R f$, and the value of this function at point x is denoted $f^R(x)$ or $(U_R f)(x)$.

To find the form of U_R, we first find the form of it corresponding to the infinitesimal rotation $R : (x_1, x_2) \mapsto (x_1 - \alpha x_2, x_2 + \alpha x_1)$, whose inverse reads $R^{-1} : (x_1, x_2) \mapsto (x_1 + \alpha x_2, x_2 - \alpha x_1)$. We have

$$f^R(x_1, x_2, x_3) = f(R^{-1}x) \tag{10.111}$$

$$= f(x_1 + \alpha x_2, x_2 - \alpha x_1, x_3) \tag{10.112}$$

$$= f(x_1, x_2, x_3) - \alpha\left(x_1 \frac{\partial f}{\partial x_2} - x_2 \frac{\partial f}{\partial x_1}\right) \tag{10.113}$$

$$= (1 - \alpha \mathcal{L}_3) f, \tag{10.114}$$

where

$$\mathcal{L}_3 = x_1 \frac{\partial}{\partial x_2} - x_2 \frac{\partial}{\partial x_1}. \tag{10.115}$$

Now, repeating the chain of reasons leading to $R = \exp(\alpha Z)$, we get

$$U(x_3, \alpha) = \exp(-\alpha \mathcal{L}_3), \tag{10.116}$$

where $\mathcal{L}_3 = x_2\,\partial_1 - x_1\,\partial_2$ is called the generator of rotations (around the x_3-axis) on functions. Using the operator $\boldsymbol{P} = -i\nabla$, we can write

$$\mathcal{L}_3 = i\,(\boldsymbol{r} \times \boldsymbol{P})_3 = i L_3, \tag{10.117}$$

which leads to

$$U(x_3, \alpha) = \exp\left(-i\alpha L_3\right). \tag{10.118}$$

We now define

$$\boldsymbol{L} = \boldsymbol{r} \times \boldsymbol{P} = -i\boldsymbol{r} \times \nabla. \tag{10.119}$$

We call it the operator of "orbital angular momentum," though it is dimensionless.

10.6.2 Vector fields

A vector field is a map that assigns a vector to any point of space. A good example is the electric field. As is shown in Fig. 10.4, the rotated field E^R at the rotated point $x' = Rx$ is not equal to $E(x)$ but is equal to $R\,E(x)$. In other words, one has to rotate the vector $E(x)$ by the same matrix R, so that $E^R(R\,x) = R\,E(x)$. Again we introduce the variable $x' = R\,x$ and solve this equation for x', which yields $x = R^{-1}x'$. Now we have $E^R(x') = R\,E\left(R^{-1}x'\right)$. Dropping the prime we get

$$E^R(x) = R\,E\left(R^{-1}x\right). \tag{10.120}$$

Let us write this equation in more detail:

$$\begin{bmatrix} E_1^R(x) \\ E_2^R(x) \\ E_3^R(x) \end{bmatrix} = \begin{bmatrix} R_{11} & R_{12} & R_{13} \\ R_{21} & R_{22} & R_{23} \\ R_{31} & R_{32} & R_{33} \end{bmatrix} \begin{bmatrix} E_1(R^{-1}x) \\ E_2(R^{-1}x) \\ E_3(R^{-1}x) \end{bmatrix}. \tag{10.121}$$

For simplicity, assume that R is a rotation around the x_3-axis. We write the 3×3 matrix $R(x_3, \alpha)$ as $R(x_3, \alpha) = \exp(-i\alpha S_3)$, and for each component $E_i(R^{-1}x)$ we write $\exp\left(-i\alpha L_3\right) E_i(x)$. Combining these, we get

$$E^R(x) = \exp\left(-i\alpha S_3\right)\exp\left(-i\alpha L_3\right) E(x). \tag{10.122}$$

Note that the operator $\exp\left(-i\alpha L_3\right)$ acts only on the arguments (x_1, x_2, x_3) of each component, while the operator $\exp\left(-i\alpha S_3\right)$, being a 3×3 matrix, has nothing to do with the arguments, but mixes the components. Therefore we can say that these two operators commute, that is,

$$\exp\left(-i\alpha S_3\right)\exp\left(-i\,\alpha L_3\right) = \exp\left(-i\,\alpha L_3\right)\exp\left(-i\,\alpha S_3\right), \tag{10.123}$$

or, in terms of generators,

$$[\,S_3,\; L_3\,] = 0. \tag{10.124}$$

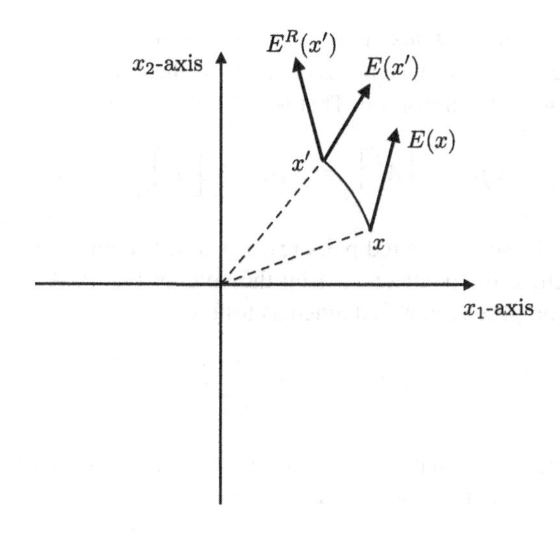

FIGURE 10.4

The vector field $\boldsymbol{E}$ assigns a vector to each point x. $\boldsymbol{E}(x)$ is the vector assigned to point $\boldsymbol{x}$, and $\boldsymbol{E}(\boldsymbol{x}')$ is the vector assigned to point $\boldsymbol{x}'$. Rotation $\boldsymbol{R}$ maps $\boldsymbol{x}$ to $\boldsymbol{x}'$ and vector $\boldsymbol{E}(\boldsymbol{x})$ to $\boldsymbol{E}^R(\boldsymbol{x}')$. Note that $\boldsymbol{E}^R(\boldsymbol{x}')$ is not the same as $\boldsymbol{E}(\boldsymbol{x})$ but is equal to $\boldsymbol{R}\,\boldsymbol{E}(\boldsymbol{x})$; that is, one has to rotate $\boldsymbol{E}(\boldsymbol{x})$ by the same matrix $\boldsymbol{R}$ to obtain $\boldsymbol{E}^R(\boldsymbol{x}')$.

Now, we define the operator $J_3 := S_3 + L_3$, in terms of which we have $E^R(x) = e^{-i\alpha J_3} E(x)$. The generalization is obvious: Defining

$$\boldsymbol{J} = \boldsymbol{L} + \boldsymbol{S}, \tag{10.125}$$

we have

$$E^R(\boldsymbol{r}) = \exp(-i\boldsymbol{\alpha} \cdot \boldsymbol{J}) E(\boldsymbol{r}). \tag{10.126}$$

In quantum mechanics $\boldsymbol{L} = (L_1, L_2, L_3)$ is called the orbital angular momentum operator, $\boldsymbol{S} = (S_1, S_2, S_3)$ is called the spin-1 angular momentum operator, and the *vector operator* $\boldsymbol{J} = \boldsymbol{L} + \boldsymbol{S}$ is called the total angular momentum operator.

10.7 The unitary groups U(2) and SU(2)

Motivation. For mathematicians, it is quite natural to study the space $\mathbb{C}^2$. For physicists, it is better to give some motivation. Here is one motivation. In the early 20th century it became obvious that quantum mechanics is a genuinely "complex" theory, meaning that the functions appearing as wave functions are complex-valued functions of (t, x, y, z). For the electron, the wave function is of the form (f, g), where f and g are two complex-valued functions of (t, x, y, z). That is, the wave function of the electron is a function with value in $\mathbb{C}^2$.

Basic definitions and notations. If A is a complex matrix, $\overline{A}$ is the matrix formed by the complex conjugate of the entries of A; A^T is the transpose of A, and $A^\dagger$ is the complex conjugate of the transpose. That is,

$$\left[\overline{A}\right]_{ij} = \overline{A}_{ij}, \qquad \left[A^T\right]_{ij} = A_{ji}, \qquad \left[A^\dagger\right]_{ij} = \overline{A}_{ji}. \tag{10.127}$$

The space $\mathbb{C}^2$ is the set of ordered pairs (z_1, z_2), where both z_1 and z_2 are complex numbers. We denote the point (z_1, z_2) by the column matrix $A = \left[\begin{smallmatrix} z_1 \\ z_2 \end{smallmatrix}\right]$. The inner product of two complex vectors is defined as follows:

$$\langle A, B\rangle = A^\dagger B = \sum_i \overline{A}_i B_i. \tag{10.128}$$

Using any complex 2×2 matrix S we can transform $A \mapsto A' = SA$ and $B \mapsto B' = SB$. The inner product of A' and B' is now

$$\left\langle A', B'\right\rangle = (SA)^\dagger (SB) = A^\dagger S^\dagger S B. \tag{10.129}$$

From this it is obvious that the inner product of *any two* complex vectors is invariant under this transformation if and only if $S^\dagger S = \mathbb{I}$.
Unitarity. The matrix S is said to be unitary if $S^\dagger S = \mathbb{I}$.

If $S^\dagger S = \mathbb{I}$, then $S^\dagger$ is the left inverse of S, so it is also its right inverse; that is, $S S^\dagger = \mathbb{I}$. It is also easy to see that if S is unitary, $S^{-1} = S^\dagger$ is also unitary. Now let S and U be two unitary matrices. We have

$$(SU)^\dagger (SU) = \left(U^\dagger S^\dagger\right)(SU) = U^\dagger \left(S^\dagger S\right) U = U^\dagger \mathbb{I} U = U^\dagger U = \mathbb{I}. \tag{10.130}$$

From these observations we see that the set of 2×2 unitary matrices form a group.
The unitary group. U(2) is the set of 2×2 unitary matrices, with the usual matrix multiplication.

It is obvious that $\det U^\dagger = \overline{\det U}$, and therefore from $\det(U^\dagger U) = \det \mathbb{I} = 1$ it follows that $|\det U| = 1$. In other words, $\det U = e^{i\alpha}$ for some real α. Physicists say that $\det U$ is a "phase."

Since $\det(SU) = \det S \cdot \det U$, it is obvious that those matrices of U(2) with $\det U = 1$ form a closed set, and therefore they are a subgroup of U(2).

10.8 Topology of SU(2)

SU(2) is the subset of U(2) with $\det S = 1$. To find the topology of SU(2), we write

$$S = \begin{bmatrix} a & b \\ c & d \end{bmatrix} \quad \Rightarrow \quad S^\dagger = \begin{bmatrix} \overline{a} & \overline{c} \\ \overline{b} & \overline{d} \end{bmatrix}. \tag{10.131}$$

We know that $\det S = ad - bc = 1$, and we know that the inverse of S is

$$S^{-1} = \begin{bmatrix} d & -b \\ -c & a \end{bmatrix}. \tag{10.132}$$

The condition of unitarity is $S^{\dagger} = S^{-1}$, which means that

$$\begin{bmatrix} \overline{a} & \overline{c} \\ \overline{b} & \overline{d} \end{bmatrix} = \begin{bmatrix} d & -b \\ -c & d \end{bmatrix}, \tag{10.133}$$

or

$$d = \overline{a}, \qquad c = -\overline{b}. \tag{10.134}$$

Therefore

$$S = \begin{bmatrix} a & b \\ -\overline{b} & \overline{a} \end{bmatrix}. \tag{10.135}$$

Now, the condition $\det S = ad - bc = 1$ reads

$$a\,\overline{a} - b\,(-\overline{b}) = |a|^2 + |b|^2 = 1. \tag{10.136}$$

Writing $a = x + iy$ and $b = z + iw$, this condition is

$$x^2 + y^2 + z^2 + w^2 = 1. \tag{10.137}$$

This is the equation of a 3D sphere.

Definition. (3D Sphere) The 3D sphere with radius a is the set

$$S_a^3 = \left\{x^2 + y^2 + z^2 + w^2 = a^2\right\} \subset \mathbb{R}^4. \tag{10.138}$$

10.8.1 Topology of U(2)

Let A be in U(2), and let $\det A = e^{2i\alpha}$. Since A is a 2×2 matrix, we have

$$\det\left(e^{-i\alpha} A\right) = e^{-2i\alpha} \det A = 1. \tag{10.139}$$

So the matrix $B = e^{-i\alpha} A$ is in SU(2). Therefore any matrix A in U(2) is of the form $e^{i\alpha} B$ for a unique B in SU(2), and it is uniquely determined by the ordered pair $(e^{i\alpha}, \boldsymbol{\beta})$, where $\boldsymbol{\beta}$ is a point in S^3. This means that as a set U(2) is the Cartesian product of S^1 and S^3. If this is not obvious, recall the meaning of Cartesian product: The Cartesian product of two sets A and B is the set of ordered pairs (a, b) where $a \in A$ and $b \in B$. So $S^1 \times S^3$ means simply the set of ordered pairs (a, b) where $a \in S^1$ and $b \in S^3$.

10.9 Exponentiating 2 x 2 anti-Hermitian matrices

Recall that a matrix Ω is Hermitian if $\Omega^\dagger = \Omega$, and if that is the case, then $(i\Omega)^\dagger = -i\Omega$. Matrices of the form $i\Omega$, where Ω is Hermitian, are named *anti-Hermitian*. Recall also that for any square matrix A the inverse of e^A is e^{-A}. The reason for this latter is that A and $-A$ commute with each other, and therefore the same reasoning that says that for complex z we have $e^z e^{-z} = 1$ now leads to $e^A e^{-A} = \mathbb{I}$. From these recollections, it is seen that if Ω is a Hermitian matrix, then $e^{i\Omega}$ is unitary: $(e^{i\Omega})^\dagger = e^{(i\Omega)^\dagger} = e^{-i\Omega}$.

Any 2×2 Hermitian matrix is a *unique, real* combination of the following four *complex* matrices:

$$\sigma_0 := \begin{bmatrix} 1 & 0 \\ 0 & 1 \end{bmatrix}, \tag{10.140}$$

$$\sigma_1 := \begin{bmatrix} 0 & 1 \\ 1 & 0 \end{bmatrix}, \quad \sigma_2 := \begin{bmatrix} 0 & -i \\ i & 0 \end{bmatrix}, \quad \sigma_3 := \begin{bmatrix} 1 & 0 \\ 0 & -1 \end{bmatrix}. \tag{10.141}$$

The proof is simply writing such a combination:

$$t\,\sigma_0 + x\,\sigma_1 + y\,\sigma_2 + z\,\sigma_3 = \begin{bmatrix} t+z & x-iy \\ x+iy & t-z \end{bmatrix}. \tag{10.142}$$

Since the set of equations $(t+z=a,\ t-z=b)$ for real (a,b) has always real solutions for $(t,\ z)$, it is obvious that the above matrix is the most general Hermitian 2×2 matrix.

We are now going to calculate the exponential of an anti-Hermitian matrix. This is done in two steps. We calculate the exponential first in the case $t = 0$ and then in the general case.

Exponentiating an anti-Hermitian traceless matrix. Define, as usual, a spherical polar coordinate system on the real space of (x, y, z) and name the coordinates $\beta = \sqrt{x^2+y^2+z^2}$, θ, and φ. The two variables θ and φ define a unit vector $\hat{\boldsymbol{n}}$. At this moment it might seem strange to name $\sqrt{x^2+y^2+z^2}$ not r but β. It soon will become clear why we have done so. As a shorthand, we define a vector with matrix elements, $\boldsymbol{\sigma} = (\sigma_1,\ \sigma_2,\ \sigma_3)$, and the real vector $\boldsymbol{\beta} = \beta\,\hat{\boldsymbol{n}}$, where $\hat{\boldsymbol{n}} = (n_1,\ n_2,\ n_3)$ is a real vector of unit length. With these notations we have

$$\boldsymbol{\beta}\cdot\boldsymbol{\sigma} = \beta\,\hat{\boldsymbol{n}}\cdot\boldsymbol{\sigma}, \qquad \hat{\boldsymbol{n}}\cdot\boldsymbol{\sigma} = \begin{bmatrix} n_3 & n_1 - in_2 \\ n_1 + in_2 & -n_3 \end{bmatrix}. \tag{10.143}$$

A simple multiplication reveals that $(\hat{\boldsymbol{n}}\cdot\boldsymbol{\sigma})^2 = \mathbb{I}$:

$$(\hat{\boldsymbol{n}}\cdot\boldsymbol{\sigma})^2 = \begin{bmatrix} n_3 & n_1 - in_2 \\ n_1 + in_2 & -n_3 \end{bmatrix}\begin{bmatrix} n_3 & n_1 - in_2 \\ n_1 + in_2 & -n_3 \end{bmatrix} \tag{10.144}$$

$$= \begin{bmatrix} n_3^2 + n_1^1 + n_2^2 & 0 \\ 0 & n_1^2 + n_2^2 + n_3^2 \end{bmatrix} = \begin{bmatrix} 1 & 0 \\ 0 & 1 \end{bmatrix}. \tag{10.145}$$

From this it follows that

$$(\hat{\boldsymbol{n}} \cdot \boldsymbol{\sigma})^n = \begin{cases} \mathbb{I} & n \text{ even}, \\ \hat{\boldsymbol{n}} \cdot \boldsymbol{\sigma} & n \text{ odd}. \end{cases} \tag{10.146}$$

It is now easy to calculate $\exp(-i\boldsymbol{\beta} \cdot \boldsymbol{\sigma})$. For simplicity of notation, let us temporarily define $X = \hat{\boldsymbol{n}} \cdot \boldsymbol{\sigma}$. We have

$$\exp(-i\beta X) = \sum_{n=0}^{\infty} \frac{(-i\beta)^n}{n!} X^n \tag{10.147}$$

$$= \mathbb{I} - i\beta X - \frac{\beta^2}{2!}\mathbb{I} + i\frac{\beta^3}{3!}X + \frac{\beta^4}{4!}\mathbb{I} - i\frac{\beta^4}{5!}X + \ldots \tag{10.148}$$

$$= \cos\beta\, \mathbb{I} - i \sin\beta\, \hat{\boldsymbol{n}} \cdot \boldsymbol{\sigma}, \tag{10.149}$$

$$e^{-i\beta\, \hat{\boldsymbol{n}} \cdot \boldsymbol{\sigma}} = \begin{bmatrix} \cos\beta - i n_3 \sin\beta & -\sin\beta\, n_2 - i \sin\beta\, n_1 \\ -i \sin\beta\, n_1 + \sin\beta\, n_2 & \cos\beta + i n_3 \sin\beta \end{bmatrix}. \tag{10.150}$$

We know that this matrix is unitary. It is easy to check that the determinant of this matrix is 1:

$$\det e^{-i\beta\, \hat{\boldsymbol{n}} \cdot \boldsymbol{\sigma}} = \cos^2\beta + n_3^2 \sin^2\beta + \sin^2\beta (n_1^2 + n_2^2) = 1. \tag{10.151}$$

Therefore this matrix is in SU(2). The question now is this: Can any matrix in SU(2) be written as $\exp(-i\beta\, \hat{\boldsymbol{n}} \cdot \boldsymbol{\sigma})$? To answer this question, we remember that any matrix in SU(2) is specified by a point on the 3D sphere of unit radius S_1^3, that is, a set of four real numbers (x_1, x_2, x_3, x_4) such that $x_1^2 + x_2^2 + x_3^2 + x_4^2 = 1$. We also recall that the unitary matrix S is of the following form:

$$S = \begin{bmatrix} x_4 - i x_3 & -x_2 - i x_1 \\ x_2 - i x_1 & x_4 + i x_3 \end{bmatrix}. \tag{10.152}$$

Comparing this with the entries of $e^{i\beta\, \hat{\boldsymbol{n}} \cdot \boldsymbol{\sigma}}$ we see that we must have

$$x_4 = \cos\beta, \tag{10.153}$$

$$x_3 = \sin\beta\, n_3 = \sin\beta\, \cos\theta, \tag{10.154}$$

$$x_1 = \sin\beta\, n_1 = \sin\beta\, \sin\theta\, \cos\varphi, \tag{10.155}$$

$$x_2 = \sin\beta\, n_2 = \sin\beta\, \sin\theta\, \sin\varphi. \tag{10.156}$$

We now note that for (β, θ, φ) in the following parameter set, the whole S_1^3 is covered:

$$0 \leq \beta \leq \pi, \qquad 0 \leq \theta \leq \pi, \qquad 0 \leq \varphi \leq 2\pi. \tag{10.157}$$

To see this, note that in this range of parameters, x_1 goes from -1 to $+1$, and so do x_2, x_3, and x_4, with all possible sign combinations covered. For example, to have x_1, x_2, and x_3 negative and x_4 positive, we must have $0 < \beta < \pi/2$, $\pi/2 < \theta < \pi$, $\pi < \varphi < 3\pi/2$.

Exponentiating the general anti-Hermitian matrix. The matrix $\sigma_0 = \mathbb{I}$ commutes with any 2×2 matrix, and we have $\exp(-i\beta_0 \mathbb{I}) = e^{-i\beta_0}\mathbb{I}$. Therefore

$$\exp(-i\beta_0 \mathbb{I} - i\boldsymbol{\beta} \cdot \boldsymbol{\sigma}) = \exp(-i\beta_0 \mathbb{I}) \exp(-i\boldsymbol{\beta} \cdot \boldsymbol{\sigma}) = e^{-i\beta_0} S(\boldsymbol{\beta}). \tag{10.158}$$

Obviously

$$\left(e^{-i\beta_0} S(\boldsymbol{\beta})\right)^\dagger \left(e^{-i\beta_0} S(\boldsymbol{\beta})\right) = S^\dagger S = \mathbb{I}, \tag{10.159}$$

showing that $e^{-i\beta_0} S(\boldsymbol{\beta})$ is in U(2). Conversely, if T is in U(2), then $\det T$ is a phase (a number of unit modulus), which can be written as $e^{-2i\beta_0}$. The matrix $e^{i\beta_0} T$ is then in SU(2) because

$$\left(e^{-i\beta_0} T^\dagger\right)\left(e^{i\beta_0} T\right) = T^\dagger T = \mathbb{I}, \tag{10.160}$$

$$\det\left(e^{i\beta_0} T\right) = e^{2i\beta_0} \det T = e^{2i\beta_0} e^{-2i\beta_0} = 1, \tag{10.161}$$

so that $e^{i\beta_0} T$ could be written as $\exp(-i\boldsymbol{\beta} \cdot \boldsymbol{\sigma})$. Therefore any T in U(2) is of the form $\exp(-i\beta_0 \sigma_0 - i\boldsymbol{\beta} \cdot \boldsymbol{\sigma})$.

10.9.1 The algebra of Pauli matrices

The three matrices

$$\sigma_1 = \begin{bmatrix} 0 & 1 \\ 1 & 0 \end{bmatrix}, \quad \sigma_2 = \begin{bmatrix} 0 & -i \\ i & 0 \end{bmatrix}, \quad \sigma_3 = \begin{bmatrix} 1 & 0 \\ 0 & -1 \end{bmatrix} \tag{10.162}$$

are called the Pauli matrices.[3] It is straightforward to check that the following relations hold:

$$\sigma_1^2 = \sigma_2^2 = \sigma_3^2 = \mathbb{I}, \tag{10.163}$$

$$\sigma_1 \sigma_2 = -\sigma_2 \sigma_1 = i\,\sigma_3, \tag{10.164}$$

$$\sigma_2 \sigma_3 = -\sigma_3 \sigma_2 = i\,\sigma_1, \tag{10.165}$$

$$\sigma_3 \sigma_1 = -\sigma_1 \sigma_3 = i\,\sigma_2. \tag{10.166}$$

All these relations can be written compactly as

$$\sigma_j \sigma_k = \delta_{jk} \mathbb{I} + i \sum_m \epsilon_{jkm} \sigma_m. \tag{10.167}$$

If $[A, B]$ stands for the commutator $AB - BA$ and $\{A, B\}$ stands for the anticommutator $AB + BA$, then the above relations can also be written as

$$\left[\sigma_j, \sigma_k\right] = 2i \sum_m \epsilon_{jkm} \sigma_m, \tag{10.168}$$

[3] After Wolfgang Pauli (1900–1958).

$$\{\sigma_j, \sigma_k\} = 2\delta_{jk}\,\mathbb{I}. \tag{10.169}$$

If $\boldsymbol{a} = (a_1, a_2, a_3)$ is a real vector, then $\boldsymbol{a}\cdot\boldsymbol{\sigma}$ is a 2×2 Hermitian matrix:

$$\boldsymbol{a}\cdot\boldsymbol{\sigma} = \begin{bmatrix} a_3 & a_1 - i\,a_2 \\ a_1 + i\,a_2 & -a_3 \end{bmatrix}. \tag{10.170}$$

We now give and prove some useful identities:

$$(\boldsymbol{a}\cdot\boldsymbol{\sigma})(\boldsymbol{b}\cdot\boldsymbol{\sigma}) = \left(\sum_j a_j\,\sigma_j\right)\left(\sum_k b_k\,\sigma_k\right) \tag{10.171}$$

$$= \sum_{j,k} a_j\,b_k\,\sigma_j\,\sigma_k \tag{10.172}$$

$$= \sum_{j,k} a_j\,b_k\left(\delta_{jk}\,\mathbb{I} + i\sum_m \epsilon_{jkm}\,\sigma_m\right) \tag{10.173}$$

$$= (\boldsymbol{a}\cdot\boldsymbol{b})\,\mathbb{I} + i\,(\boldsymbol{a}\times\boldsymbol{b})\cdot\boldsymbol{\sigma}. \tag{10.174}$$

Taking the trace of both sides we get

$$\mathrm{tr}(\boldsymbol{a}\cdot\boldsymbol{\sigma}\,\boldsymbol{b}\cdot\boldsymbol{\sigma}) = 2\,\boldsymbol{a}\cdot\boldsymbol{b}. \tag{10.175}$$

We leave it to the reader to prove that the following relation also holds:

$$\mathrm{tr}(\boldsymbol{a}\cdot\boldsymbol{\sigma}\,\boldsymbol{b}\cdot\boldsymbol{\sigma}\,\boldsymbol{c}\cdot\boldsymbol{\sigma}) = 2\,i\,\boldsymbol{a}\times\boldsymbol{b}\cdot\boldsymbol{c}. \tag{10.176}$$

10.9.2 Lie algebras $\mathfrak{u}(2)$ and $\mathfrak{su}(2)$

We have seen that any matrix in SU(2) can be written, in a unique way, as $\exp(-i\boldsymbol{\beta}\cdot\boldsymbol{\sigma})$. The matrices $\boldsymbol{\sigma} = (-i\sigma_1, -i\sigma_2, -i\sigma_3)$ are called the generators of the Lie group SU(2). But let us do a change of parameter. Instead of β, let us write everything in terms of $\alpha = 2\,\beta$. The range of α is $[0, 2\pi]$. We therefore have

$$S(\alpha, \hat{\boldsymbol{n}}) = \exp\left(-i\,\boldsymbol{\alpha}\cdot\frac{\boldsymbol{\sigma}}{2}\right). \tag{10.177}$$

Remark. In physics the Hermitian matrices

$$\boldsymbol{S} = \frac{1}{2}\boldsymbol{\sigma} \tag{10.178}$$

without the $-i$ factor are called the generators of the Lie group SU(2).

Adding the matrix $\frac{1}{2}\,\sigma_0 = \frac{1}{2}\,\mathbb{I}$, we get $\mathfrak{u}(2)$, which is the Lie algebra of the group U(2). The basic commutation relations are thus exactly as those of the $\mathfrak{so}(3)$ given in (10.103). We have

$$\left[S_j, S_k\right] = i\,\epsilon_{jkn}\,S_n. \tag{10.179}$$

Note that in $\mathfrak{u}(2)$ there is an element ($\frac{i}{2}\,\mathbb{I}$) which commutes with every element of the algebra (any 2×2 anti-Hermitian matrix). In a Lie algebra an element is said to be in the center (or is a central element) if it commutes with any element of the Lie algebra. The zero of any Lie algebra has this property—it commutes with everything! Now, for $\mathfrak{su}(2)$, the center consists of the single element 0, but in $\mathfrak{u}(2)$ the center is a 1D subspace—the elements of the form $\beta_0\,\mathbb{I}$, which form a line in $\mathfrak{u}(2)$.

Remark. The Lie group SU(2), although defined as a subset of complex matrices, is a real group. Note that it is the real sphere S^3 as a subset of $\mathbb{R}^4$, and all the parameters (β, θ, φ) are real numbers.

10.10 Action of U(2) and SU(2) on $\mathbb{R}^3$

Let us recollect some simple facts. First, if X is a Hermitian matrix and T is a unitary matrix, then $TXT^\dagger$ is also Hermitian, because

$$(TXT^\dagger)^\dagger = (T^\dagger)^\dagger X^\dagger T^\dagger = TXT^\dagger. \tag{10.180}$$

Besides, if X is traceless, then $TXT^\dagger$ is also traceless. This follows from the cyclic property of trace, namely, $\mathrm{tr}(ABC) = \mathrm{tr}(CAB)$, from which it follows that $\mathrm{tr}(TXT^\dagger) = \mathrm{tr}(T^\dagger TX) = \mathrm{tr}\,X$. We also know that to any real vector $\boldsymbol{a}$ we can associate a 2×2 traceless Hermitian matrix $\boldsymbol{a} \cdot \boldsymbol{\sigma}$. Therefore given T in U(2) the transformation

$$\boldsymbol{a} \cdot \boldsymbol{\sigma} \mapsto T\,\boldsymbol{a} \cdot \boldsymbol{\sigma}\,T^\dagger \tag{10.181}$$

is in fact a mapping from $\mathbb{R}^3$ to $\mathbb{R}^3$. We also know that any T in U(2) is of the form $T = e^{-i\beta_0} S$, where β_0 is real and S is in SU(2). Now observe that

$$TXT^\dagger = (e^{-i\beta_0}\,S)X(e^{-i\beta_0}\,S)^\dagger = e^{-i\beta_0}\,e^{i\beta_0}\,SXS^\dagger = SXS^\dagger. \tag{10.182}$$

That is, the action of T is the same as the action of S. In other words, the action of S is the same as the action of $e^{-i\beta_0} S$.

What we are saying is very simple. To see that, let us consider a simple example. Consider two real numbers β_0 and β. The following matrix is in U(2):

$$S = \begin{bmatrix} e^{-i(\beta_0+\beta)} & 0 \\ 0 & e^{-i(\beta_0-\beta)} \end{bmatrix} = e^{-i\beta_0} \begin{bmatrix} e^{-i\beta} & 0 \\ 0 & e^{i\beta} \end{bmatrix}. \tag{10.183}$$

Let $\boldsymbol{r} = (x, y, z)$. We then have

$$X := \boldsymbol{r} \cdot \boldsymbol{\sigma} = \begin{bmatrix} z & x - iy \\ x + iy & -z \end{bmatrix}, \tag{10.184}$$

which is Hermitian, as expected. Now let us calculate

$$SXS^\dagger = e^{-i\beta_0}\begin{bmatrix} e^{-i\beta} & 0 \\ 0 & e^{i\beta} \end{bmatrix}\begin{bmatrix} z & x-iy \\ x+iy & -z \end{bmatrix}\begin{bmatrix} e^{i\beta} & 0 \\ 0 & e^{-i\beta} \end{bmatrix} e^{i\beta_0} \quad (10.185)$$

$$= \begin{bmatrix} z & e^{-2i\beta}(x-iy) \\ e^{2i\beta}(x+iy) & -z \end{bmatrix} =: X'. \quad (10.186)$$

This matrix is Hermitian and traceless, as expected. We can write this matrix X' in the form $X' = \boldsymbol{r}' \cdot \boldsymbol{\sigma}$ for the real vector $\boldsymbol{r}'$ with components (x', y', z'):

$$X' = \boldsymbol{r}' \cdot \boldsymbol{\sigma} = \begin{bmatrix} z' & x'-iy' \\ x'+iy' & -z' \end{bmatrix}. \quad (10.187)$$

Comparing these two expressions of X' we see that x', y', and z' are solutions to the following linear system:

$$z' = z, \quad (10.188)$$

$$x' + iy' = e^{2i\beta}(x+iy), \quad (10.189)$$

$$x' - iy' = e^{-2i\beta}(x-iy). \quad (10.190)$$

The solutions is very simple:

$$\begin{cases} x' = x\cos(2\beta) - y\sin(2\beta), \\ y' = x\sin(2\beta) + y\cos(2\beta), \\ z' = z. \end{cases} \quad (10.191)$$

This is simply rotation by angle 2β around the z-axis. By defining $\alpha = 2\beta$, $\hat{\boldsymbol{n}} = (0, 0, 1)$, and $\boldsymbol{\alpha} = \alpha\,\hat{\boldsymbol{n}}$, the above transformation is nothing but $R(\boldsymbol{\alpha})$. The generalization is this.

Theorem. *For* $S(\boldsymbol{\alpha}) = \exp\left(-\frac{i}{2}\boldsymbol{\alpha}\cdot\boldsymbol{\sigma}\right)$ *in* SU(2), *the transformation*

$$\boldsymbol{r}\cdot\boldsymbol{\sigma} \mapsto \boldsymbol{r}'\cdot\boldsymbol{\sigma} = S\,(\boldsymbol{r}\cdot\boldsymbol{\sigma})S^\dagger \quad (10.192)$$

is the rotation $R(\boldsymbol{\alpha}) \in$ SO(3).

Proof. We first write a useful formula to take x_j out of $\boldsymbol{r}\cdot\boldsymbol{\sigma}$:

$$x_j = \frac{1}{2}\operatorname{tr}(\boldsymbol{r}\cdot\boldsymbol{\sigma}\,\sigma_j). \quad (10.193)$$

Remembering $\operatorname{tr}(\sigma_j\sigma_k) = 2\delta_{jk}$, the proof of this identity is simple:

$$\frac{1}{2}\operatorname{tr}(\boldsymbol{r}\cdot\boldsymbol{\sigma}\,\sigma_j) = \frac{1}{2}\operatorname{tr}\left(\sum_k x_k\,\sigma_k\,\sigma_j\right) = \frac{1}{2}\sum_k x_k \operatorname{tr}(\sigma_k\,\sigma_j) \quad (10.194)$$

$$= \frac{1}{2} \sum_k x_k (2\,\delta_{jk}) = x_j. \tag{10.195}$$

Now we calculate. The transformation is $\boldsymbol{r} \cdot \boldsymbol{\sigma} \mapsto \boldsymbol{r}' \cdot \boldsymbol{\sigma} = S\,\boldsymbol{r} \cdot \boldsymbol{\sigma}\, S^\dagger$, and we want to calculate x'_j:

$$x'_j = \frac{1}{2} \operatorname{tr}\left(S\,\boldsymbol{r} \cdot \boldsymbol{\sigma}\, S^\dagger\, \sigma_j \right). \tag{10.196}$$

Using the cyclic property of the trace, $\operatorname{tr}(ABCD) = \operatorname{tr}(BCDA)$,

$$x'_j = \frac{1}{2} \operatorname{tr}\left(\boldsymbol{r} \cdot \boldsymbol{\sigma}\, S^\dagger\, \sigma_j\, S \right). \tag{10.197}$$

Now we use Eq. (10.150) on page 223 and compute $S^\dagger\, \sigma_j\, S$:

$$\begin{aligned} \sigma'_j &= S^\dagger\, \sigma_j\, S \qquad (10.198) \\ &= \sum_{k,m} \left(\cos\frac{\alpha}{2} + i\,\sin\frac{\alpha}{2}\, n_k\, \sigma_k \right) \sigma_j \left(\cos\frac{\alpha}{2} - i\,\sin\frac{\alpha}{2}\, n_m\, \sigma_m \right) \\ &= \cos^2\frac{\alpha}{2}\, \sigma_j + i \sum_k n_k\, \sin\frac{\alpha}{2}\, \cos\frac{\alpha}{2}\, [\sigma_k, \sigma_j] + \sin^2\frac{\alpha}{2} \sum_{k,m} n_k\, n_m\, \sigma_k\, \sigma_j\, \sigma_m. \end{aligned}$$

We now use

$$\sigma_k\, \sigma_j\, \sigma_m = \delta_{jm}\, \sigma_k - \delta_{km}\, \sigma_j + \delta_{jk}\, \sigma_m + i\, \epsilon_{kjm} \tag{10.199}$$

to get

$$\begin{aligned} \sigma'_j &= \sum_k \left[\cos\alpha\, \delta_{jk} - \sin\alpha \sum_m \epsilon_{jmk}\, n_m + (1 - \cos\alpha)\, n_j\, n_k \right] \sigma_k \qquad (10.200) \\ &= \sum_k \left[\delta_{jk} + (1 - \cos\alpha)(n_j\, n_k - \delta_{jk}) - \sin\alpha \sum_m \epsilon_{jmk}\, n_m \right] \sigma_k. \qquad (10.201) \end{aligned}$$

If we look at the form of the *rotation* $R(\boldsymbol{\alpha})$ given in Eq. (10.97) on page 214, we see that

$$\sigma'_j := S^\dagger\, \sigma_j\, S = \sum_k R_{jk}\, \sigma_k. \tag{10.202}$$

We already know that the determinant of R is $+1$ (since it is a rotation, not an inversion).

Remark. This identity is remembered by saying that *the Pauli matrices transform (or rotate) as the components of a vector.*

From this transformation of the Pauli matrices we conclude that

$$\begin{aligned} x'_j &= \frac{1}{2}\operatorname{tr}\left(\boldsymbol{r}\cdot\boldsymbol{\sigma}\,\sigma'_j\right) \\ &= \frac{1}{2}\sum_k \operatorname{tr}\left(\boldsymbol{r}\cdot\boldsymbol{\sigma}\,R_{jk}\,\sigma_k\right) \\ &= \sum_k R_{jk}\frac{1}{2}\operatorname{tr}\left(\boldsymbol{r}\cdot\boldsymbol{\sigma}\,\sigma_k\right) \\ &= \sum_k R_{jk}\,x_k. \end{aligned} \tag{10.203}$$

This completes the proof. □

The function $f : \mathrm{SU}(2) \to \mathrm{SO}(3)$ defined in the previous theorem (Eq. (10.192)) has the following properties:

1. $f(\mathbb{I}) = \mathbb{I}$, that is, the rotation corresponding to the identity $\mathbb{I}$ in SU(2) is the identity of SO(3).
2. $f(S_1\,S_2) = f(S_1)\,f(S_2)$. The proof is simple: $f(S_1\,S_2)$ is the rotation corresponding to the transformation $(S_1\,S_2)X(S_1\,S_2)^\dagger$, where $X = \boldsymbol{r}\cdot\boldsymbol{\sigma}$. Now we have

$$(S_1\,S_2)\,X\,(S_1\,S_2)^\dagger = S_1\,S_2\,X\,S_2^\dagger\,S_1^\dagger. \tag{10.204}$$

   Given the matrix X, first $S_2\,X\,S_2^\dagger$ acts, which corresponds to $f(S_2)$. After that $S_1\,X'\,S_1^\dagger$ acts, which corresponds to $f(S_1)$. First $f(S_2)$, then $f(S_1)$, which means $f(S_1)\,f(S_2)$.
3. $f(S) = f(-S)$, that is, both S and $-S$ are being mapped to the same rotation. This is simply because $(-1)^2 = 1$. Also note that since S is a 2×2 matrix, $\det(-S) = \det S = 1$.

This means that SO(3) is homomorphic to SU(2). However, the homomorphism $f :$ $\mathrm{SU}(2) \to \mathrm{SO}(3)$ is not one to one, it is two to one! Because $f(S) = f(-S)$,

$$f : \mathrm{SU}(2) \to \mathrm{SO}(3) \qquad f : \begin{matrix} S \\ -S \end{matrix} \rightrightarrows R.$$

Therefore *the groups* SU(2) *and* SO(3) *are not isomorphic*. In other words, there is no homomorphism from SO(3) to SU(2)—one cannot unambiguously send a rotation into a special unitary matrix.

Covering group. The group SU(2) is said to be the covering group of SO(3). Since the homomorphism f is two to one, it is said to be the *double cover* of SO(3).

10.10.1 Relations of the topologies of SU(2) and SO(3)

To specify the unit vector $\hat{\boldsymbol{n}} = (n_1, n_2, n_3)$ on the unit sphere S_1^2, we use polar coordinates (θ, φ). We have

$$n_3 = \cos\theta, \quad n_1 = \sin\theta\, \cos\varphi, \quad n_2 = \sin\theta\, \sin\varphi. \tag{10.205}$$

From these formulas it is seen that sending $\hat{\boldsymbol{n}} \mapsto -\hat{\boldsymbol{n}}$, the antipodal map, is the following:

$$\text{antipodal map of } S_1^2: \quad (\theta, \varphi) \mapsto (\pi - \theta, \varphi + \pi), \tag{10.206}$$

$$\begin{bmatrix} \sin\theta\, \cos\varphi \\ \sin\theta\, \sin\varphi \\ \cos\theta \end{bmatrix} \longmapsto \begin{bmatrix} \sin(\pi - \theta)\, \cos(\varphi + \pi) \\ \sin(\pi - \theta)\, \sin(\varphi + \pi) \\ \cos(\pi - \theta) \end{bmatrix} = \begin{bmatrix} -\sin\theta\, \cos\varphi \\ -\sin\theta\, \sin\varphi \\ -\cos\theta \end{bmatrix}. \tag{10.207}$$

Looking at the polar angle coordinates (β, θ, φ) of $\mathbb{R}^4$ given in (10.153)–(10.156) on page 223, we see that here also the antipodal map is given by

$$\text{antipodal map on } S^3: \quad (\beta, \theta, \varphi) \mapsto (\pi - \beta, \pi - \theta, \varphi + \pi). \tag{10.208}$$

For the group SU(2), we used the parameter $\alpha = 2\beta$. Since β is in the interval $[0, \pi]$, the angle α is in the interval $[0, 2\pi]$, and the antipodal map reads $(\alpha, \theta, \varphi) \mapsto (2\pi - \alpha, \pi - \theta, \varphi + \pi)$. Now (rewriting Eq. (10.150) of page 223) we look at the matrix $S(\alpha, \theta, \varphi) = \exp(-\frac{i}{2}\alpha\, \hat{\boldsymbol{n}} \cdot \boldsymbol{\sigma})$:

$$S(\alpha, \theta, \varphi) = \begin{bmatrix} \cos\frac{\alpha}{2} - i\, n_3 \sin\frac{\alpha}{2} & -\sin\frac{\alpha}{2}\, n_2 - i\, \sin\frac{\alpha}{2}\, n_1 \\ -i\, \sin\frac{\alpha}{2}\, n_1 + \sin\frac{\alpha}{2}\, n_2 & \cos\frac{\alpha}{2} + i\, n_3 \sin\frac{\alpha}{2} \end{bmatrix}. \tag{10.209}$$

We see that

$$S(2\pi - \alpha, \pi - \theta, \varphi + \pi) = -S(\alpha, \theta, \varphi). \tag{10.210}$$

Changing $\alpha \mapsto \alpha + \pi$, this can be written in the following equivalent form:

$$S(\pi - \alpha, \pi - \theta, \varphi + \pi) = -S(\pi + \alpha, \theta, \varphi). \tag{10.211}$$

For rotations, however, we have

$$R(2\pi - \alpha, \pi - \theta, \varphi + \pi) = +R(\alpha, \theta, \varphi), \tag{10.212}$$

or equivalently (by changing $\alpha \mapsto \alpha + \pi$),

$$R(\pi - \alpha, \pi - \theta, \varphi + \pi) = +R(\alpha + \pi, \theta, \varphi). \tag{10.213}$$

This can be seen from Eq. (10.97) on page 214. The geometric interpretation of identity (10.213) for rotations is obvious: A rotation by $\pi + \alpha$ around any vector $\hat{\boldsymbol{n}}$ is equivalent to a rotation by $\pi - \alpha$ around the antipodal direction $-\hat{\boldsymbol{n}}$. For example,

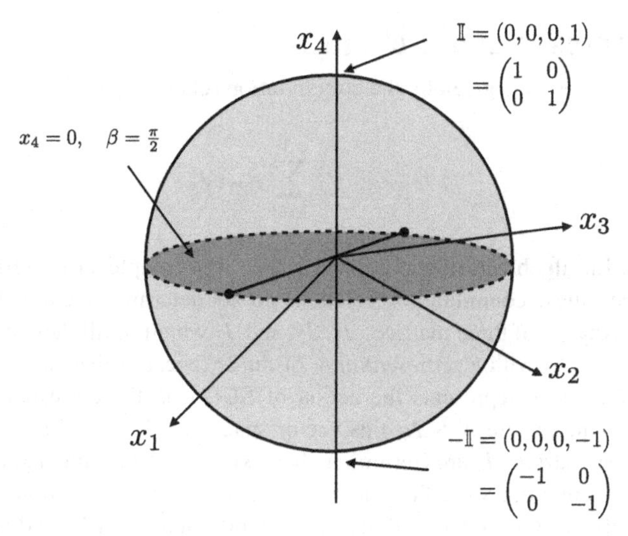

FIGURE 10.5 Topology of SU(2) and SO(3).

SU(2) is the 3D sphere $x_1^2 + x_2^2 + x_3^2 + x_4^2 = 1$. The hyperplane $x_4 = 0$ cuts the 3D sphere in two parts. SO(3) is either the "northern" hemisphere $x_4 \geq 0$ or the "southern" hemisphere $x_4 \leq 0$. To get SO(3), one should identify the antipodal points. Identifying the antipodal points on the "equator" $x_4 = 0$ (two are shown as black points) causes the hemispheres to become the real projective space $\mathbb{R}\mathrm{P}^3$.

a rotation by 190 degrees around $(0, 0, 1)$ is equivalent to a rotation by 10 degrees around $(0, 0, -1)$. As a consequence, the rotation by $\alpha = \pi$ around $\hat{\boldsymbol{n}}$ is equivalent to the rotation by the same angle $\alpha = \pi$ around $-\hat{\boldsymbol{n}}$. And, finally, a rotation by $\alpha = 2\pi$ around any vector $\hat{\boldsymbol{n}}$ is equal to the identity map $\mathbb{I}$. Therefore because of (10.213) to specify all rotations, each one only once, we must consider vectors $\alpha\,\hat{\boldsymbol{n}}$ in $\mathbb{R}^3$, with $\alpha \in [0, \pi]$, and we must identify the antipodal points on the surface, that is, $\pi\,\hat{\boldsymbol{n}}$ and $-\pi\,\hat{\boldsymbol{n}}$ are the same rotations. This is exactly $\mathbb{R}\mathrm{P}^3$ introduced on page 216.

However, because of (10.211) to specify all SU(2) matrices, we must consider all vectors $\alpha\,\hat{\boldsymbol{n}}$ with $\alpha \in [0, 2\pi]$ and with all the points of the surface, that is, $2\pi\,\hat{\boldsymbol{n}}$, identified—since they are all the same point $\left[\begin{smallmatrix} -1 & 0 \\ 0 & -1 \end{smallmatrix}\right] \in \mathrm{SU}(2)$. This is the 3D sphere.

The geometric picture, shown in Fig. 10.5, is this: SU(2) is the sphere $x_1^2 + x_2^2 + x_3^2 + x_4^2 = 1$ with polar coordinates given in (10.153)–(10.156). The plane $x_4 = 0$, given by $\beta = \frac{\pi}{2}$ or $\alpha = \pi$, intersects this 3D sphere in the "equator," which is a 2D sphere. SO(3) is the upper 3D "hemisphere" with the antipodal points on the "equator" identified. The equator represents rotations by π along various directions, and the identity of the antipodal points of the equator is the manifestation of (10.213). The "north" pole of the sphere, $(0, 0, 0, 1)$, is the identity matrix $\mathbb{I}$, and the "south" pole is the matrix $-\mathbb{I} \in \mathrm{SU}(2)$.

10.11 Representations of $\mathfrak{su}(2)$

The Lie algebra $\mathfrak{su}(2)$ is given by the commutation relations

$$[J_n, J_m] = i \sum_{k=1}^{3} \epsilon_{nmk} J_k. \tag{10.214}$$

Although the Lie algebra itself was defined as the set of complex, Hermitian, traceless 2×2 matrices, these commutations relations do not determine the dimension of the J_n matrices. Any set of three matrices J_1, J_2, and J_3 which fulfill these commutation relations provide us with a *representation* of $\mathfrak{su}(2)$, because then we can construct $\exp(-i\,\boldsymbol{\alpha} \cdot \boldsymbol{J})$, which represents the action of SU(2) on the corresponding vector space. If the matrices are $N \times N$, this vector space is $\mathbb{C}^N$, or $\mathbb{R}^N$ if $-i\,J_n$ are real matrices. If the matrices J_n are Hermitian, then $\exp(-i\,\boldsymbol{\alpha} \cdot \boldsymbol{J})$ is unitary, and we have a unitary representation. Since the generators specify the representation, we speak of representation $\{J_n\}$. Given two N-dimensional representations $\{J_n\}$ and $\{K_n\}$, a good question is: Are they different? To answer this question we must have a notion of equivalence.

Equivalent representations. We say two representations $\{J_n\}$ and $\{K_n\}$ are equivalent if there is a nonsingular matrix S such that $S^{-1} J_n S = K_n$ for $n = 1, 2, 3$. Let us explore some examples. Consider the following sets of 3×3 matrices:

$$A_1 = \begin{bmatrix} 0 & 0 & 0 \\ 0 & 0 & -i \\ 0 & i & 0 \end{bmatrix}, \quad A_2 = \begin{bmatrix} 0 & 0 & i \\ 0 & 0 & 0 \\ -i & 0 & 0 \end{bmatrix}, \quad A_3 = \begin{bmatrix} 0 & -i & 0 \\ i & 0 & 0 \\ 0 & 0 & 0 \end{bmatrix},$$

$$B_1 = \begin{bmatrix} 0 & 1 & 0 \\ 1 & 0 & 1 \\ 0 & 1 & 0 \end{bmatrix}, \quad B_2 = \frac{1}{\sqrt{2}} \begin{bmatrix} 0 & -i & 0 \\ i & 0 & -i \\ 0 & i & 0 \end{bmatrix}, \quad B_3 = \begin{bmatrix} 1 & 0 & 0 \\ 0 & 0 & 0 \\ 0 & 0 & -1 \end{bmatrix},$$

$$C_1 = \begin{bmatrix} 0 & 1 & 0 \\ 1 & 0 & 0 \\ 0 & 0 & 0 \end{bmatrix}, \quad C_2 = \frac{1}{2} \begin{bmatrix} 0 & -i & 0 \\ i & 0 & 0 \\ 0 & 0 & 0 \end{bmatrix}, \quad C_3 = \frac{1}{2} \begin{bmatrix} 1 & 0 & 0 \\ 0 & -1 & 0 \\ 0 & 0 & 0 \end{bmatrix},$$

$$D_1 = \frac{1}{2} \begin{bmatrix} 0 & c & -s \\ c & 0 & 0 \\ -s & 0 & 0 \end{bmatrix}, \quad D_2 = \frac{i}{2} \begin{bmatrix} 0 & c & -s \\ -c & 0 & 0 \\ s & 0 & 0 \end{bmatrix},$$

$$D_3 = \frac{1}{2} \begin{bmatrix} 1 & 0 & 0 \\ 0 & -c^2 & s\,c \\ 0 & s\,c & -s^2 \end{bmatrix},$$

where for brevity we have used the notation

$$c := \cos\theta, \qquad s := \sin\theta, \tag{10.215}$$

θ being a fixed real number. The matrices $\{A_n\}$ are the generators of SO(3) given on page 215, and we already know that they satisfy $\mathfrak{su}(2)$. For the other sets, by simple matrix multiplications, which we leave to the reader, it is seen that they also satisfy the Lie algebra $\mathfrak{su}(2)$. Now, a very natural question is this: Are these four representations all different? To investigate this question, let us do a little calculation and calculate $J^2 = \sum_n (J_n)^2$ and $\operatorname{tr} J^2$ for these four representations. We have

$$A^2 = (A_1)^2 + (A_2)^2 + (A_3)^2 = 2\begin{bmatrix} 1 & 0 & 0 \\ 0 & 1 & 0 \\ 0 & 0 & 1 \end{bmatrix}, \qquad \operatorname{tr} A^2 = 6, \tag{10.216}$$

$$B^2 = (B_1)^2 + (B_2)^2 + (B_3)^2 = 2\begin{bmatrix} 1 & 0 & 0 \\ 0 & 1 & 0 \\ 0 & 0 & 1 \end{bmatrix}, \qquad \operatorname{tr} B^2 = 6, \tag{10.217}$$

$$C^2 = (C_1)^2 + (C_2)^2 + (C_3)^2 = \frac{3}{4}\begin{bmatrix} 1 & 0 & 0 \\ 0 & 1 & 0 \\ 0 & 0 & 0 \end{bmatrix}, \qquad \operatorname{tr} C^2 = \frac{3}{2}, \tag{10.218}$$

$$D^2 = (D_1)^2 + (D_2)^2 + (D_3)^2 = \frac{3}{4}\begin{bmatrix} 1 & 0 & 0 \\ 0 & c^2 & s\,c \\ 0 & s\,c & s^2 \end{bmatrix}, \qquad \operatorname{tr} D^2 = \frac{3}{2}. \tag{10.219}$$

We are sure that representation $\{A_n\}$ is not equivalent to representation $\{C_n\}$ or representation $\{D_n\}$, because if two representations are equivalent, then $\operatorname{tr} J^2$ must be the same for the two. Here is the proof:

$$\operatorname{tr}\left[\left(S^{-1} J_n S\right)\left(S^{-1} J_n S\right)\right] = \operatorname{tr}\left(S^{-1} J_n J_n S\right) = \operatorname{tr}(J_n J_n). \tag{10.220}$$

Casimir of $\mathfrak{su}(2)$. The matrix $J^2 = \sum_n (J_n)^2$, which is not in $\mathfrak{su}(2)$, has a very important property: It commutes with every element of $\mathfrak{su}(2)$. The proof is based on the identity $[A\,B,\, C] = A\,[B,\, C] + [A,\, C]\,B$ and the fact that if S_{nk} is symmetric in nk, then $\sum_{n,k} \epsilon_{nmk}\, S_{nk} = 0$. We have

$$\left[J^2,\, J_m\right] = \left[\sum_n J_n\, J_n,\, J_m\right] = \sum_n [J_n\, J_n,\, J_m] \tag{10.221}$$

$$= \sum_n \{J_n\,[J_n,\, J_m] + [J_n,\, J_m]\,J_n\} \tag{10.222}$$

$$= \sum_{n,k} \{i\,\epsilon_{nmk}\, J_n\, J_k + i\,\epsilon_{nmk}\, J_k\, J_n\} \tag{10.223}$$

$$= \sum_{n,k} i\,\epsilon_{nmk} \underbrace{(J_n J_k + J_k J_n)}_{\text{symmetric}} = 0. \tag{10.224}$$

Casimir operators. Operators (matrices) that commute with all vectors of a Lie algebra are called Casimir[4] operators.

Now let us return to our four 3D representations of $\mathfrak{su}(2)$, which we call, for obvious reasons, representations A, B, C, and D. We see that for representations A and B the Casimir is a multiple of unity, but for representations C and D it is not. Representations C and D are not "genuine" 3D representations, by which we mean "they are decomposable." Let us see what this means.

Direct sum of vector spaces. Consider two vector spaces V and W. The Cartesian product $V \times W$ can be considered a vector space. Saying that $(\boldsymbol{a}, \boldsymbol{b})$ is in $V \times W$ means that $\boldsymbol{a}$ is in V and $\boldsymbol{b}$ is in W. Now, if $(\boldsymbol{a}, \boldsymbol{b})$ and $(\boldsymbol{a}', \boldsymbol{b}')$ are two elements of $V \times W$, we can add them naturally:

$$(\boldsymbol{a}, \boldsymbol{b}) + (\boldsymbol{a}', \boldsymbol{b}') = (\boldsymbol{a} + \boldsymbol{a}', \boldsymbol{b} + \boldsymbol{b}'). \tag{10.225}$$

Moreover, we can multiply them by numbers, thus

$$\lambda(\boldsymbol{a}, \boldsymbol{b}) = (\lambda\,\boldsymbol{a}, \lambda\,\boldsymbol{b}). \tag{10.226}$$

This vector space $V \times W$ is called the direct sum of V and W and is denoted by $V \oplus W$. The dimension of $V \oplus W$ is the sum of the dimensions of V and W:

$$\dim(V \oplus W) = \dim V + \dim W. \tag{10.227}$$

Now we return to our representations C and D. For representation C, we note that all basis matrices, C_1, C_2, and C_3, are "block diagonal":

$$C_1 = \frac{1}{2}\left[\begin{array}{cc|c} 0 & 1 & 0 \\ 1 & 0 & 0 \\ \hline 0 & 0 & 0 \end{array}\right], \quad C_2 = \frac{1}{2}\left[\begin{array}{cc|c} 0 & -i & 0 \\ i & 0 & 0 \\ \hline 0 & 0 & 0 \end{array}\right], \quad C_3 = \frac{1}{2}\left[\begin{array}{cc|c} 1 & 0 & 0 \\ 0 & -1 & 0 \\ \hline 0 & 0 & 0 \end{array}\right].$$

This means that the 3D vector space V_3 on which these matrices act is actually a direct sum of a 2D vector space V_2 and a 1D vector space V_1:

$$V_3 = V_2 \oplus V_1. \tag{10.228}$$

The action of C_n on sector V_2 is simply the action of the Pauli matrices, while the action of C_n on sector V_1 is simply "no action," or the "trivial action."

Decomposable representation. A D-dimensional representation $\{J_n\}$ is said to be decomposable if the D-dimensional vector space V_D on which the generators act is the direct sum of two vector spaces $V_N \oplus V_{D-N}$, such that the action of all J_ns on V leaves the sectors V_N and V_{D-N} invariant, that is, they do not mix these two sectors.

[4] After Hendrik Casimir (1909–2000).

Reducible representation. A D-dimensional representation is said to be reducible if it contains a nontrivial invariant subspace. By a nontrivial subspace, we mean some subspace other than V_D itself and the zero[5] subspace.

Note that decomposability is stronger than reducibility. If a representation is decomposable, it surely is reducible. The matrices of a decomposable representation are block diagonal, and the matrices of a reducible representation are block upper triangular, something like

$$\left[\begin{array}{cc|c} a & b & e \\ c & d & f \\ \hline 0 & 0 & h \end{array}\right]. \tag{10.229}$$

Under the action of this matrix, the subspace $(x, y, 0)$ is invariant, but the subspace $(0, 0, z)$ is not invariant. The action on $(x, y, 0)$ is determined by the submatrix $\left[\begin{smallmatrix} a & b \\ c & d \end{smallmatrix}\right]$.

Up to this point, with respect to our four representations A, B, C, and D, we know that representation C is decomposable. What about representation D? The answer is that it is equivalent to representation C. Actually, we have used the unitary matrix

$$S = \begin{bmatrix} 1 & 0 & 0 \\ 0 & c & s \\ 0 & -s & c \end{bmatrix}, \qquad c := \cos\theta, \quad s := \sin\theta, \tag{10.230}$$

to construct representation D from representation C, by the formula $D_n = S^{-1} C_n S$. Representations A and B are also equivalent (see Worked Example 44 on page 238). But how can one find out whether two given representations are equivalent or not? Here we make use of the representation theory of $\mathfrak{su}(2)$. We now state the basic assertions of this theory without proof. For the proof of these assertions we refer the reader to other texts.

The representation theory of $\mathfrak{su}(2)$. Up to unitary equivalence, for any integer $n \geq 1$, the Lie algebra $\mathfrak{su}(2)$ has one and only one n-dimensional irreducible representation. An n-dimensional representation is irreducible if and only if

$$J^2 = \frac{n^2 - 1}{4} \mathbb{I}, \tag{10.231}$$

or, defining $s = (n-1)/2$, if and only if $J^2 = s(s+1)\mathbb{I}$. Note that $s \geq 0$ is an integer if n is odd and a half-integer if n is even. The representation is called *the spin-s representation*.

Notation. The spin-s representation is denoted by $\mathbf{n}$, where $n = 2s + 1$ is the dimension of the representation. For example, the spin-$\frac{1}{2}$ representation is denoted by $\mathbf{2}$.

[5] The zero subspace is the subspace consisting only of the origin.

Spin-s representation of $\mathfrak{su}(2)$. It is customary to choose the basis of the vector space V_n such that the matrix S_3 is diagonal, that is, the basis vectors are chosen to be eigenvectors of S_3. The eigenvalues of S_3 are denoted by m, and the eigenvector corresponding to m (in representation s) is denoted by $|s, m\rangle$. In the usual quantum mechanical notation this is written as

$$S_3 \, |s, m\rangle = m \, |s, m\rangle . \tag{10.232}$$

The eigenvalues of S_3 are the numbers

$$m \in \{-s, -s+1, \ldots, s-1, s\} . \tag{10.233}$$

To specify the representation fully, we must specify S_1 and S_2, or, equivalently, we can specify the following two non-Hermitian matrices:

$$S_+ = S_1 + i\, S_2, \tag{10.234}$$

$$S_- = S_1 - i\, S_2. \tag{10.235}$$

These two matrices are given once their action on the basis vectors $\{|s, m\rangle\}$ is given; the actions are as follows:

$$S_+ \, |s, m\rangle = \sqrt{(s-m)(s+m+1)} \, |s, m+1\rangle , \tag{10.236}$$

$$S_- \, |s, m\rangle = \sqrt{(s+m)(s-m+1)} \, |s, m-1\rangle . \tag{10.237}$$

Note that for $m = s$ we have $S_+ \, |s, +s\rangle = 0$, and for $m = -s$ we have $S_- \, |s, -s\rangle = 0$. The above relations can be inverted to get S_1 and S_2 in terms of $S_\pm$ as follows:

$$S_1 = \frac{1}{2} (S_- + S_+) , \tag{10.238}$$

$$S_2 = \frac{i}{2} (S_- - S_+) . \tag{10.239}$$

Let us explore three examples.

Example 1. Representation **1**. The first example is $s = 0$, which means that $n = 2s + 1 = 1$. This is a 1D representation. Here we have only one basis vector, $|0, 0\rangle$, which we can denote by a 1×1 column matrix [1]. The matrix S_z is the 1×1 matrix [0], sending $|0, 0\rangle$ to zero. The matrices $S_\pm$ also are the same [0]. So $\exp(i\, \boldsymbol{\alpha} \cdot \mathbf{S}) = 1$. This is called the trivial representation.

Example 2. Representation **2**. Our next example is $s = \frac{1}{2}$, which means that $n = 2 \cdot \frac{1}{2} + 1 = 2$. This is a 2D representation. Now the basis vectors are

$$\left|\tfrac{1}{2}, +\tfrac{1}{2}\right\rangle = \begin{bmatrix} 1 \\ 0 \end{bmatrix} =: |+\rangle , \tag{10.240}$$

$$\left|\tfrac{1}{2},-\tfrac{1}{2}\right\rangle=\begin{bmatrix}0\\1\end{bmatrix}=:|-\rangle. \tag{10.241}$$

The diagonal matrix S_3 is as follows:

$$S_z=\frac{1}{2}\begin{bmatrix}1&0\\0&-1\end{bmatrix}. \tag{10.242}$$

The matrices $S_\pm$ are given by the relations

$$S_+|+\rangle=0, \qquad S_+|-\rangle=|+\rangle, \tag{10.243}$$

$$S_-|+\rangle=|-\rangle, \qquad S_-|-\rangle=0. \tag{10.244}$$

Therefore in the basis $\{|+\rangle,|-\rangle\}$ we have

$$S_+=\begin{bmatrix}0&1\\0&0\end{bmatrix}, \qquad S_-=\begin{bmatrix}0&0\\1&0\end{bmatrix}, \tag{10.245}$$

$$S_1=\frac{1}{2}(S_-+S_+)=\frac{1}{2}\begin{bmatrix}0&1\\1&0\end{bmatrix}, \tag{10.246}$$

$$S_2=\frac{i}{2}(S_--S_+)=\frac{1}{2}\begin{bmatrix}0&-i\\i&0\end{bmatrix}. \tag{10.247}$$

We see that these S_ns are the same generators of SU(2) we are familiar with.

Example 3. Representation **3**. For $s=1$, we have $n=3$, and if we define

$$|1,+1\rangle=\begin{bmatrix}1\\0\\0\end{bmatrix}, \qquad |1,0\rangle=\begin{bmatrix}0\\1\\0\end{bmatrix}, \qquad |1,-1\rangle=\begin{bmatrix}0\\0\\1\end{bmatrix}, \tag{10.248}$$

we get

$$S_3=\begin{bmatrix}1&0&0\\0&0&0\\0&0&-1\end{bmatrix}, \tag{10.249}$$

$$S_+=\sqrt{2}\begin{bmatrix}0&1&0\\0&0&1\\0&0&0\end{bmatrix}, \qquad S_-=\sqrt{2}\begin{bmatrix}0&0&0\\1&0&0\\0&1&0\end{bmatrix}, \tag{10.250}$$

$$S_1=\frac{1}{\sqrt{2}}\begin{bmatrix}0&1&0\\1&0&1\\0&1&0\end{bmatrix}, \qquad S_2=\frac{1}{\sqrt{2}}\begin{bmatrix}0&-i&0\\i&0&-i\\0&i&0\end{bmatrix}. \tag{10.251}$$

This is representation A of page 232. From Worked Example 44 (page 238) we know that it is equivalent to the Lie algebra of SO(3).

Worked Example 44. Consider the following two 3D representations of $\mathfrak{su}(2)$:

$$A_1 = \begin{bmatrix} 0 & 0 & 0 \\ 0 & 0 & -i \\ 0 & i & 0 \end{bmatrix}, \; A_2 = \begin{bmatrix} 0 & 0 & i \\ 0 & 0 & 0 \\ -i & 0 & 0 \end{bmatrix}, \; A_3 = \begin{bmatrix} 0 & -i & 0 \\ i & 0 & 0 \\ 0 & 0 & 0 \end{bmatrix},$$

$$B_1 = \frac{1}{\sqrt{2}} \begin{bmatrix} 0 & 1 & 0 \\ 1 & 0 & 1 \\ 0 & 1 & 0 \end{bmatrix}, \; B_2 = \frac{1}{\sqrt{2}} \begin{bmatrix} 0 & -i & 0 \\ i & 0 & -i \\ 0 & i & 0 \end{bmatrix}, \; B_3 = \begin{bmatrix} 1 & 0 & 0 \\ 0 & 0 & 0 \\ 0 & 0 & -1 \end{bmatrix}.$$

We now prove that these two representations are unitarily equivalent. To do this, we find the matrix U such that $B_n = U^\dagger A_n U$. To find U, we first note that B_3 is diagonal. Multiply the equation $B_3 = U^\dagger A_3 U$ on the left by U to get $A_3 U = U B_3$, or explicitly,

$$\begin{bmatrix} 0 & -i & 0 \\ i & 0 & 0 \\ 0 & 0 & 0 \end{bmatrix} \begin{bmatrix} U_{11} & U_{12} & U_{13} \\ U_{21} & U_{22} & U_{23} \\ U_{31} & U_{32} & U_{33} \end{bmatrix} = \begin{bmatrix} U_{11} & U_{12} & U_{13} \\ U_{21} & U_{22} & U_{23} \\ U_{31} & U_{32} & U_{33} \end{bmatrix} \begin{bmatrix} 1 & 0 & 0 \\ 0 & 0 & 0 \\ 0 & 0 & -1 \end{bmatrix}$$
$$= \begin{bmatrix} U_{11} & 0 & -U_{13} \\ U_{21} & 0 & -U_{23} \\ U_{31} & 0 & -U_{33} \end{bmatrix}.$$

Therefore the columns of U must be eigenvectors of A_3, corresponding respectively to 1, 0, and -1. Let us check that the eigenvalues of A_3 are in fact 0 and ± 1. The eigenvalues are the roots of the characteristic equation

$$\det(A_3 - \lambda \mathbb{I}) = -\lambda \begin{vmatrix} -\lambda & i \\ -i & -\lambda \end{vmatrix} = \lambda\,(\lambda^2 - 1) = \lambda\,(\lambda - 1)\,(\lambda + 1). \qquad (10.252)$$

It is easy to check that the following column vectors are the most general "normalized" eigenvectors of A_3 (by normalized we mean that their norm is equal to 1):

$$\lambda = +1 \leftrightarrow \frac{1}{\sqrt{2}} \begin{bmatrix} e^{i\alpha} \\ i\,e^{i\alpha} \\ 0 \end{bmatrix}, \quad \lambda = 0 \leftrightarrow \begin{bmatrix} 0 \\ 0 \\ e^{i\gamma} \end{bmatrix}, \quad \lambda = -1 \leftrightarrow \frac{1}{\sqrt{2}} \begin{bmatrix} i\,e^{i\beta} \\ e^{i\beta} \\ 0 \end{bmatrix}.$$

So the matrix U we are searching for has the form

$$U = \begin{bmatrix} b\,e^{i\alpha} & 0 & i\,b\,e^{i\beta} \\ i\,b\,e^{i\alpha} & 0 & b\,e^{i\beta} \\ 0 & e^{i\gamma} & 0 \end{bmatrix}, \quad \text{where } b := \frac{1}{\sqrt{2}}. \qquad (10.253)$$

Now we must calculate $U^\dagger A_1 U$ and solve the equation $B_1 = U^\dagger A_1 U$. After a straightforward calculation we see that this condition is fulfilled if the following two conditions are satisfied:

$$\begin{aligned} \gamma - \alpha &= \pi, \\ \gamma - \beta &= \tfrac{\pi}{2}, \end{aligned} \qquad \text{which means} \qquad \begin{aligned} \beta &= \alpha + \tfrac{\pi}{2}, \\ \gamma &= \alpha + \pi. \end{aligned} \tag{10.254}$$

By straightforward calculation it is seen that by these same conditions on the phases α, β, and γ, the other equation $U^\dagger A_2 U = B_2$ is also satisfied. A suitable choice for these phases is $\alpha = -\pi/2$, $\beta = 0$, and $\gamma = \pi/2$, which leads to

$$U = \begin{bmatrix} -i\,b & 0 & i\,b \\ b & 0 & b \\ 0 & i & 0 \end{bmatrix}, \quad \text{where } b := \frac{1}{\sqrt{2}}. \tag{10.255}$$

The complex conjugate representation $\bar{2}$. The matrices $J_n = -\frac{i}{2}\sigma_n$ satisfy the following commutation relation:

$$[\,J_n,\, J_m\,] = \epsilon_{nmk}\, J_k. \tag{10.256}$$

Now let us consider the complex conjugate matrices $\overline{J}_n$. Since the Pauli matrices are Hermitian, we have $\bar{\sigma}_n = \sigma_n^T$, where T denotes transposition. It is easy to see that

$$\overline{-i\,\sigma_1} = i\,\sigma_1^T = i\,\sigma_1, \quad \overline{-i\,\sigma_2} = i\,\sigma_2^T = -i\,\sigma_2, \quad \overline{-i\,\sigma_3} = i\,\sigma_3^T = i\,\sigma_3. \tag{10.257}$$

From this it follows that the matrices $\overline{-i\,\sigma_n}$ satisfy the same algebra as $-i\,\sigma_n$:

$$\left[\overline{-i\,\sigma_1},\, \overline{-i\,\sigma_2}\right] = [\,i\,\sigma_1,\, -i\,\sigma_2\,] = 2i\,\sigma_3 = 2\left(\overline{-i\,\sigma_3}\right), \tag{10.258}$$

$$\left[\overline{-i\,\sigma_2},\, \overline{-i\,\sigma_3}\right] = [\,-i\,\sigma_2,\, i\,\sigma_3\,] = 2i\,\sigma_1 = 2\left(\overline{-i\,\sigma_1}\right), \tag{10.259}$$

$$\left[\overline{-i\,\sigma_3},\, \overline{-i\,\sigma_1}\right] = [\,i\,\sigma_3,\, i\,\sigma_1\,] = -2i\,\sigma_2 = 2\left(\overline{-i\,\sigma_2}\right). \tag{10.260}$$

Therefore

$$\left[\,\bar{J}_n,\, \bar{J}_m\,\right] = \epsilon_{nmk}\, \bar{J}_k. \tag{10.261}$$

Thus we have two 2D representations of $\mathfrak{su}(2)$, $\{-i\,\sigma_n/2\}$ and $\left\{\overline{-i\,\sigma_n}/2\right\}$. We know that these two representations are equivalent, because both are 2D and both are irreducible (since they satisfy $J^2 = \frac{3}{4}\,\mathbb{I}$). The matrix connecting these two representations is $S = -i\sigma_2 = \left[\begin{smallmatrix} 0 & -1 \\ 1 & 0 \end{smallmatrix}\right]$ and we have $S^{-1}\,(-i\,\sigma_n)\,S = \overline{-i\,\sigma_n}$. The proof consists of simple matrix multiplications, which are left to the reader.

10.12 Representations of $\mathfrak{so}(3)$

As a Lie algebra, $\mathfrak{so}(3)$ is the same as $\mathfrak{su}(2)$. The groups SO(3) and SU(2) are however not the same. As we have seen earlier, SU(2) is the double cover of SO(3).

Bosonic and fermionic representations. Irreducible representations of $\mathfrak{su}(2)$ are divided into two categories:

- Representations with integer spin s, which are odd-dimensional. These are called *bosonic* or *tensorial* representations.
- Representations with half-integer spin s, which are even-dimensional. These are called *fermionic* or *spinorial* representations.

Recall that for SU(2), in the expression $\exp(-i\alpha\hat{\boldsymbol{n}}\cdot\boldsymbol{S})$, the parameter α has the domain $[0, 2\pi]$, while for SO(3) it has the domain $[0, \pi]$ (see §10.9.2 on page 225). Consider the spin-$\frac{1}{2}$ representation, and let us look at $\exp(\mp i\pi S_3)$, where $S_3 = \frac{1}{2}\sigma_3$. We have $-i\, S_3 = \frac{1}{2}\left[\begin{smallmatrix} -i & 0 \\ 0 & +i \end{smallmatrix}\right]$ and $\exp(\pm\frac{1}{2}i\pi) = \pm i$; thus

$$\exp(-i\,\pi\, S_3) = \begin{bmatrix} -i & 0 \\ 0 & +i \end{bmatrix}, \qquad \exp(+i\,\pi\, S_3) = \begin{bmatrix} +i & 0 \\ 0 & -i \end{bmatrix}. \tag{10.262}$$

These two matrices are not the same, so this representation does not fulfill $R(\pi, \hat{z}) = R(\pi, -\hat{z})$ (see (10.213) on page 230), and therefore it is not a representation of SO(3). Since the eigenvalues of S_3 in half-integer-spin representations of $\mathfrak{su}(2)$ are all half-integers, the obvious generalization of the above statement is true for all half-integer-spin representations.

Note. By exponentiating a fermionic representation of $\mathfrak{su}(2)$ one gets an even-dimensional representation of SU(2), but this is not a true representation of SO(3), because such matrices do not fulfill $R(\pi, \hat{\boldsymbol{n}}) = R(\pi, -\hat{\boldsymbol{n}})$.

10.13 Tensors and index notation

Consider a vector space V of dimension N with basis $B_V = \{\boldsymbol{E}_i\}$. A vector $\boldsymbol{A}$ in V could be written as a linear combination of the basis vectors. We write it as

$$\boldsymbol{A} = \sum_{j=1}^{N} A^j\, \boldsymbol{E}_j. \tag{10.263}$$

The convention is to write the index of basis vectors as subscript and the index of the components of a vector as superscript. We can drop the summation by stating the convention that whenever an index is repeated, once as a subscript and once as a superscript, one should sum on that index.

Now consider a linear map $T : V \to V$. This map is represented by a matrix which depends on the basis. In basis B_V the matrix of T is defined by the following relation:

$$T(\boldsymbol{E}_j) = T^i{}_j\, \boldsymbol{E}_i. \tag{10.264}$$

Note that the summation (understood from our convention) is on the *first* index of ${T^i}_j$. Now consider the action of T on a vector $\boldsymbol{A} = A^j\, \boldsymbol{E}_j$:

$$T(\boldsymbol{A}) = T(A^j\, \boldsymbol{E}_j) = A^j\, T(\boldsymbol{E}_j) = A^j\, {T^i}_j\, \boldsymbol{E}_i = \left({T^i}_j\, A^j\right) \boldsymbol{E}_i. \tag{10.265}$$

Therefore if we write the vector $\boldsymbol{A}$ as a column matrix with entries A^i, then the column matrix representing $\boldsymbol{A}' = T(\boldsymbol{A})$ is obtained by usual matrix multiplication:

$$A'^i = {T^i}_j\, A^j, \qquad \text{linear transformation.} \tag{10.266}$$

Now consider another basis $\tilde{B}_V = \left\{\tilde{\boldsymbol{E}}_i\right\}$ for the same vector space V. We can write vectors $\tilde{\boldsymbol{E}}_i$ as linear combinations of B_V. In this way a matrix S is defined:

$$\tilde{\boldsymbol{E}}_i = {S^k}_i\, \boldsymbol{E}_k. \tag{10.267}$$

If $\boldsymbol{A} = A^k\, \boldsymbol{E}_k$ is a vector in V, *the same vector* can be written in basis B'_V:

$$\boldsymbol{A} = \tilde{A}^i\, \tilde{\boldsymbol{E}}_i = \tilde{A}^i\, {S^k}_i\, \boldsymbol{E}_k = A^k\, \boldsymbol{E}_k, \tag{10.268}$$

from which we get

$$A^k = {S^k}_i\, \tilde{A}^i, \qquad \text{change of basis.} \tag{10.269}$$

Now let us find the matrix of the linear transformation $T : V \to V$ in the basis $\tilde{B}_V$, which we denote by $\tilde{T}$. To find that, we must find $T(\tilde{\boldsymbol{E}}_i)$ and write it as ${\tilde{T}^i}_j\, \tilde{\boldsymbol{E}}_i$:

$$T(\tilde{\boldsymbol{E}}_i) = T\left({S^n}_i\, \boldsymbol{E}_n\right) = {S^n}_i\, T(\boldsymbol{E}_n) = {S^n}_i\, {T^k}_n\, \boldsymbol{E}_k. \tag{10.270}$$

Now this must be equal to

$$T(\tilde{\boldsymbol{E}}_i) = {\tilde{T}^n}_i\, \tilde{\boldsymbol{E}}_n = {\tilde{T}^n}_i\, {S^k}_n\, \boldsymbol{E}_k. \tag{10.271}$$

Comparing the last equality above we get

$${T^k}_n\, {S^n}_i = {S^k}_n\, {\tilde{T}^n}_i, \tag{10.272}$$

or in matrix notation, $T\,S = S\,\tilde{T}$, which is usually written as

$$\tilde{T} = S^{-1}\, T\, S. \tag{10.273}$$

This is called a similarity transformation. We say that the matrix representing the transformation T in the basis $\tilde{B}_V$ is related by this similarity transformation to the matrix representing the same transformation in the basis B_V.

The reverse similarity transformation appears in a different context. Let V and W be two vector spaces having the same dimension, and let $S : V \to W$ be a linear map (not a change of basis). If a linear transformation $T : V \to V$ is given, the linear transformation $\tilde{T} = S \circ T \circ S^{-1}$ is a linear transformation acting on W:

$$\begin{array}{ccc} V & \xleftarrow{S^{-1}} & W \\ {\scriptstyle T}\downarrow & & \downarrow{\scriptstyle \tilde{T}} \\ V & \xrightarrow{S} & W \end{array} \qquad \tilde{T} = S \circ T \circ S^{-1}. \tag{10.274}$$

10.14 Tensor product

Let V be a vector space of dimension N with basis $B_V = \{\boldsymbol{E}_1, \boldsymbol{E}_2, \ldots, \boldsymbol{E}_N\}$, and let W be a vector space of dimension M with basis $B_W = \{\boldsymbol{F}_1, \boldsymbol{F}_2, \ldots, \boldsymbol{F}_M\}$. We construct another vector space $V \otimes W$ with dimension NM whose basis is the set $B_{V\otimes W} = \{\boldsymbol{E}_i \otimes \boldsymbol{F}_\mu\}$. We have used different sets of indices to emphasize that the dimensions of the vector spaces, and thus the ranges of the indices, may be different.

In the basis B_V, the linear map $S : V \to V$ is represented by the square matrix $S^i{}_j$, defined by $S(\boldsymbol{E}_j) = S^i{}_j\, \boldsymbol{E}_i$. Similarly, the linear map $T : W \to W$ is represented by a square matrix T^μ_ν defined by $T(\boldsymbol{F}_\nu) = T^\mu_\nu\, \boldsymbol{F}_\mu$.

We can construct the linear map $S \otimes T$ which acts on $V \otimes W$. As any linear transformation, it is specified once its action on the basis vectors is specified, and the natural action on the basis vectors is

$$\boldsymbol{E}_j \otimes \boldsymbol{F}_\nu \mapsto S^i{}_j\, T^\mu_\nu\, \boldsymbol{E}_i \otimes \boldsymbol{F}_\mu, \tag{10.275}$$

or, in matrix (index) notation,

$$[S \otimes T]^{i\mu}_{j\nu} = S^i{}_j\, T^\mu_\nu. \tag{10.276}$$

This means that $(S \otimes T)(\boldsymbol{a} \otimes \boldsymbol{b}) = S(\boldsymbol{a}) \otimes T(\boldsymbol{b})$.

Usually we order the set $\{\boldsymbol{E}_i \otimes \boldsymbol{F}_\mu\}$ by first running μ and then i; for example, the usual ordered basis for $V \otimes V$, if $\dim V = 2$, is

$$B_{V\otimes V} = \{\boldsymbol{E}_1 \otimes \boldsymbol{E}_1,\ \boldsymbol{E}_1 \otimes \boldsymbol{E}_2,\ \boldsymbol{E}_2 \otimes \boldsymbol{E}_1,\ \boldsymbol{E}_2 \otimes \boldsymbol{E}_2\}. \tag{10.277}$$

For the special case of $\mathbb{C}^2$, if we use the standard real basis vectors $\boldsymbol{E}_1 = \left[\begin{smallmatrix}1\\0\end{smallmatrix}\right]$ and $\boldsymbol{E}_2 = \left[\begin{smallmatrix}0\\1\end{smallmatrix}\right]$, we get

$$\boldsymbol{E}_{11} = \begin{bmatrix}1\\0\\0\\0\end{bmatrix}, \quad \boldsymbol{E}_{12} = \begin{bmatrix}0\\1\\0\\0\end{bmatrix}, \quad \boldsymbol{E}_{21} = \begin{bmatrix}0\\0\\1\\0\end{bmatrix}, \quad \boldsymbol{E}_{21} = \begin{bmatrix}0\\0\\0\\1\end{bmatrix}, \tag{10.278}$$

where $\boldsymbol{E}_{ij} = \boldsymbol{E}_i \otimes \boldsymbol{E}_j$. In this basis the matrix representing $S \otimes T$ is given by the so-called tensor or Kronecker[6] product:

$$S \otimes T = \left[S^i{}_m T^j{}_n \right] = \begin{bmatrix} S^1{}_1 T^1{}_1 & S^1{}_1 T^1{}_2 & S^1{}_2 T^1{}_1 & S^1{}_2 T^1{}_2 \\ S^1{}_1 T^2{}_1 & S^1{}_1 T^2{}_2 & S^1{}_2 T^2{}_1 & S^1{}_2 T^2{}_2 \\ S^2{}_1 T^1{}_1 & S^2{}_1 T^1{}_2 & S^2{}_2 T^1{}_1 & S^2{}_2 T^1{}_2 \\ S^2{}_1 T^2{}_1 & S^2{}_1 T^2{}_2 & S^2{}_2 T^2{}_1 & S^2{}_2 T^2{}_2 \end{bmatrix}. \tag{10.279}$$

This could be remembered by the following compact form, in which T is the 2×2 matrix $\left[T^i{}_j \right]$:

$$S \otimes T = \begin{bmatrix} S^1{}_1 T & S^1{}_2 T \\ S^2{}_1 T & S^2{}_2 T \end{bmatrix}. \tag{10.280}$$

As special cases we have

$$S \otimes \mathbb{I} = \begin{bmatrix} S^1{}_1 \mathbb{I} & S^1{}_2 \mathbb{I} \\ S^2{}_1 \mathbb{I} & S^2{}_2 \mathbb{I} \end{bmatrix}, \qquad \mathbb{I} \otimes T = \begin{bmatrix} T & \mathbb{O} \\ \mathbb{O} & T \end{bmatrix}, \tag{10.281}$$

where $\mathbb{O}$ is the zero matrix.

It can be shown that the Kronecker product has the following properties[7]:

$$\det S \otimes \mathbb{I} = (\det S)^2, \qquad \det(\mathbb{I} \otimes T) = (\det T)^2. \tag{10.282}$$

The generalization is

$$\det(A \otimes \mathbb{I}_{M \times M}) = (\det A)^M, \tag{10.283}$$

$$\det(\mathbb{I}_{N \times N} \otimes B) = (\det B)^N. \tag{10.284}$$

Finally, since $A \otimes B = (A \otimes \mathbb{I})(\mathbb{I} \otimes B)$, we see that

$$\det A \otimes B = (\det A)^M (\det B)^N, \qquad M = \dim B, \quad N = \dim A. \tag{10.285}$$

The tensor product representation. Let A and A' be two linear transformations on V, and let B and B' be two linear transformations on W. Observe that

$$(A \otimes B)(A' \otimes B') = (A\,A') \otimes (B\,B'). \tag{10.286}$$

From this, it follows that

$$(A \otimes \mathbb{I} + \mathbb{I} \otimes B)(A' \otimes \mathbb{I} + \mathbb{I} \otimes B') = A\,A' \otimes \mathbb{I} + A \otimes B' \\ + A' \otimes B + \mathbb{I} \otimes B\,B',$$

[6] After Leopold Kronecker (1823–1891).

[7] The proof is simple for $\mathbb{I} \otimes T$, which in this basis is block diagonal. For $S \otimes \mathbb{I}$ one should use elementary row and column transformations (reshuffling rows and columns) to transform it into a block diagonal form.

$$(A' \otimes \mathbb{I} + \mathbb{I} \otimes B')(A \otimes \mathbb{I} + \mathbb{I} \otimes B) = A' A \otimes \mathbb{I} + A' \otimes B + A \otimes B' + \mathbb{I} \otimes B' B,$$

and from these two equations it follows that

$$\left[A \otimes \mathbb{I} + \mathbb{I} \otimes B,\ A' \otimes \mathbb{I} + \mathbb{I} \otimes B'\right] = \left[A,\ A'\right] \otimes \mathbb{I} + \mathbb{I} \otimes \left[B,\ B'\right]. \tag{10.287}$$

This last identity has a very important consequence.

Theorem. *If $\{A_n\}$ and $\{B_n\}$ are two representations of a Lie algebra $\mathfrak{g}$ with structure constants $C^k{}_{ij}$, then the matrices*

$$S_n := A_n \otimes \mathbb{I} + \mathbb{I} \otimes B_n \tag{10.288}$$

are a representation of $\mathfrak{g}$ on $V \otimes W$.

The proof is simple:

$$\begin{aligned}\left[S_i,\ S_j\right] &= \left[A_i,\ A_j\right] \otimes \mathbb{I} + \mathbb{I} \otimes \left[B_i,\ B_j\right] \\ &= C^k{}_{ij}\,(A_k \otimes \mathbb{I} + \mathbb{I} \otimes B_k) \\ &= C^k{}_{ij}\, S_k.\end{aligned} \tag{10.289}$$

Observation. $A_n \otimes \mathbb{I}$ and $\mathbb{I} \otimes B_n$ commute, because

$$(A_n \otimes \mathbb{I})(\mathbb{I} \otimes B_n) = A_n \otimes B_n = (\mathbb{I} \otimes B_n)(A_n \otimes \mathbb{I}). \tag{10.290}$$

Therefore

$$\exp(\alpha\, A_n \otimes \mathbb{I} + \mathbb{I} \otimes B_n\, \alpha) = \exp(\alpha\, A_n) \otimes \exp(\alpha\, B_n). \tag{10.291}$$

Remark. Physicists usually do not use the notation of tensor product $\otimes$. For example, in quantum mechanics one usually writes $\boldsymbol{J} = \boldsymbol{L} + \boldsymbol{S}$, where $\boldsymbol{S}$ is the spin, $\boldsymbol{L}$ is the orbital angular momentum, and $\boldsymbol{J}$ is the total angular momentum. The operators $\boldsymbol{L}$ and $\boldsymbol{S}$ act on different objects—the former on the arguments of the wave functions, the latter on the spinor components. In tensor notation it is written as $\boldsymbol{J} = \boldsymbol{L} \otimes \mathbb{I} + \mathbb{I} \otimes \boldsymbol{S}$ (remember that $[\, L_n,\ S_m\,] = 0$). If we are aware of the fact that $\boldsymbol{L}$ and $\boldsymbol{S}$ act on different objects, then not writing the tensor product makes formulas look simpler.

Worked Example 45. Decomposing $\mathbf{2} \otimes \overline{\mathbf{2}}$ of $\mathfrak{su}(2)$. As is customary in physics, we write the Lie algebra of $\mathfrak{su}(2)$ as $[\, S_n,\ S_m\,] = i\,\epsilon_{nmk}\, S_k$. For the defining representation $S_n = \frac{1}{2}\sigma_n$, and for the complex conjugate representation $S_n = -\frac{1}{2}\sigma_n^T$. Both these representations are 2D. It is customary to name and denote them by $\mathbf{2}$ and $\overline{\mathbf{2}}$. On the tensor product space $\mathbf{2} \otimes \overline{\mathbf{2}}$, we have

$$S_n = \tfrac{1}{2}\left(\sigma_n \otimes \mathbb{I} - \mathbb{I} \otimes \sigma_n^T\right). \tag{10.292}$$

It is straightforward to find these matrices and their squares,

$$S_1 = \tfrac{1}{2}\begin{bmatrix} -\sigma_1 & \mathbb{I} \\ \mathbb{I} & -\sigma_1 \end{bmatrix}, \qquad (S_1)^2 = \tfrac{1}{2}\begin{bmatrix} \mathbb{I} & -\sigma_1 \\ -\sigma_1 & \mathbb{I} \end{bmatrix}, \tag{10.293}$$

$$S_2 = \tfrac{1}{2}\begin{bmatrix} \sigma_2 & -i\,\mathbb{I} \\ i\,\mathbb{I} & \sigma_2 \end{bmatrix}, \qquad (S_2)^2 = \frac{1}{2}\begin{bmatrix} \mathbb{I} & -i\,\sigma_2 \\ i\,\sigma_2 & \mathbb{I} \end{bmatrix}, \tag{10.294}$$

$$S_3 = \begin{bmatrix} 0 & 0 & 0 & 0 \\ 0 & 1 & 0 & 0 \\ 0 & 0 & -1 & 0 \\ 0 & 0 & 0 & 0 \end{bmatrix}, \qquad (S_3)^2 = \begin{bmatrix} 0 & 0 & 0 & 0 \\ 0 & 1 & 0 & 0 \\ 0 & 0 & 1 & 0 \\ 0 & 0 & 0 & 0 \end{bmatrix}, \tag{10.295}$$

and finally the Casimir

$$S^2 = \begin{bmatrix} 1 & 0 & 0 & -1 \\ 0 & 2 & 0 & 0 \\ 0 & 0 & 2 & 0 \\ -1 & 0 & 0 & 1 \end{bmatrix}, \qquad \operatorname{tr} S^2 = 6. \tag{10.296}$$

Since the Casimir is not a multiple of the unit matrix, this representation of $\mathfrak{su}(2)$ is reducible. Remember that for **3** (the spin-1 representation) we have $\operatorname{tr} S^2 = 3 \times 1\,(1+1) = 6$, while for **2** we have $\operatorname{tr} S^2 = 2 \times \frac{1}{2}(\frac{1}{2}+1) = \frac{3}{2}$. For $\mathbf{2} \oplus \mathbf{2}$ we have $\operatorname{tr} S^2 = 2 \cdot \frac{3}{2} = 3$, and for **1** we have $\operatorname{tr} S^2 = 0$. We conclude that the representation $\mathbf{2} \otimes \bar{\mathbf{2}}$, for which $\operatorname{tr} S^2 = 6$, is the direct sum of **1** and **3**. The subspace V_1 on which **1** acts is called the singlet, the subspace V_3 on which **3** acts is called the triplet. We denote these by writing

$$\mathbf{2} \otimes \bar{\mathbf{2}} = \mathbf{1} \oplus \mathbf{3}. \tag{10.297}$$

From the form of the matrix S_3 it is clear that $\boldsymbol{E}_{12} = \boldsymbol{E}_1 \otimes \boldsymbol{E}_2$ is a normalized eigenvector of S_3, with eigenvalue 1. To find the other basis vectors of V_3 we should apply S_-. The matrices $S_\pm = S_1 \pm i\,S_2$ are as follows:

$$S_- = \begin{bmatrix} 0 & -1 & 0 & 0 \\ 0 & 0 & 0 & 0 \\ 1 & 0 & 0 & -1 \\ 0 & 1 & 0 & 0 \end{bmatrix}, \qquad S_+ = \begin{bmatrix} 0 & 0 & 1 & 0 \\ -1 & 0 & 0 & 1 \\ 0 & 0 & 0 & 0 \\ 0 & 0 & -1 & 0 \end{bmatrix}. \tag{10.298}$$

Applying these matrices to $\boldsymbol{E}_{12} = (0, 1, 0, 0)$ is easy, and we must remember that

$$S_-\,|1,1\rangle = \sqrt{2}\,|1,0\rangle\,, \quad S_-\,|1,0\rangle = \sqrt{2}\,|1,-1\rangle\,, \quad S_-\,|1,-1\rangle = 0. \tag{10.299}$$

In this way we can find the V_3 subspace. The result is

$$|1,+1\rangle = \boldsymbol{E}_1 \otimes \boldsymbol{E}_2, \tag{10.300}$$

$$|1,\,0\rangle = \frac{1}{\sqrt{2}}(-\boldsymbol{E}_1 \otimes \boldsymbol{E}_1 + \boldsymbol{E}_2 \otimes \boldsymbol{E}_2)\,, \tag{10.301}$$

$$|1,-1\rangle = -\boldsymbol{E}_2 \otimes \boldsymbol{E}_1. \tag{10.302}$$

The subspace V_1 is a 1D eigenspace of S_3 with eigenvalue 0. It is easy to see that $\boldsymbol{E}_1 \otimes \boldsymbol{E}_2 + \boldsymbol{E}_2 \otimes \boldsymbol{E}_1$ is such a vector. We normalize it and write it as

$$|0,0\rangle = \frac{1}{\sqrt{2}}(\boldsymbol{E}_1 \otimes \boldsymbol{E}_1 + \boldsymbol{E}_2 \otimes \boldsymbol{E}_2). \tag{10.303}$$

Remark. In the books of quantum mechanics the tensor product $\mathbf{2} \otimes \mathbf{2}$ is decomposed into $\mathbf{3} \oplus \mathbf{1}$, because the motivation is the structure of the state space of two particles, which could be identical. In the above we have decomposed $\mathbf{2} \otimes \overline{\mathbf{2}}$. The reader is recommended to compare these two decompositions. For the moment, note that the procedures are similar, but the forms of the generators and the basis vectors are different. Especially note that in the decomposition of $\mathbf{2} \otimes \mathbf{2}$, the singlet state $|0,0\rangle$ is antisymmetric in $1 \leftrightarrow 2$, and $|1,0\rangle$, belonging to the triplet, is symmetric. In the decomposition of $\mathbf{2} \otimes \overline{\mathbf{2}}$, however, the singlet $|0,0\rangle$ *seems to be symmetric*, and the state $|1,0\rangle$ belonging to the triplet *seems to be antisymmetric*. But note that now, since $\overline{\mathbf{2}}$ is different from $\mathbf{2}$, symmetry and antisymmetry are not well defined.

10.15 The Euclidean groups

The symmetry group of the Euclidean space, denoted by E(3), is the set of all rotations and translations. This means transformations of the form

$$X \mapsto X' = RX + A, \tag{10.304}$$

where R is an orthogonal matrix and A is a vector (a column matrix). So the transformation is specified by the pair $g = (R, A)$. Consider two successive Euclidean transformations:

$$g_1 = (R_1, A_1) : X \mapsto X' = R_1 X + A_1, \tag{10.305}$$

$$g_2 = (R_2, A_2) : X' \mapsto X'' = R_2 X' + A_2. \tag{10.306}$$

The combination of these two transformations is

$$\begin{aligned} g_2 \circ g_1(X) = g_2(g_1(X)) &= R_2(R_1 X + A_1) + A_2 \\ &= R_2 R_1 X + (R_2 A_1 + A_2), \end{aligned} \tag{10.307}$$

meaning that

$$g_2 \circ g_1 = (R_2 R_1, R_2 A_1 + A_2). \tag{10.308}$$

It is easy to see that $(\mathbb{I}, 0)$, where $\mathbb{I}$ is the unit matrix and 0 is the zero vector, is the identity of this action. It is also obvious that this group contains the following two subgroups:

- The subset of pure translations with $g = (\mathbb{I}, A)$. This is an Abelian subgroup.
- The subset of pure rotations with $g = (R, 0)$.

Jargon. The mathematical term for this structure is *semidirect product*. For example, it is said that the Euclidean group E(2) is the semidirect product of the rotation group O(2) and the translation group $\mathbb{R}^2$. The symbol $\ltimes$ is used, as follows:

$$\mathrm{E}(2) = \mathrm{O}(2) \ltimes \mathbb{R}^2. \tag{10.309}$$

Another notation is IO(2), which is called the *inhomogeneous* orthogonal group. The term inhomogeneous refers to the "inhomogeneous" term A in (10.304). The group $\mathrm{ISO}(2) = \mathrm{SO}(2) \ltimes \mathbb{R}^2$ is the combination of all rotations and translations (no inversions).

***A matrix representation of the Euclidean group* E(2).** Consider the 3D space with Cartesian coordinates (x_1, x_2, x_3). The set $x_3 = a$ on this space is a plane which we denote Π_a. Now consider the following matrix:

$$G = \begin{pmatrix} R_{11} & R_{12} & \alpha_1 \\ R_{21} & R_{22} & \alpha_2 \\ 0 & 0 & 1 \end{pmatrix}. \tag{10.310}$$

Let us investigate the action of this matrix on Π_a:

$$\begin{pmatrix} x_1' \\ x_2' \\ a \end{pmatrix} = \begin{pmatrix} R_{11} & R_{12} & \alpha_1 \\ R_{21} & R_{22} & \alpha_2 \\ 0 & 0 & 1 \end{pmatrix} \begin{pmatrix} x_1 \\ x_2 \\ a \end{pmatrix}. \tag{10.311}$$

Note that this transformation maps the plane Π_a into itself, but the points of Π_a are transformed according to

$$\begin{cases} x_1' = R_{11} x_1 + R_{12} x_2 + \alpha_1 a, \\ x_2' = R_{21} x_1 + R_2 2 x_2 + \alpha_2 a, \end{cases} \tag{10.312}$$

which is exactly the action of E(2) on the (x_1, x_2)-plane.

Remark. Note that a is a fixed parameter with dimensions of length, and α_1 and α_2 are dimensionless, so the combinations $\alpha_i a$ have dimensions of length, as expected. One could use the convention $a = 1$, that is, consider the plane $x_3 = 1$. In that case the parameters of translation are simply α_i and have the dimensions of length.

Let us investigate the situation when R is an infinitesimal rotation. In this case we have

$$\begin{cases} x_1' = x_1 - \theta\, x_2 + \alpha_1 a, \\ x_2' = \theta\, x_1 + x_2 + \alpha_2 a. \end{cases} \tag{10.313}$$

Thus the matrix G in this case is

$$G = \begin{pmatrix} 1 & -\theta & \alpha_1 \\ \theta & 1 & \alpha_2 \\ 0 & 0 & 1 \end{pmatrix} = \mathbb{I} + \begin{pmatrix} 0 & -\theta & \alpha_1 \\ \theta & 0 & \alpha_2 \\ 0 & 0 & 0 \end{pmatrix}. \tag{10.314}$$

We now introduce the generators of ISO(2),

$$\mathcal{S}_3 = \begin{pmatrix} 0 & -1 & 0 \\ 1 & 0 & 0 \\ 0 & 0 & 0 \end{pmatrix}, \quad P_1 = \begin{pmatrix} 0 & 0 & 1 \\ 0 & 0 & 0 \\ 0 & 0 & 0 \end{pmatrix}, \quad P_2 = \begin{pmatrix} 0 & 0 & 0 \\ 0 & 0 & 1 \\ 0 & 0 & 0 \end{pmatrix}. \tag{10.315}$$

By simple matrix multiplications, which we leave to the reader, it is easy to see that

$$[\mathcal{S}_3, P_1] = +P_2, \tag{10.316}$$

$$[\mathcal{S}_3, P_2] = -P_1, \tag{10.317}$$

$$[P_1, P_2] = 0. \tag{10.318}$$

We already know that $\exp(\theta \mathcal{S}_3)$ is the rotation around the x_3-axis. The exponential $\exp(\alpha_1 P_1 + \alpha_2 P_2)$ is easily calculated once we realize that P_1 and P_2 commute with each other and

$$P_1^2 = P_2^2 = P_1 P_2 = 0. \tag{10.319}$$

Therefore

$$\exp(\alpha_1 P_1 + \alpha_2 P_2) = \mathbb{I} + \alpha_1 P_1 + \alpha_2 P_2 = \begin{pmatrix} 1 & 0 & \alpha_1 \\ 0 & 1 & \alpha_2 \\ 0 & 0 & 1 \end{pmatrix}. \tag{10.320}$$

The action of this matrix on Π_a is a pure translation of $(\alpha_1 a, \alpha_2 a)$.

The Lie algebra $\mathfrak{iso}(3)$**.** The generalization to the Euclidean group E(3) is obvious. The Euclidean transformation is

$$X \to X' = RX + A, \tag{10.321}$$

where now R is in O(3) and $A^T = (\alpha_1 a, \alpha_2 a, \alpha_3 a)$ is represented by

$$\begin{pmatrix} x_1 \\ x_2 \\ x_3 \\ a \end{pmatrix} \mapsto \begin{pmatrix} x_1' \\ x_2' \\ x_3' \\ a \end{pmatrix} = \begin{pmatrix} R_{11} & R_{12} & R_{13} & \alpha_1 \\ R_{21} & R_{22} & R_{23} & \alpha_2 \\ R_{31} & R_{32} & R_{33} & \alpha_3 \\ 0 & 0 & 0 & 1 \end{pmatrix} \begin{pmatrix} x_1 \\ x_2 \\ x_3 \\ a \end{pmatrix}. \tag{10.322}$$

ISO(3) is 6D: three parameters to specify rotations (R) and three parameters to specify translations (A). The Lie algebra $\mathfrak{iso}(3)$ thus has six generators:

$$\mathcal{S}_1 = \begin{pmatrix} 0 & 0 & 0 & 0 \\ 0 & 0 & -1 & 0 \\ 0 & 1 & 0 & 0 \\ 0 & 0 & 0 & 0 \end{pmatrix}, \qquad P_1 = \begin{pmatrix} 0 & 0 & 0 & 1 \\ 0 & 0 & 0 & 0 \\ 0 & 0 & 0 & 0 \\ 0 & 0 & 0 & 0 \end{pmatrix}, \tag{10.323}$$

$$\mathcal{S}_2 = \begin{pmatrix} 0 & 0 & 1 & 0 \\ 0 & 0 & 0 & 0 \\ -1 & 0 & 0 & 0 \\ 0 & 0 & 0 & 0 \end{pmatrix}, \qquad P_2 = \begin{pmatrix} 0 & 0 & 0 & 0 \\ 0 & 0 & 0 & 1 \\ 0 & 0 & 0 & 0 \\ 0 & 0 & 0 & 0 \end{pmatrix}, \tag{10.324}$$

$$\mathcal{S}_3 = \begin{pmatrix} 0 & -1 & 0 & 0 \\ 1 & 0 & 0 & 0 \\ 0 & 1 & 0 & 0 \\ 0 & 0 & 0 & 0 \end{pmatrix}, \qquad P_3 = \begin{pmatrix} 0 & 0 & 0 & 0 \\ 0 & 0 & 0 & 0 \\ 0 & 0 & 0 & 1 \\ 0 & 0 & 0 & 0 \end{pmatrix}. \tag{10.325}$$

The commutation relations are as follows:

$$[\,\mathcal{S}_n,\, \mathcal{S}_m\,] = \epsilon_{nmk}\, \mathcal{S}_k, \tag{10.326}$$

$$[\,\mathcal{S}_n,\, P_m\,] = \epsilon_{nmk}\, P_k, \tag{10.327}$$

$$[\,P_n,\, P_m\,] = 0. \tag{10.328}$$

Relation to the Galilean group. The Galilean transformations are transformations of the form

$$\begin{pmatrix} x_1 \\ x_2 \\ x_3 \\ t \end{pmatrix} \to \begin{pmatrix} x_1' \\ x_2' \\ x_3' \\ t' \end{pmatrix} = \begin{pmatrix} R_{11} & R_{12} & R_{13} & v_1 \\ R_{21} & R_{22} & R_{23} & v_2 \\ R_{31} & R_{32} & R_{33} & v_3 \\ 0 & 0 & 0 & 1 \end{pmatrix} \begin{pmatrix} x_1 \\ x_2 \\ x_3 \\ t \end{pmatrix}, \tag{10.329}$$

where R is an orthogonal matrix and (v_1, v_2, v_3) is the velocity vector. The Lie algebra of this group is the same as $\mathfrak{iso}(3)$. If the matrix G is considered to act on the Π_a space, it represents a Euclidean transformation; if it is considered to act on the (x_1, x_2, x_3, t) space (which is called the Galilean space-time), G represents a Galilean transformation.

Remark. In the literature the term Galilean group may refer to one of the following two groups:

- The six-parameter group of all rotations and Galilean boosts. The Lie algebra of this group is isomorphic to $\mathfrak{iso}(3)$, and it is this group that was discussed in the preceding lines. The generators of this group are usually denoted by $\mathcal{S}_1$, $\mathcal{S}_2$, and $\mathcal{S}_3$, generating rotations, and $\mathcal{K}_1$, $\mathcal{K}_2$, and $\mathcal{K}_3$, generating Galilean boosts.

- The 10-parameter group of all rotations, Galilean boosts, *and* spatial and temporal translations. This group is the symmetry group of the space-time in the nonrelativistic classical and quantum mechanics. The structure of this 10D Lie algebra is usually studied by the $v \ll c$ limit of the Poincaré group as we will see in §11.8.

Guide to Literature. For a brief text about Lie algebras see Duistermaat and Kolk (2004, pp. 169–171). For a more detailed exposition see Kosmann-Schwarzbach (2010, pp. 47–55) or Lipkin (1965).

Rotations are usually taught in textbooks on classical mechanics. A good example is Goldstein (1980, pp. 128–187).

In the physics curriculum representation theory of $\mathfrak{su}(2)$ and SO(3) is usually covered in the courses on quantum mechanics, so the subject can be found in almost any textbook on quantum mechanics. Some standard texts are Sakurai (1985, pp. 187–195) and Greiner and Müller (1994, pp. 37–57). The approach of Schwinger (2001, pp. 149–181) is original and worthy of study. A concise and informative text is Fuchs and Schweigert (1997, pp. 17–22).

Sexl and Urbantke (1992, pp. 169–228) give an almost complete treatment of the subject.

References

Duistermaat, J.J., Kolk, J.A.C., 2004. Multidimensional Real Analysis I. Differentiation. Cambridge University Press, Cambridge.
Fuchs, J., Schweigert, C., 1997. Symmetries, Lie Algebras and Representations. A Graduate Course for Physicists. Cambridge University Press, Cambridge.
Goldstein, H., 1980. Classical Mechanics, 2nd edition. Addison-Wesley, Reading, MA.
Greiner, W., Müller, B., 1994. Quantum Mechanics II: Symmetries. Springer-Verlag, Berlin.
Kosmann-Schwarzbach, Y., 2010. Groups and Symmetries. From Finite Groups to Lie Groups. Springer-Verlag, New York.
Lipkin, H.J., 1965. Lie Groups for Pedestrians. North-Holland, Amesterdam.
Sakurai, J.J., 1985. Modern Quantum Mechanics. Addison-Wesley, Redwood City, CA.
Schwinger, J., 2001. Quantum Mechanics. Symbolism of Atomic Measurements. Springer-Verlag, Berlin.
Sexl, R.U., Urbantke, H.K., 1992. Relativity, Groups, Particles. Special Relativity and Relativistic Symmetry in Field and Particle Physics. Springer-Verlag, Wien.

CHAPTER

The Lorentz group 11

In this chapter we introduce the 4D Minkowski space-time and the Lorentz group. We then study the mathematical structure of the Lorentz group. This mathematical structure is of fundamental importance in our current theories of space-time, both at very small scales appropriate to high-energy physics and at larger scales important in our theories of gravitation.

In 1908, Hermann Minkowski, who once was Einstein's mathematics teacher while Einstein was a student at ETH Zürich,[1] delivered an address at the 80th Assembly of German Natural Scientists and Physicians, in Cologne, in which the geometric structure of Einstein's special relativity was explained. Ever since, all physicists and mathematicians appreciate the importance of this geometric structure. Minkowski used complex coordinates $w = i\,c\,t$ so that $-c^2 t^2 + x^2 + y^2 + z^2$ appear as $w^2 + x^2 + y^2 + z^2$, but nowadays we use real coordinates by using a metric which is explained in this chapter.

Guide to Literature. For an English translation of Minkowski's address with valuable notes by A. Sommerfeld, see Einstein et al. (1952, pp. 73–96).

11.1 The Minkowski space-time

Let c be the velocity of light in vacuum. We call the following symmetric matrix the Minkowski metric:

$$H = \left[\eta_{\mu\nu}\right] = \begin{bmatrix} \eta_{00} & \eta_{01} & \eta_{02} & \eta_{03} \\ \eta_{10} & \eta_{11} & \eta_{12} & \eta_{13} \\ \eta_{20} & \eta_{21} & \eta_{22} & \eta_{23} \\ \eta_{30} & \eta_{31} & \eta_{32} & \eta_{33} \end{bmatrix} = \begin{bmatrix} -\varsigma\, c^2 & 0 & 0 & 0 \\ 0 & \varsigma & 0 & 0 \\ 0 & 0 & \varsigma & 0 \\ 0 & 0 & 0 & \varsigma \end{bmatrix}. \qquad (11.1)$$

Here ς is either 1, for the so-called *space-like* convention, or -1, for the so-called *time-like* convention. Since both conventions have been used by excellent textbooks and great physicists, it is better if the reader can use both of the conventions.

Remark. In this book we use the convention $x^0 = t$, so we have $\eta_{00} = -\varsigma\, c^2$, $\eta^{00} = -\varsigma\, c^{-2}$. Usually in textbooks authors define $x^0 = c\,t$. This gives $\eta_{00} = \eta^{00} = -\varsigma$.

---

[1] Eidgenössische Technische Hochschule Zürich.

A Mathematical Approach to Special Relativity. https://doi.org/10.1016/B978-0-32-399708-9.00021-X

The reason we have chosen to define $x^0 = t$ is that first, it is then clear that $\eta_{\mu\nu}$ is not the same as $\eta^{\mu\nu}$, and second, now we can study the $c \to \infty$ case and see how that process *contracts*[2] the Lorentz group to the Galilean group.

The inverse of this H is the following matrix:

$$H^{-1} = \left[\eta^{\mu\nu}\right] = \begin{bmatrix} \eta^{00} & \eta^{01} & \eta^{02} & \eta^{03} \\ \eta^{10} & \eta^{11} & \eta^{12} & \eta^{13} \\ \eta^{20} & \eta^{21} & \eta^{22} & \eta^{23} \\ \eta^{30} & \eta^{31} & \eta^{32} & \eta^{33} \end{bmatrix} = \begin{bmatrix} -\varsigma\, c^{-2} & 0 & 0 & 0 \\ 0 & \varsigma & 0 & 0 \\ 0 & 0 & \varsigma & 0 \\ 0 & 0 & 0 & \varsigma \end{bmatrix}. \tag{11.2}$$

Let

$$A = \begin{bmatrix} A^0 \\ A^1 \\ A^2 \\ A^3 \end{bmatrix} \quad \text{and} \quad B = \begin{bmatrix} B^0 \\ B^1 \\ B^2 \\ B^3 \end{bmatrix} \tag{11.3}$$

be two vectors in $\mathbb{R}^4$. We define the Minkowski (or Lorentz, or relativistic) inner product $A \cdot B$ as follows:

$$A \cdot B = A^T\, H\, B, \tag{11.4}$$

where A^T is the transpose of the column vector A. It is easy to see that we have

$$A \cdot B = \eta_{\mu\nu}\, A^\mu\, B^\nu \tag{11.5}$$

$$= \varsigma \left(-c^2\, A^0\, B^0 + A^1\, B^1 + A^2\, B^2 + A^3\, B^3\right). \tag{11.6}$$

We say two vectors A and B are orthogonal if $A \cdot B = 0$. This inner product has the following properties:

1. It is commutative, that is, $A \cdot B = B \cdot A$.
2. It is bilinear, that is, for real numbers λ and μ,

$$(\lambda\, A) \cdot (\mu\, B) = (\lambda\, \mu)(A \cdot B). \tag{11.7}$$

3. If $A \cdot B = 0$ is always true for any vector B, then $A = 0$.
4. $A \cdot A$ can be positive, zero, or negative.

Signature. Consider a nonsingular matrix S, and define

$$X' = S\, X. \tag{11.8}$$

[2] See §11.5 and the Guide to Literature at the end of it.

Since $X = S^{-1} X'$ and $X^T = X'^T \left(S^{-1}\right)^T$, we have

$$X \cdot Y = X^T H Y = X'^T \left(S^{-1}\right)^T H S^{-1} Y' = X'^T H' Y', \tag{11.9}$$

where

$$H' = \left(S^{-1}\right)^T H S^{-1} \qquad \text{or} \qquad H = S^T H' S. \tag{11.10}$$

Such a transformation is called *congruence*—it is not a similarity transformation! H and H' are said to be congruent. Consider the following S:

$$S = \begin{bmatrix} \alpha^{-1} & 0 & 0 & 0 \\ 0 & \beta^{-1} & 0 & 0 \\ 0 & 0 & \gamma^{-1} & 0 \\ 0 & 0 & 0 & \delta^{-1} \end{bmatrix}, \qquad S^{-1} = \begin{bmatrix} \alpha & 0 & 0 & 0 \\ 0 & \beta & 0 & 0 \\ 0 & 0 & \gamma & 0 \\ 0 & 0 & 0 & \delta \end{bmatrix}. \tag{11.11}$$

It is easy to see that for H given in (11.1), we have

$$H' = \varsigma \begin{bmatrix} -\alpha^2 c^2 & 0 & 0 & 0 \\ 0 & \beta^2 & 0 & 0 \\ 0 & 0 & \gamma^2 & 0 \\ 0 & 0 & 0 & \delta^2 \end{bmatrix}. \tag{11.12}$$

From this, it is seen that the absolute values of the diagonal elements of H are not important—they can be changed by a change of basis. However, the signs of these entries do not change (as far as we consider real transformation matrices S). Therefore if $\varsigma = 1$, the space-like convention, the metric has three positive eigenvalues and one negative one. If $\varsigma = -1$, the time-like convention, the metric has three negative eigenvalues and one positive one. We may say that the metric has the signature $(-\varsigma, \varsigma, \varsigma, \varsigma)$. However, mathematicians define the *signature* to be the sum of these ± 1 s, that is, H has the signature 2ς.

Definition. The space $\mathbb{R}^4$ endowed with inner product $A \cdot B = A^T H B$, where $H = \varsigma\ \mathrm{diag}(-c^2, 1, 1, 1)$, is called the Minkowski space-time, and we denote it by $\mathbb{M}$.

11.2 The Lorentz group

The linear transformation $L : \mathbb{M} \to \mathbb{M}$ is said to be a symmetry of $\mathbb{M}$ if the inner product $A \cdot B = A^T H B$ is left invariant under its action. In other words, $L : \mathbb{M} \to \mathbb{M}$ is a symmetry if the inner product of two vectors A and B is the same as the inner product of the vectors $L A$ and $L B$, that is, if

$$(L A) \cdot (L B) = A \cdot B. \tag{11.13}$$

It is easy to find the condition over L for this to be true:

$$(L\,A)\cdot(L\,B) = (L\,A)^T\; H\;(L\,B) \tag{11.14}$$

$$= \left(A^T\,L^T\right)\,H\,(L\,B) \tag{11.15}$$

$$= A^T\,\left(L^T\,H\,L\right)\,B. \tag{11.16}$$

We see that $(L\,A)\cdot(L\,B) = A\cdot B$ is true for any A and B if and only if

$$L^T\,H\,L = H. \tag{11.17}$$

Let us write the same condition in index notation:

$$(L\,A)\cdot(L\,B) = \eta_{\mu\nu}\,(L\,A)^\mu\,(L\,B)^\nu \tag{11.18}$$

$$= \eta_{\mu\nu}\,L^\mu{}_\alpha\,A^\alpha\,L^\nu{}_\beta\,B^\beta \tag{11.19}$$

$$= \left(L^\mu{}_\alpha\,L^\nu{}_\beta\,\eta_{\mu\nu}\right)A^\alpha\,B^\beta, \tag{11.20}$$

$$A\cdot B = \eta_{\alpha\beta}\,A^\alpha\,B^\beta. \tag{11.21}$$

We see that $(L\,A)\cdot(L\,B) = A\cdot B$ is true for any A and B if and only if

$$L^\mu{}_\alpha\,L^\nu{}_\beta\,\eta_{\mu\nu} = \eta_{\alpha\beta}. \tag{11.22}$$

The set of matrices with the property $L^T\,H\,L = H$ form a group, which means that:

1. If M and N are two such matrices, their product MN is also such a matrix.
2. The unit matrix $\mathbb{I}$ is such a matrix.
3. If M is such a matrix, M^{-1} is also such a matrix.

Here is the proof. First, if M and N are two such matrices, then $L = M\,N$ is also such a matrix, because

$$L^T\,H\,T = (M\,N)^T\;H\,(M\,N) = \left(N^T\,M^T\right)H\,(M\,N) \tag{11.23}$$

$$= N^T\left(M^T\,H\,M\right)N = N^T\,H\,N = H. \tag{11.24}$$

Second, the unit matrix $\mathbb{I}$ is also such a matrix, because $\mathbb{I}^T H\mathbb{I} = H$. It remains to prove that if the matrix L satisfies $L^T\,H\,L = H$, then the inverse of L, L^{-1}, also satisfies the same relation. The proof is simple if we multiply the relation $L^T\,H\,L = H$ from the left by H^{-1}, which shows that $L^{-1} = H^{-1}\,L^T\,H$. We now calculate

$$\left(L^{-1}\right)^T\,H\,L^{-1} = \left(H^{-1}\,L^T\,H\right)^T\,H\,\left(H^{-1}\,L^T\,H\right) \tag{11.25}$$

$$= H^T L \left(H^{-1}\right)^T H H^{-1} L^T H \tag{11.26}$$

$$= H L H^{-1} H H^{-1} L^T H \tag{11.27}$$

$$= H L \left(H^{-1} L^T H\right) = H L L^{-1} = H. \tag{11.28}$$

In the above calculation we have used the symmetry of H and H^{-1}, which means that $H = H^T$ and $\left(H^{-1}\right)^T = H^{-1}$.

The group of matrices L with property $L^T H L = H$ is called the Lorentz group and is denoted by $\mathrm{O}(1,3)$. In this notation O is for *orthogonal*, and $(1,3)$ reminds us of the fact that of the four diagonal entries of H, one is negative and three are positive in the space-like convention, while in the time-like convention, one is positive and three are negative.

Remark. The group $\mathrm{O}(1,3)$ depends on the velocity of light, c. For a moment, let us denote it by $\mathrm{O}_c(1,3)$. The groups $\mathrm{O}_{c'}(1,3)$ and $\mathrm{O}_c(1,3)$ are isomorphic (see §11.5 on page 266). So usually we forget about c, or we choose units so that $c = 1$.

11.2.1 Some important subgroups of $\mathrm{O}(1,3)$

***The special subgroup* $\mathbf{SO(1,3)}$.** The determinant of H is $-c^2$. Now from $L^T H L = H$ we conclude that $\det L^T \det H \det L = \det H$. Using $\det L^T = \det L$, we get

$$(\det L)^2 = 1 \qquad \Rightarrow \qquad \det L = \pm 1. \tag{11.29}$$

If M and N are both members of $\mathrm{O}(1,3)$, both with determinant 1, then $L = M N$ is also in $\mathrm{O}(1,3)$ with determinant 1. Therefore the set

$$\mathrm{SO}(1,3) = \{L \in \mathrm{O}(1,3) \quad | \quad \det L = 1\} \tag{11.30}$$

forms a subgroup of $\mathrm{O}(1,3)$. (Remember that $\det \mathbb{I} = 1$.) The letter S in $\mathrm{SO}(1,3)$ is the abbreviation for *special*, which means "determinant 1."

***The orthochronous subgroup* $\mathbf{O^{\uparrow}(1,3)}$.** Let sgn be the sign function, giving the sign of the real argument:

$$\operatorname{sgn}(x) = \begin{cases} -1 & \text{if} \quad x < 0, \\ 0 & \text{if} \quad x = 0, \\ 1 & \text{if} \quad x > 1. \end{cases} \tag{11.31}$$

Define

$$\tau(L) = \operatorname{sgn}\left(L^0{}_0\right). \tag{11.32}$$

Acting on matrices $\mathrm{O}(1,3)$, this function has a property similar to that of the determinant, that is,

$$\tau(L M) = \tau(L)\, \tau(M). \tag{11.33}$$

To prove, consider two matrices L and M in $O(1,3)$ and their product $L = M\,N$, and define

$$\boldsymbol{L} := \left(L^0{}_1,\, L^0{}_2,\, L^0{}_3\right), \tag{11.34}$$

$$\boldsymbol{M} := \left(M^1{}_0,\, M^2{}_0,\, M^3{}_0\right), \tag{11.35}$$

$$L^2 := \boldsymbol{L}\cdot\boldsymbol{L} = \sum_{i=1}^{3}\left(L^0{}_i\right)^2, \tag{11.36}$$

$$M^2 := \boldsymbol{M}\cdot\boldsymbol{M} = \sum_{j=1}^{3}\left(M^j{}_0\right)^2, \tag{11.37}$$

$$\ell := L^0{}_0, \tag{11.38}$$

$$m := M^0{}_0. \tag{11.39}$$

Now if we set $\mu = \nu = 0$ in $\eta^{\mu\nu} = L^\mu{}_\alpha\, L^\nu{}_\beta\, \eta^{\alpha\beta}$ we get

$$\frac{1}{c^2} = \frac{1}{c^2}\left(L^0{}_0\right)^2 - \sum_{i=1}^{3}\left(L^0{}_i\right)^2, \tag{11.40}$$

which means

$$\frac{1}{c^2} = \frac{\ell^2}{c^2} - L^2, \tag{11.41}$$

or

$$L^2 = \frac{1}{c^2}\left(\ell^2 - 1\right), \tag{11.42}$$

which shows two inequalities:

$$|\ell| \geq 1 \tag{11.43}$$

since $L^2 \geq 0$ and

$$|L| \leq \frac{1}{c}\,|\ell|\,. \tag{11.44}$$

Note also that if $\ell^2 = (L^0{}_0)^2 = 1$, then $L^2 = 0$, which means that $L^0{}_i = 0$.

Similarly, if in $\eta_{\mu\nu} = M^\alpha{}_\mu\, M^\beta{}_\nu\, \eta_{\alpha\beta}$ we set $\mu = \nu = 0$, we get

$$c^2 = c^2 m^2 - M^2, \tag{11.45}$$

from which it follows that

$$M^2 = c^2(m^2 - 1), \tag{11.46}$$

and therefore the following two inequalities:

$$M \leq c\,|m|, \qquad m \geq 1. \tag{11.47}$$

Note also that if $m^2 = (M^0{}_0)^2 = 1$, then $M^2 = 0$, which means that $M^i{}_0 = 0$.

Before proceeding, let us observe an important property of the matrix $L^\mu{}_\nu$: if $(L^0{}_0)^2 = 1$, then $\ell = 0$ so that $\boldsymbol{L} = 0$ and therefore both $L^0{}_i$s and $L^i{}_0$s vanish, which means that if $L^0{}_0$ is either 1 or -1, then the matrix $L^\mu{}_\nu$ is of the form

$$L = \begin{bmatrix} \pm 1 & 0 & 0 & 0 \\ 0 & A_{11} & A_{12} & A_{13} \\ 0 & A_{21} & A_{22} & A_{23} \\ 0 & A_{31} & A_{32} & A_{33} \end{bmatrix}. \tag{11.48}$$

The condition $L^T\, H\, L = H$ is now satisfied if $A^T\, A = \mathbb{I}$, that is, if A is an orthogonal matrix.

Let us return to our proof that $\tau(L\,M) = \tau(L) \cdot \tau(M)$. From the inequalities proven above, it is obvious that

$$|\boldsymbol{L} \cdot \boldsymbol{M}| \leq |\boldsymbol{L}|\;|\boldsymbol{M}| \leq \ell\, m. \tag{11.49}$$

We calculate $N^0{}_0$:

$$N^0{}_0 = L^0{}_\alpha\, M^\alpha{}_0 = L^0{}_0\, M^0{}_0 + L^0{}_i\, M^i{}_0 = \ell\, m + \boldsymbol{L} \cdot \boldsymbol{M}. \tag{11.50}$$

Since $|\boldsymbol{L} \cdot \boldsymbol{M}| \leq \ell\, m$, adding $\boldsymbol{L} \cdot \boldsymbol{M}$ to $\ell\, m$ cannot change the sign of $\ell\, m$. Therefore the sign of $N^0{}_0$ is the same as the sign of $\ell\, m$. This means that

$$\operatorname{sgn}\left(N^0{}_0\right) = \operatorname{sgn}(\ell\, m) = \operatorname{sgn}\left(L^0{}_0\, M^0{}_0\right), \tag{11.51}$$

or in our notation

$$\tau(L\,M) = \tau(L)\,\tau(M). \tag{11.52}$$

It is now obvious that those matrices in $\mathrm{O}(1,3)$ with $\tau(L) = 1$ form a subgroup of $\mathrm{O}(1,3)$ (note that the unit matrix, $\mathbb{I}$, has this property). This subset is called the orthochronous subgroup and is denoted by $\mathrm{O}^\uparrow(1,3)$.

Remark. The word *orthochronous* means that the transformation preserves the direction of time. To see this, consider the case $\tau(L) = -1$. Then the time component of the transformation $X' = L\,X$ is something like $t' = -\alpha^2 t + f(\boldsymbol{r})$ for some real α. Now consider a sequence of events $E_1, E_2, E_3, \ldots$, all happening at the same $\boldsymbol{r}$, such that

$$t_1 \leq t_2 \leq t_3 \leq \ldots. \tag{11.53}$$

For the same sequence of events, we have

$$t_1' = -\alpha^2\, t_1 \geq t_2' = -\alpha^2\, t_2 \geq t_3' = -\alpha^2\, t_3 \geq \ldots. \tag{11.54}$$

This means that if $\tau(L) = -1$, then the transformation $X \to X' = LX$, from frame K to frame K', is such that the order of events in K' has the reverse order compared to the sequence of events in K.

The special orthochronous subgroup $\mathbf{SO^{\uparrow}(1,3)}$**.** The intersection of $SO(1,3)$ and $O^{\uparrow}(1,3)$ is also a subgroup. It is called the *special orthochronous orthogonal* group and is denoted by $SO^{\uparrow}(1,3)$.

Time reversal and parity operators. Consider the following two matrices:

$$T = \begin{bmatrix} -1 & 0 \\ 0 & \mathbb{I} \end{bmatrix} = \begin{bmatrix} -1 & 0 & 0 & 0 \\ 0 & 1 & 0 & 0 \\ 0 & 0 & 1 & 0 \\ 0 & 0 & 0 & 1 \end{bmatrix}, \tag{11.55}$$

$$P = \begin{bmatrix} 1 & 0 \\ 0 & -\mathbb{I} \end{bmatrix} = \begin{bmatrix} 1 & 0 & 0 & 0 \\ 0 & -1 & 0 & 0 \\ 0 & 0 & -1 & 0 \\ 0 & 0 & 0 & -1 \end{bmatrix}. \tag{11.56}$$

It is easy to check that these two matrices are both in $O(1,3)$. It is also clear that

$$\det T = \det P = -1, \qquad \tau(T) = -1, \qquad \tau(P) = 1. \tag{11.57}$$

Besides, T and P commute, that is, $TP = PT$. Note also that $TP = -\mathbb{I}$.

We can now prove the following theorem. Any matrix in $O(1,3)$ is of one of the following four forms, in which L is a matrix in $SO^{\uparrow}(1,3)$:

$$L, \quad LT, \quad LP, \quad -L = LTP. \tag{11.58}$$

The proof is simple. If $\det M = \tau(M) = 1$, then there is nothing to prove. If $\det M = -1$ and $\tau(M) = 1$, then define $L = MP$, for which $\det L = +1$ and $\tau(L) = \tau(M) \cdot \tau(P) = 1$. Obviously now $M = LP$. If $\det M = 1$ and $\tau(M) = -1$, then define $L = -M$, for which $\det L = (-1)^4 \det M = 1$ and $\tau(L) = \tau(-\mathbb{I}) \cdot \tau(M) = +1$. Obviously now $M = -L$. If $\det M = \tau(M) = -1$, then define $L = MT$, for which $\det L = \det M \cdot \det T = +1$ and $\tau(L) = \tau(M) \cdot \tau(T) = +1$. Now $M = LT$. All this is contained in the following table.

$M =$	L	$-L$	LP	LT
$\det(M)$	+1	+1	−1	−1
$\tau(M)$	+1	−1	+1	−1

11.3 The $SO^{\uparrow}(1,3)$ matrices

As in the case of rotation matrices, we try to find matrices Ω, such that $L = e^{\Omega}$ is a Lorentz transformation. Expanding e^{Ω} up to first-order terms, we get $L = \mathbb{I} + \Omega$, so $L^T = \mathbb{I} + \Omega^T$. Now the condition $L^T\,H\,L = H$ becomes (up to first-order terms) $H + \Omega^T\,H + H\,\Omega = H$, or simply $\Omega^T\,H + H\,\Omega = 0$. Since H is a symmetric matrix, we have $\Omega^T\,H = \Omega^T\,H^T = (H\,\Omega)^T$. Therefore the condition on Ω is $(H\,\Omega)^T + H\,\Omega = 0$, which is saying that $H\,\Omega$ must be a 4×4 antisymmetric matrix. So we have

$$H\,\Omega = \varsigma \begin{bmatrix} 0 & -\alpha_1 & -\alpha_2 & -\alpha_3 \\ \alpha_1 & 0 & \theta_3 & -\theta_2 \\ \alpha_2 & -\theta_3 & 0 & \theta_1 \\ \alpha_3 & \theta_2 & -\theta_1 & 0 \end{bmatrix}. \tag{11.59}$$

We want to find e^{Ω}. Before that, let us prove that if $\Omega^T\,H = -H\,\Omega$, then e^{Ω} is in $O(1,3)$. To see this, first note that $\Omega^T = -H\,\Omega\,H^{-1}$. Now recall that $\exp(H\,A\,H^{-1}) = H\,e^A\,H^{-1}$, the proof of which is simple.[3] Therefore we have

$$\left(e^{\Omega}\right)^T\,H\,e^{\Omega} = e^{\Omega^T}\,H\,e^{\Omega} = e^{-H\,\Omega\,H^{-1}}\,H\,e^{\Omega} = H\,e^{-\Omega}\,H^{-1}\,H\,e^{\Omega} = H. \tag{11.60}$$

Using the identity $\det e^A = e^{\operatorname{tr} A}$, the proof of which is also simple,[4] we now prove that $\det e^{\Omega} = 1$. To do that, just note that $\Omega = -H^{-1}\,\Omega^T\,H$, from which it follows that $\operatorname{tr}\Omega = -\operatorname{tr}\Omega^T = -\operatorname{tr}\Omega$ or $\operatorname{tr}\Omega = 0$. Therefore e^{Ω} is in $SO(1,3)$. It can also be shown that $\tau(e^{\Omega}) = 1$, and therefore e^{Ω} is in $SO^{\uparrow}(1,3)$. Let us postpone the proof (see Eq. (11.73) and the note after that).

If we multiply $H\,\Omega$ on the left by H^{-1}, we get

$$\Omega = \begin{bmatrix} 0 & c^{-2}\alpha_1 & c^{-2}\alpha_2 & c^{-2}\alpha_3 \\ \alpha_1 & 0 & -\theta_3 & \theta_2 \\ \alpha_2 & \theta_3 & 0 & -\theta_1 \\ \alpha_3 & -\theta_2 & \theta_1 & 0 \end{bmatrix}. \tag{11.61}$$

Now we define

$$\mathcal{T}_1 = \begin{bmatrix} 0 & c^{-2} & 0 & 0 \\ 1 & 0 & 0 & 0 \\ 0 & 0 & 0 & 0 \\ 0 & 0 & 0 & 0 \end{bmatrix}, \qquad \mathcal{S}_1 = \begin{bmatrix} 0 & 0 & 0 & 0 \\ 0 & 0 & 0 & 0 \\ 0 & 0 & 0 & -1 \\ 0 & 0 & 1 & 0 \end{bmatrix}, \tag{11.62}$$

---

[3] Just recall that $(H\,A\,H^{-1})^n = H\,A^n\,H^{-1}$, which follows by just writing $(H\,A\,H^{-1})^n$. For example, $(H\,A\,H^{-1})^2 = (H\,A\,H^{-1})(H\,A\,H^{-1}) = HA^2\,H^{-1}$. Now write $\exp(H\,A\,H^{-1})$ as a series: $\exp(H\,A\,H^{-1}) = \mathbb{I} + H\,A\,H^{-1} + \frac{1}{2!}H\,A^2\,H^{-1} + \cdots = H\,e^A\,H^{-1}$.

[4] The proof of this identity is simple if the matrix A is either symmetric or antisymmetric, because then there exists a complex matrix U such that $U\,A\,U^{-1}$ is a diagonal matrix, the diagonal elements being the eigenvalues of A denoted by λ_i. Now $\exp(U\,A\,U^{-1}) = \operatorname{diag}(e^{\lambda_1}, e^{\lambda_2}, e^{\lambda_3})$. The determinant of this diagonal matrix is $\exp(\sum_i \lambda_i) = \exp(\operatorname{tr}(U\,A\,U^{-1})) = \exp(\operatorname{tr} A)$. Besides, we can use the identity $\exp(U\,A\,U^{-1}) = \exp(U)\,\exp(A)\,\exp(U^{-1})$ to show that $\det\exp(U\,A\,U^{-1}) = \det\exp(A)$. If A is real symmetric, it is Hermitian. If A is real antisymmetric, $i\,A$ is Hermitian. In either case A can be diagonalized by a complex matrix U.

$$\mathcal{T}_2 = \begin{bmatrix} 0 & 0 & c^{-2} & 0 \\ 0 & 0 & 0 & 0 \\ 1 & 0 & 0 & 0 \\ 0 & 0 & 0 & 0 \end{bmatrix}, \qquad \mathcal{S}_2 = \begin{bmatrix} 0 & 0 & 0 & 0 \\ 0 & 0 & 0 & 1 \\ 0 & 0 & 0 & 0 \\ 0 & -1 & 0 & 0 \end{bmatrix}, \tag{11.63}$$

$$\mathcal{T}_3 = \begin{bmatrix} 0 & 0 & 0 & c^{-2} \\ 0 & 0 & 0 & 0 \\ 0 & 0 & 0 & 0 \\ 1 & 0 & 0 & 0 \end{bmatrix}, \qquad \mathcal{S}_3 = \begin{bmatrix} 0 & 0 & 0 & 0 \\ 0 & 0 & -1 & 0 \\ 0 & 1 & 0 & 0 \\ 0 & 0 & 0 & 0 \end{bmatrix}, \tag{11.64}$$

$$\boldsymbol{\mathcal{T}} = (\mathcal{T}_1, \mathcal{T}_2, \mathcal{T}_3), \qquad \boldsymbol{\mathcal{S}} = (\mathcal{S}_1, \mathcal{S}_2, \mathcal{S}_3), \tag{11.65}$$

$$\boldsymbol{\alpha} = (\alpha_1, \alpha_2, \alpha_3), \qquad \boldsymbol{\theta} = (\theta_1, \theta_2, \theta_3). \tag{11.66}$$

Using these definitions, we usually write $\Omega = \boldsymbol{\alpha} \cdot \boldsymbol{\mathcal{T}} + \boldsymbol{\theta} \cdot \boldsymbol{\mathcal{S}}$. Note that $\mathcal{S}_i$ is antisymmetric. If $c = 1$, then $\mathcal{T}_i$s is symmetric, but we prefer to work with these asymmetric forms.

11.3.1 Pure boosts

Let α denote the magnitude of $\boldsymbol{\alpha}$ and $\hat{\boldsymbol{n}}$ the unit vector in the direction of $\boldsymbol{\alpha}$, so that $\boldsymbol{\alpha} = \alpha(n_1, n_2, n_3)$; and, for simplicity, let us consider the case $\boldsymbol{\theta} = 0$. Now $\Omega = \alpha A$, where

$$A = \begin{bmatrix} 0 & c^{-2}n_1 & c^{-2}n_2 & c^{-2}n_3 \\ n_1 & 0 & 0 & 0 \\ n_2 & 0 & 0 & 0 \\ n_3 & 0 & 0 & 0 \end{bmatrix}. \tag{11.67}$$

Computing A^2 is straightforward:

$$A^2 = \frac{1}{c^2} B, \quad \text{where} \quad B = \begin{bmatrix} 1 & 0 & 0 & 0 \\ 0 & n_1^2 & n_1 n_2 & n_1 n_3 \\ 0 & n_2 n_1 & n_2^2 & n_2 n_3 \\ 0 & n_3 n_1 & n_3 n_2 & n_3^2 \end{bmatrix}. \tag{11.68}$$

Now, one more matrix multiplication, which is quite easy, leads to

$$A^3 = \frac{1}{c^2} \begin{bmatrix} 0 & c^{-2}n_1 & c^{-2}n_2 & c^{-2}n_3 \\ n_1 & 0 & 0 & 0 \\ n_2 & 0 & 0 & 0 \\ n_3 & 0 & 0 & 0 \end{bmatrix} = \frac{1}{c^2} A. \tag{11.69}$$

Summarizing, we have

$$\Omega = \alpha A, \quad \Omega^2 = \frac{\alpha^2}{c^2} B, \quad \Omega^3 = \frac{\alpha^3}{c^2} A = \frac{\alpha^3}{c^3} (c A), \quad \Omega^4 = \frac{\alpha^4}{c^4} A, \quad \ldots. \tag{11.70}$$

Now we can calculate e^{Ω}:

$$\begin{aligned}
e^{\Omega} &= e^{\alpha A} = \sum_{n=0}^{\infty} \frac{\alpha^n}{n!} A^n \\
&= \mathbb{I} + \alpha A + \frac{1}{2!}\frac{\alpha^2}{c^2} B + \frac{1}{3!}\frac{\alpha^3}{c^2} A + \frac{1}{4!}\frac{\alpha^4}{c^4} B + \dots \\
&= (\mathbb{I} - B) + B\left(1 + \frac{1}{2!}\frac{\alpha^2}{c^2} + \frac{1}{4!}\frac{\alpha^4}{c^4} + \dots\right) + cA\left(\frac{\alpha}{c} + \frac{1}{3!}\frac{\alpha^3}{c^3} + \dots\right) \\
&= \mathbb{I} - B + B\cosh\frac{\alpha}{c} + c\,A\sinh\frac{\alpha}{c} \\
&= \mathbb{I} + \left(\cosh\frac{\alpha}{c} - 1\right) B + \left(\sinh\frac{\alpha}{c}\right) c\,A.
\end{aligned} \tag{11.71}$$

Instead of the parameters $\boldsymbol{\alpha} = \alpha\,\hat{\boldsymbol{n}}$, let us define $v = c\,\tanh(\alpha/c)$ and $\boldsymbol{v} = v\,\hat{\boldsymbol{n}}$. Writing $\sinh(\alpha/c)$ and $\cosh(\alpha/c)$ in terms of v, we get

$$\cosh\frac{\alpha}{c} = \gamma(v) = \frac{1}{\sqrt{1-\frac{v^2}{c^2}}}, \qquad \sinh\frac{\alpha}{c} = \frac{v}{c}\,\gamma(v). \tag{11.72}$$

Inserting these in the expression for $L = e^{\Omega}$, we get

$$\begin{aligned}
L = e^{\Omega} &= \mathbb{I} + (\gamma(v) - 1)\,B + v\,\gamma(v) A \\
&= \mathbb{I} + (\gamma(v) - 1)\begin{bmatrix} 1 & 0 & 0 & 0 \\ 0 & n_1^2 & n_1 n_2 & n_1 n_3 \\ 0 & n_2 n_1 & n_2^2 & n_2 n_3 \\ 0 & n_3 n_1 & n_3 n_2 & n_3^2 \end{bmatrix} + \gamma(v)\,v \begin{bmatrix} 0 & \frac{n_1}{c^2} & \frac{n_2}{c^2} & \frac{n_3}{c^2} \\ n_1 & 0 & 0 & 0 \\ n_2 & 0 & 0 & 0 \\ n_3 & 0 & 0 & 0 \end{bmatrix} \\
&= \begin{bmatrix} \gamma & c^{-2}\gamma\,v_1 & c^{-2}\gamma\,v_2 & c^{-2}\gamma\,v_3 \\ \gamma\,v_1 & 1 + (\gamma-1)n_1^2 & (\gamma-1)n_1 n_2 & (\gamma-1)n_1 n_3 \\ \gamma\,v_2 & (\gamma-1)n_2 n_1 & 1+(\gamma-1)n_2^2 & (\gamma-1)n_2 n_3 \\ \gamma\,v_3 & (\gamma-1)n_3 n_1 & (\gamma-1)n_3 n_2 & 1+(\gamma-1)n_3^2 \end{bmatrix}.
\end{aligned} \tag{11.73}$$

Note that for $\hat{\boldsymbol{n}} = (1,0,0)$ this is the same L that was introduced and studied in Chapter 3. Also, note that the sign of $L^0{}_0$ is positive, so this matrix is in fact in SO$^{\uparrow}(1,3)$.

Matrices of the above form ($L = e^{\boldsymbol{\alpha}\cdot\boldsymbol{\mathcal{T}}}$) are called *pure boosts*. To simplify the discussion, we now introduce a notation:

$$\mathcal{B}(\boldsymbol{v}) = \exp(\boldsymbol{\alpha}\cdot\boldsymbol{\mathcal{T}}), \quad \text{where} \quad \boldsymbol{v} := \hat{\boldsymbol{n}}\,c\,\tanh\frac{\alpha}{c}. \tag{11.74}$$

For later use, let us write formulas for $[\mathcal{B}(\boldsymbol{v})]^{\mu}{}_{\nu}$:

$$[\mathcal{B}(\boldsymbol{v})]^0{}_0 = \gamma(v), \tag{11.75}$$

$$[\mathcal{B}(\boldsymbol{v})]^i{}_0 = c^{-2}\gamma(v)\, v_i, \tag{11.76}$$
$$[\mathcal{B}(\boldsymbol{v})]^0{}_j = \gamma(v)\, v_j, \tag{11.77}$$
$$[\mathcal{B}(\boldsymbol{v})]^i{}_j = \delta^i_j + (\gamma(v) - 1) n_i\, n_j, \quad \text{where} \quad n_i = \frac{v_i}{v}. \tag{11.78}$$

11.3.2 Rotations

We now study matrices of the form $e^{\boldsymbol{\theta}\cdot\boldsymbol{S}}$, which we call *(pure) rotations*. To find the form of them we set $\boldsymbol{\alpha} = 0$, $\theta = \theta\,\hat{\boldsymbol{n}}$, $\hat{\boldsymbol{n}} = (n_1, n_2, n_3)$ and calculate the powers of Ω. We have

$$\Omega = \theta\, A, \quad A = \begin{bmatrix} 0 & 0 & 0 & 0 \\ 0 & 0 & -n_3 & n_2 \\ 0 & n_3 & 0 & -n_1 \\ 0 & -n_2 & n_1 & 0 \end{bmatrix}. \tag{11.79}$$

The computation is an exact repetition of the one performed in §10.4.2 (page 213). The result reads

$$e^{\boldsymbol{\theta}\cdot\boldsymbol{S}} = L^\mu{}_\nu = \begin{bmatrix} 1 & 0 & 0 & 0 \\ 0 & R_{11} & R_{12} & R_{13} \\ 0 & R_{21} & R_{22} & R_{23} \\ 0 & R_{31} & R_{32} & R_{33} \end{bmatrix}, \tag{11.80}$$

where the 3×3 matrix R_{ij} is the same matrix found in §10.4.2:

$$R_{ij} = \delta_{ij} + (1 - \cos\theta)\left(n_i\, n_j - \delta_{ij}\right) - \sin\theta\, \epsilon_{ijk}\, n_k. \tag{11.81}$$

Note that $e^{\theta\,\hat{\boldsymbol{n}}\cdot\boldsymbol{S}}$ also belongs to $\mathrm{SO}^\uparrow(1,3)$, because $L^0{}_0 = 1$ is positive. We introduce the notation

$$\mathcal{R}(\boldsymbol{\theta}) := \exp(\boldsymbol{\theta}\cdot\boldsymbol{S}), \tag{11.82}$$

which means

$$[\mathcal{R}(\boldsymbol{\theta})]^0{}_0 = 1, \tag{11.83}$$
$$[\mathcal{R}(\boldsymbol{\theta})]^i{}_0 = [\mathcal{R}(\boldsymbol{\theta})]^0{}_i = 0, \tag{11.84}$$
$$[\mathcal{R}(\boldsymbol{\theta})]^i{}_j = R_{ij} = \delta_{ij} + (1 - \cos\theta)(n_i\, n_j - \delta_{ij}) - \sin\theta\, \epsilon_{ijk}\, n_k. \tag{11.85}$$

11.3.3 General transformations

We now prove the following theorem.

Theorem. *Any matrix L in $\mathrm{SO}^\uparrow(1,3)$ could be written as $L = \mathcal{B}(\boldsymbol{v})\ \mathcal{R}(\boldsymbol{\theta})$, and this so-called decomposition is unique; that is, $\boldsymbol{v}$ and $\boldsymbol{\theta}$ are unique.*

To prove this theorem, consider the transformation $x^\mu = L^\mu{}_\nu x'^\nu$. The origin of coordinate system K' is defined by $x'^1 = x'^2 = x'^3 = 0$, for any value $t' = x'^0$. Writing $x^\mu = L^\mu{}_\nu x'^\nu$ we get

$$t = x^0 = L^0{}_\nu x'^\nu = L^0{}_0 t', \tag{11.86}$$

$$x^1 = L^i{}_\nu x'^\nu = L^i{}_0 t', \tag{11.87}$$

from which it follows that $x^i/t = L^i{}_0/L^0{}_0$, which means that the origin of the coordinate system of K' is moving with velocity $\boldsymbol{v} = \frac{1}{L^0{}_0}\left(L^1{}_0, L^2{}_0, L^3{}_0\right)$. Let us denote the components of this vector by (v_1, v_2, v_3) and calculate

$$\frac{1}{\gamma^2(v)} = 1 - \frac{v^2}{c^2} = 1 - \frac{\sum(L^i{}_0)^2}{c^2\,(L^0{}_0)^2} = \frac{c^2\,(L^0{}_0)^2 - \sum(L^i{}_0)^2}{c^2(L^0{}_0)^2} \tag{11.88}$$

$$= \frac{c^2}{c^2(L^0{}_0)^2}. \tag{11.89}$$

In the last step above we have used equation $M^2 = c^2(m^2 - 1)$ on page 256. We also have

$$[\mathcal{B}(-\boldsymbol{v})L]^0{}_0 = [\mathcal{B}(-\boldsymbol{v})]^0{}_0\, L^0{}_0 + [\mathcal{B}(-\boldsymbol{v})]^0{}_i\, L^i{}_0 \tag{11.90}$$

$$= \gamma(v)\, L^0{}_0 - c^{-2}\,\gamma(v)\, v_i\, L^i{}_0 \tag{11.91}$$

$$= \gamma(v)\, L^0{}_0 \left(1 - c^{-2} \sum v_i\, v_i\right) = 1. \tag{11.92}$$

Earlier (page 257) we have seen that if $L^0{}_0 = 1$, then L has the form $\left(\begin{smallmatrix}1 & 0\\ 0 & A\end{smallmatrix}\right)$, where A is a 3×3 orthogonal matrix. Since $\det L = 1$, we have $\det A = 1$ and A is a rotation. Therefore $\mathcal{B}(-\boldsymbol{v})\, L = \mathcal{R}$. Solving for L we get $L = \mathcal{B}(\boldsymbol{v})\, \mathcal{R}$. This decomposition is by construction unique, because starting from L, the velocity vector $\boldsymbol{v}$ is uniquely determined, then $\mathcal{B}(\boldsymbol{v})$ is uniquely determined, and then $\mathcal{R} = \mathcal{B}(-\boldsymbol{v})\, L$ is uniquely determined.

11.3.4 Lie algebra of the Lorentz group

Lorentz matrices are real. They can be expressed as e^Ω for real Ω. We have studied the form of Ω, and we found that $H\,\Omega$ is real antisymmetric. Because of the conventions generally used in quantum mechanics, it is customary to define imaginary generators instead of real ones. This is done by multiplying the real generators by imaginary i. So we change the generators as follows:

$$\begin{cases} S_n = i\,\mathcal{S}_n, \\ T_n = i\,\mathcal{T}_n, \end{cases} \quad \text{which means} \quad \begin{cases} \mathcal{S}_n = -i\,S_n, \\ \mathcal{T}_n = -i\,T_n. \end{cases} \tag{11.93}$$

We therefore have

$$\boldsymbol{\alpha} \cdot \boldsymbol{\mathcal{T}} + \boldsymbol{\theta} \cdot \boldsymbol{\mathcal{S}} = -i\,(\boldsymbol{\alpha} \cdot \boldsymbol{T} + \boldsymbol{\theta} \cdot \boldsymbol{S}), \tag{11.94}$$

$$L = e^{\alpha\cdot\mathcal{T}+\theta\cdot\mathcal{S}} = e^{-i\,(\alpha\cdot T+\theta\cdot S)}. \tag{11.95}$$

Using the form of matrices $\mathcal{S}$ and $\mathcal{T}$ given on page 259, it is straightforward to check that the following commutations relations hold:

$$\begin{cases} [\,\mathcal{S}_n,\,\mathcal{S}_m\,] = \epsilon_{nmk}\,\mathcal{S}_k, \\ [\,\mathcal{S}_n,\,\mathcal{T}_m\,] = \epsilon_{nmk}\,\mathcal{T}_k, \\ [\,\mathcal{T}_n,\,\mathcal{T}_m\,] = -c^{-2}\,\epsilon_{nmk}\,\mathcal{S}_k, \end{cases} \qquad \begin{cases} [\,S_n,\,S_m\,] = i\,\epsilon_{nmk}\,S_k, \\ [\,S_n,\,T_m\,] = i\,\epsilon_{nmk}\,T_k, \\ [\,T_n,\,T_m\,] = -i\,c^{-2}\,\epsilon_{nmk}\,S_k. \end{cases} \tag{11.96}$$

Thomas rotation. In §3.11 we saw that Lorentz boosts do not commute. We now see the same phenomenon in terms of the commutation relations of the generators, $\mathcal{T}_1$, $\mathcal{T}_2$, and $\mathcal{T}_3$. For example, consider the following sequence of boosts: first, boost in x_1 with velocity v_1, then boost in x_2 with velocity v_2, then boost in x with velocity $-v_1$, and finally boost in x_2 with velocity $-v_2$. If $|v_1|$, $|v_2| \ll c$, then the combined transformation is given by

$$\begin{aligned} L &= (\mathbb{I} - v_2\,\mathcal{T}_2)\,(\mathbb{I} - v_1\,\mathcal{T}_1)\,(\mathbb{I} + v_2\,\mathcal{T}_2)\,(\mathbb{I} + v_1\,\mathcal{T}_1) \\ &= \mathbb{I} - \frac{v_1\,v_2}{c^2}\,[\,\mathcal{T}_1,\,\mathcal{T}_2\,] \\ &= \mathbb{I} + \frac{v_1\,v_2}{c^2}\,\mathcal{S}_3. \end{aligned}$$

This is a small rotation around the x_3-axis. We also see that in the Galilean limit $c \to \infty$ this Thomas rotation vanishes.

11.3.5 Representations of $\mathfrak{so}^{\uparrow}(1,3)$

Let us define $A = \frac{1}{2}\,(S + i\,c\,T)$ and $B = \frac{1}{2}\,(S - i\,c\,T)$, that is,

$$A_n = \frac{1}{2}\,(S_n + i\,c\,T_n), \tag{11.97}$$

$$B_n = \frac{1}{2}\,(S_n - i\,c\,T_n). \tag{11.98}$$

By straightforward calculation it is seen that

$$[\,A_n,\,A_m\,] = i\,\epsilon_{nmj}\,A_j, \tag{11.99}$$

$$[\,B_n,\,B_m\,] = i\,\epsilon_{nmj}\,B_j, \tag{11.100}$$

$$[\,A_n,\,B_m\,] = 0. \tag{11.101}$$

Therefore in this basis the algebra is decomposed into two independent $\mathfrak{su}(2)$.

Let us now write the inverse relations:

$$S_n = B_n + A_n, \tag{11.102}$$

$$T_n = i\,c^{-1}\,(B_n - A_n)\,. \tag{11.103}$$

How should we interpret these equations, knowing that As and Bs commute? The answer is simple: We invoke the mathematics of tensor product of vector spaces and write

$$B_n \to B_n \otimes \mathbb{I}, \qquad A_n \to \mathbb{I} \otimes A_n, \tag{11.104}$$

which means

$$S_n = B_n \otimes \mathbb{I} + \mathbb{I} \otimes A_n, \tag{11.105}$$

$$T_n = i\,c^{-1}\,(B_n \otimes \mathbb{I} - \mathbb{I} \otimes A_n)\,. \tag{11.106}$$

We know that an irreducible representation of $\mathfrak{su}(2)$ is specified by the dimension of the representation. We use the symbol $\mathbf{n}$ for the irreducible representation of $\mathfrak{su}(2)$ and $\bar{\mathbf{n}}$ for the complex conjugate representation.

Picking two representations $\mathbf{n}_1$ and $\mathbf{n}_2$ provides us with a representation of the Lorentz group.

Example: the representation $(\bar{\mathbf{2}}, \mathbf{2})$. For this representation we have

$$B_n = \frac{\sigma_n}{2} \otimes \mathbb{I}, \qquad A_n = -\frac{1}{2}\mathbb{I} \otimes \sigma_n^T\,. \tag{11.107}$$

Writing tensor products explicitly we get

$$B_1 = \frac{1}{2}\begin{bmatrix} 0&0&1&0\\ 0&0&0&1\\ 1&0&0&0\\ 0&1&0&0 \end{bmatrix}, \quad B_2 = \frac{1}{2}\begin{bmatrix} 0&0&-i&0\\ 0&0&0&-i\\ i&0&0&0\\ 0&i&0&0 \end{bmatrix}, \quad B_3 = \frac{1}{2}\begin{bmatrix} 1&0&0&0\\ 0&1&0&0\\ 0&0&-1&0\\ 0&0&0&-1 \end{bmatrix}, \tag{11.108}$$

$$A_1 = \frac{1}{2}\begin{bmatrix} 0&-1&0&0\\ -1&0&0&0\\ 0&0&0&-1\\ 0&0&-1&0 \end{bmatrix}, \quad A_2 = \frac{1}{2}\begin{bmatrix} 0&-i&0&0\\ i&0&0&0\\ 0&0&0&-i\\ 0&0&i&0 \end{bmatrix}, \quad A_3 = \frac{1}{2}\begin{bmatrix} -1&0&0&0\\ 0&1&0&0\\ 0&0&-1&0\\ 0&0&0&1 \end{bmatrix}. \tag{11.109}$$

From these forms it is seen that

$$S_1 = \frac{1}{2}\begin{bmatrix} 0&-1&1&0\\ -1&0&0&1\\ 1&0&0&-1\\ 0&1&-1&0 \end{bmatrix}, \quad S_2 = \frac{1}{2}\begin{bmatrix} 0&-i&-i&0\\ i&0&0&-i\\ i&0&0&-i\\ 0&i&i&0 \end{bmatrix}, \quad S_3 = \begin{bmatrix} 0&0&0&0\\ 0&1&0&0\\ 0&0&-&0\\ 0&0&0&0 \end{bmatrix}, \tag{11.110}$$

$$T_1 = \frac{i}{2c}\begin{bmatrix} 0&1&1&0\\ 1&0&0&1\\ 1&0&0&1\\ 0&1&1&0 \end{bmatrix}, \quad T_2 = \frac{i}{2c}\begin{bmatrix} 0&1&-1&0\\ -1&0&0&-1\\ 1&0&0&1\\ 0&1&-1&0 \end{bmatrix}, \quad T_3 = \frac{i}{c}\begin{bmatrix} 1&0&0&0\\ 0&0&0&0\\ 0&0&0&0\\ 0&0&0&-1 \end{bmatrix}. \tag{11.111}$$

Note that S_ns are exactly those we studied in Worked Example 45 (page 244). As we saw there, $\{S_1, S_2, S_3\}$ is the decomposable representation $\mathbf{3} \oplus \mathbf{1}$. We leave it to the reader to show that this representation of $\mathfrak{so}^{\uparrow}(1,3)$ is equivalent to the defining representation of (11.62).

Remark. Looking at the form of $\mathcal{T}_n = i\, T_n$ (Eq. (11.62) on page 259), we see that $\lim_{c\to\infty} T_n$ is well defined, but $\lim_{c\to\infty} (c\, T_n)$ does not exist, because one entry of $c\, T_n$ goes to infinity. Therefore these elements, $\boldsymbol{A}$ and $\boldsymbol{B}$, are not suitable for going to the $c \to \infty$ limit. To study this limit, the generators $\boldsymbol{S}$ and $\boldsymbol{T}$ are suitable.

11.4 Topology of the Lorentz group

Any point L in $\mathrm{SO}^{\uparrow}(1,3)$ has the following unique presentation:

$$L = \exp(\boldsymbol{\alpha} \cdot \boldsymbol{\mathcal{S}} + \boldsymbol{\beta} \cdot \boldsymbol{\mathcal{T}}), \tag{11.112}$$

where $\boldsymbol{\alpha} = (\alpha_1, \alpha_2, \alpha_3)$ specifies a rotation and $\boldsymbol{\beta} = (\beta_1, \beta_2, \beta_3)$ specifies a boost. We already know that $\boldsymbol{\alpha}$ is a point in $\mathbb{RP}^3$. The vector $\boldsymbol{\beta}$ is in $\mathbb{R}^3$. Therefore topologically we can say that

$$\mathrm{SO}^{\uparrow}(1,3) \sim \mathbb{RP}^3 \times \mathbb{R}^3. \tag{11.113}$$

We already know that $\mathrm{O}(1,3)$ is four copies of $\mathrm{SO}^{\uparrow}(1,3)$, which could be written as

$$\mathrm{O}(1,3) \sim \mathbb{RP}^3 \times \mathbb{R}^3 \times \{0,1,2,3\}. \tag{11.114}$$

11.5 Dependence on the velocity of light

Remember that the Minkowski metric depends on the velocity of light, c:

$$H_c = \left[\eta^c_{\mu\nu}\right] = \begin{pmatrix} -\varsigma c^2 & 0 & 0 & 0 \\ 0 & \varsigma & 0 & 0 \\ 0 & 0 & \varsigma & 0 \\ 0 & 0 & 0 & \varsigma \end{pmatrix}, \qquad c > 0. \tag{11.115}$$

By definition, the group $\mathrm{O}(1,3)$ is the group of 4×4 matrices L satisfying $L^T H_c L = H_c$. So let us temporarily denote it by $\mathrm{O}_c(1,3)$.

Theorem. *The groups* $\mathrm{O}_c(1,3)$ *and* $\mathrm{O}_{c'}(1,3)$ *are isomorphic.*

Proof. Define the symmetric matrix

$$S = \begin{bmatrix} \dfrac{c'}{c} & 0 & 0 & 0 \\ 0 & 1 & 0 & 0 \\ 0 & 0 & 1 & 0 \\ 0 & 0 & 0 & 1 \end{bmatrix}. \tag{11.116}$$

By simple matrix multiplications we see that $S\, H_c\, S = H_{c'}$. Now let L be a matrix in $\mathrm{O}_c(1,3)$, and define

$$M = S^{-1}\, L\, S. \tag{11.117}$$

We see that

$$\begin{aligned} M^T\, H_{c'}\, M &= \left(S^{-1}\, L\, S\right)^T\, H_{c'} \left(S^{-1}\, L\, S\right) = S\, L^T\, \left(S^{-1}\, H_{c'}\, S^{-1}\right)\, L\, S \\ &= S\, L^T\, H_c\, L\, S = S\, H_c\, S = H_{c'}, \end{aligned} \tag{11.118}$$

showing that M is a matrix in $\mathrm{O}_{c'}(1,3)$. It is obvious that the map $L \mapsto M$ is bijective (invertible) and that $\mathbb{I} \mapsto \mathbb{I}$. It is also easy to see that

$$\left(S^{-1}\, L_1\, S\right)\left(S^{-1}\, L_2\, S\right) = S^{-1}\, (L_1\, L_2)\, S. \tag{11.119}$$

Therefore the mapping $L \to S^{-1}\, L\, S$ is a homomorphism. It is also invertible, so it is an isomorphism. □

The Galilean limit. The subgroup $\mathrm{SO}_c^{\uparrow}(1,3)$ is a Lie group, with real generators given in (11.62), which we write here again:

$$\mathcal{T}_1 = \begin{bmatrix} 0 & c^{-2} & 0 & 0 \\ 1 & 0 & 0 & 0 \\ 0 & 0 & 0 & 0 \\ 0 & 0 & 0 & 0 \end{bmatrix}, \qquad \mathcal{S}_1 = \begin{bmatrix} 0 & 0 & 0 & 0 \\ 0 & 0 & 0 & 0 \\ 0 & 0 & 0 & -1 \\ 0 & 0 & 1 & 0 \end{bmatrix},$$

$$\mathcal{T}_2 = \begin{bmatrix} 0 & 0 & c^{-2} & 0 \\ 0 & 0 & 0 & 0 \\ 1 & 0 & 0 & 0 \\ 0 & 0 & 0 & 0 \end{bmatrix}, \qquad \mathcal{S}_2 = \begin{bmatrix} 0 & 0 & 0 & 0 \\ 0 & 0 & 0 & 1 \\ 0 & 0 & 0 & 0 \\ 0 & -1 & 0 & 0 \end{bmatrix},$$

$$\mathcal{T}_3 = \begin{bmatrix} 0 & 0 & 0 & c^{-2} \\ 0 & 0 & 0 & 0 \\ 0 & 0 & 0 & 0 \\ 1 & 0 & 0 & 0 \end{bmatrix}, \qquad \mathcal{S}_3 = \begin{bmatrix} 0 & 0 & 0 & 0 \\ 0 & 0 & -1 & 0 \\ 0 & 1 & 0 & 0 \\ 0 & 0 & 0 & 0 \end{bmatrix}.$$

The commutation relations are

$$[\,\mathcal{S}_n,\, \mathcal{S}_m\,] = \epsilon_{nmk}\, \mathcal{S}_k, \tag{11.120}$$

$$[\,\mathcal{T}_n,\, \mathcal{T}_m\,] = -c^{-2} \epsilon_{nmk}\, \mathcal{S}_k, \tag{11.121}$$

$$[\,\mathcal{S}_n,\, \mathcal{T}_m\,] = -\epsilon_{nmk}\, \mathcal{T}_k. \tag{11.122}$$

It is obvious that as $c \to \infty$, the group $\mathrm{SO}_c^{\uparrow}(1,3)$ approaches the Galilean group, with commuting boosts. To see this, note that for $c \to \infty$, the matrices $\mathcal{T}_i$ have the property

$$\mathcal{T}_i \mathcal{T}_j = \mathbb{O}, \quad \forall i, j, \tag{11.123}$$

where $\mathbb{O}$ is the zero matrix. Now, expanding the exp function we have

$$\exp(\boldsymbol{v} \cdot \boldsymbol{\mathcal{T}}) = \mathbb{I} + \boldsymbol{v} \cdot \boldsymbol{\mathcal{T}} + \mathbb{O} = \begin{pmatrix} 1 & 0 & 0 & 0 \\ v_1 & 1 & 0 & 0 \\ v_2 & 0 & 1 & 0 \\ v_3 & 0 & 0 & 1 \end{pmatrix}. \tag{11.124}$$

This means the transformation for $c \to \infty$ is

$$t = t', \tag{11.125}$$

$$x = x' + v_1 t', \tag{11.126}$$

$$y = y' + v_2 t', \tag{11.127}$$

$$z = z' + v_3 t', \tag{11.128}$$

which is the usual Galilean transformation.

Guide to Literature. The procedure mentioned above is technically called *contraction* and was first studied by Inonu and Wigner (1953).

11.6 Transforming the fields

11.6.1 Scalar field

A function $f : \mathbb{M} \to \mathbb{R}$ is called a Lorentz scalar field if under a Lorentz transformation $x \mapsto Lx$, it transforms to f^L with the property $f^L(Lx) = f(x)$. It is easy (as we saw in §10.6.1) to show that this means

$$f^L(x) = f(L^{-1}x). \tag{11.129}$$

We now search for an operator U_L such that $f^L = U_L f$. Repeating what we did in §10.6.1 (page 217) it is easy to show that for $L = \mathcal{B}(\boldsymbol{v}) = e^{\boldsymbol{\alpha} \cdot \boldsymbol{\mathcal{T}}}$, the operator U_L has the form $e^{-\boldsymbol{\alpha} \cdot \boldsymbol{\mathcal{M}}}$, where $\boldsymbol{\mathcal{M}} = (\mathcal{M}_1, \mathcal{M}_2, \mathcal{M}_3)$ and

$$\mathcal{M}_i = c^{-2} x^i \frac{\partial}{\partial t} + t \frac{\partial}{\partial x^i}. \tag{11.130}$$

If $\mathcal{L} = \mathcal{R}(\boldsymbol{\theta}) = e^{\boldsymbol{\theta} \cdot \boldsymbol{\mathcal{S}}}$, then U_L has the form $e^{-\boldsymbol{\theta} \cdot \boldsymbol{\mathcal{L}}}$, where $\boldsymbol{\mathcal{L}} = (\mathcal{L}_1, \mathcal{L}_2, \mathcal{L}_3)$, with

$$\mathcal{L}_i = \epsilon_{ijk} x^j \frac{\partial}{\partial x^k}, \tag{11.131}$$

$$\mathcal{L}_1 = y \frac{\partial}{\partial z} - z \frac{\partial}{\partial y}, \qquad \mathcal{M}_1 = x \frac{1}{c^2} \frac{\partial}{\partial t} + t \frac{\partial}{\partial x}, \tag{11.132}$$

$$\mathcal{L}_2 = z\frac{\partial}{\partial x} - x\frac{\partial}{\partial z}, \qquad \mathcal{M}_2 = y\frac{1}{c^2}\frac{\partial}{\partial t} + t\frac{\partial}{\partial y}, \tag{11.133}$$

$$\mathcal{L}_3 = x\frac{\partial}{\partial y} - y\frac{\partial}{\partial x}, \qquad \mathcal{M}_3 = z\frac{1}{c^2}\frac{\partial}{\partial t} + t\frac{\partial}{\partial z}. \tag{11.134}$$

It is easy to see that

$$[\mathcal{L}_n, \mathcal{L}_m] = -\epsilon_{nmk}\mathcal{L}_k, \tag{11.135}$$

$$[\mathcal{L}_n, \mathcal{M}_m] = -\epsilon_{nmk}\mathcal{M}_k, \tag{11.136}$$

$$[\mathcal{M}_n, \mathcal{M}_m] = c^{-2}\epsilon_{nmk}\mathcal{L}_k. \tag{11.137}$$

Note the signs in these formulas! Now define

$$\begin{cases} L_n = -i\,\mathcal{L}_n, \\ M_n = -i\,\mathcal{M}_n, \end{cases} \quad \text{which means} \quad \begin{cases} \mathcal{L}_n = i\,L_n, \\ \mathcal{M}_n = i\,M_n. \end{cases} \tag{11.138}$$

For these operators we have

$$[L_n, L_m] = i\,\epsilon_{nmk}L_k, \tag{11.139}$$

$$[L_n, M_m] = i\,\epsilon_{nmk}M_k, \tag{11.140}$$

$$[M_n, M_m] = -i\,c^{-2}\epsilon_{nmk}M_k.$$

These commutators are exactly the same as those of the set $\{S_n, T_n\}$.

In terms of these generators the operator U_L defined by $U_L f = f^L$ reads

$$U_L = e^{-i(\boldsymbol{\alpha}\cdot\boldsymbol{M}+\boldsymbol{\theta}\cdot\boldsymbol{L})}. \tag{11.141}$$

11.6.2 A 4-vector field

A 4-vector field is a function assigning to each point of $\mathbb{M}$ a 4-vector. Let $A(x)$ be a 4-vector field. In coordinate system K it is usually specified by $A^\mu(x^0, x^1, x^2, x^3)$. Now let L be a proper orthochronous transformation, that is, a matrix in $\mathrm{SO}^\uparrow(1,3)$. Generalizing what we saw in §10.6.2 (page 218) we can say that under L the 4-vector field $A(x)$ transforms as

$$A^L(x) = L\,A(L^{-1}x). \tag{11.142}$$

Now, $A^L(x) = L\,U_L A(x)$, where

$$L = e^{-i(\boldsymbol{\alpha}\cdot\boldsymbol{T}+\boldsymbol{\theta}\cdot\boldsymbol{S})}, \tag{11.143}$$

$$A(L^{-1}x) = e^{-i(\boldsymbol{\alpha}\cdot\boldsymbol{M}+\boldsymbol{\theta}\cdot\boldsymbol{L})}A(x), \tag{11.144}$$

from which it follows that $A^L(x) = \tilde{U}_L A(x)$, where

$$\tilde{U}_L = e^{-i(\boldsymbol{\alpha}\cdot\boldsymbol{T}+\boldsymbol{\theta}\cdot\boldsymbol{S})}e^{-i(\boldsymbol{\alpha}\cdot\boldsymbol{M}+\boldsymbol{\theta}\cdot\boldsymbol{L})}. \tag{11.145}$$

Note that $\tilde{U}_L$ is the combination (multiplication) of two operators: $\tilde{U}_L = L\,U_L$. But these two operators are in a sense completely independent: L is a 4×4 matrix acting on the indices of A^μ, while U_L is a differential operator acting on the argument of $A^\mu(x^0, x^1, x^2, x^3)$. We can now *define* generators

$$\boldsymbol{J} = \boldsymbol{S} + \boldsymbol{L}, \tag{11.146}$$

$$\boldsymbol{K} = \boldsymbol{T} + \boldsymbol{M}. \tag{11.147}$$

The addition in these two formulas is the addition of a vector with real components, $\boldsymbol{S}$ or $\boldsymbol{L}$, and a vector with differential operator components, $\boldsymbol{L}$ or $\boldsymbol{M}$. Physicists are used to such simplified symbolizations, but mathematicians are not. They prefer to write these formulas as

$$J_n = S_n \otimes \mathbb{I} + \mathbb{I} \otimes L_n, \tag{11.148}$$

$$K_n = T_n \otimes \mathbb{I} + \mathbb{I} \otimes M_n. \tag{11.149}$$

See the notation of tensor product in §10.14.

Remember that $\{J_n, K_n\}$ fulfill the same commutation relations:

$$[\,J_n,\, J_m\,] = \epsilon_{nmk}\, J_k, \tag{11.150}$$

$$[\,J_n,\, K_m\,] = \epsilon_{nmk}\, K_k, \tag{11.151}$$

$$[\,K_m,\, K_m\,] = -\epsilon_{nmk}\, J_k. \tag{11.152}$$

11.7 Relation to SL(2, ℂ)

We denote the set of complex 2×2 matrices by $M(2, \mathbb{C})$ and the set of nonsingular 2×2 matrices—with nonvanishing determinant—by $\mathrm{GL}(2, \mathbb{C})$. Obviously $\mathrm{GL}(2, \mathbb{C})$ together with matrix multiplication is a group. The set $\mathrm{SL}(2, \mathbb{C})$ is the set of 2×2 matrices with determinant equal to 1, which is also a group.

A matrix $e^\Omega := \exp(\Omega)$ is in $\mathrm{SL}(2, \mathbb{C})$ if and only if $\operatorname{tr}\Omega = 0$. Remember that the Pauli matrices are

$$\sigma_1 = \begin{bmatrix} 0 & 1 \\ 1 & 0 \end{bmatrix}, \quad \sigma_2 = \begin{bmatrix} 0 & -i \\ i & 0 \end{bmatrix}, \quad \sigma_3 = \begin{bmatrix} 1 & 0 \\ 0 & -1 \end{bmatrix}. \tag{11.153}$$

The matrix Ω could be written in the form

$$\Omega = \sum_{i=1}^{3} a_i\, \sigma_i, \tag{11.154}$$

where a_1, a_2, and a_3 are three complex numbers, equivalent to six real numbers. It is more convenient to define

$$\boldsymbol{S} = \frac{1}{2}\boldsymbol{\sigma}, \qquad \boldsymbol{K} = c^{-1}\frac{i}{2}\boldsymbol{\sigma} \tag{11.155}$$

and write Ω as follows:

$$\Omega = -i\boldsymbol{\theta} \cdot \boldsymbol{S} - i\boldsymbol{v} \cdot \boldsymbol{K}, \tag{11.156}$$

where $\boldsymbol{\theta}$ and $\boldsymbol{v}$ are real. It is now easy to verify that the following commutation relations hold:

$$[\, S_n,\, S_m\,] = i\, \epsilon_{nmk}\, S_k, \tag{11.157}$$
$$[\, S_n,\, T_m\,] = i\, \epsilon_{nmk}\, T_k, \tag{11.158}$$
$$[\, T_n,\, T_m\,] = -c^{-2} i\, \epsilon_{nmk}\, S_k. \tag{11.159}$$

This is nothing but the $\mathfrak{so}(1,3)$ algebra.

11.7.1 The group PSL(2, ℂ)

If g is a 2×2 matrix, $\det(-g) = \det(g)$. Therefore $\det(g) = 1$ if and only if $\det(-g) = 1$, which means that g is in SL(2, ℂ) if and only if $-g$ is in SL(2, ℂ).

We consider the set $\{g, -g\}$ as a single element of a set which we call PSL(2, ℂ), where P is for projective. We endow PSL(2, ℂ) with the following multiplication:

$$\{g, -g\}\, \{h, -h\} = \{g\, h, -g\, h\}. \tag{11.160}$$

It is easy to see that all the group requirements are fulfilled. For example, the unit element of this set is $\{\mathbb{I}, -\mathbb{I}\}$.

Notation. The notation would be far more easy if we denoted $\{g, -g\}$ by g and $\{g, -g\}\{h, -h\}$ by $g\,h$. One must only remember that PSL(2, ℂ) is nothing but SL(2, ℂ), with antipodal points, g and $-g$, considered to be equivalent.

11.7.2 Minkowski space-time and 2 × 2 Hermitian matrices

To the event (t, x, y, z) we can associate the complex 2×2 matrix

$$X = \begin{bmatrix} c\,t + z & x - i\,y \\ x + i\,y & c\,t - z \end{bmatrix}. \tag{11.161}$$

This Hermitian matrix is uniquely determined by the event (t, x, y, z), and vice versa; that is, any Hermitian 2×2 matrix uniquely determines an event (t, x, y, z). It is easy to see that

$$\det X = c^2\, t^2 - (x^2 + y^2 + z^2) = -\varsigma\, \eta_{\mu\nu}\, x^\mu\, x^\nu. \tag{11.162}$$

Now let g be a 2×2 complex matrix, and define

$$X' = g\, X\, g^\dagger. \tag{11.163}$$

We have

$$(X')^\dagger = \left(g\, X\, g^\dagger\right)^\dagger = \left(g^\dagger\right)^\dagger X^\dagger\, g^\dagger = g\, X\, g^\dagger = X', \tag{11.164}$$

from which it follows that X' is Hermitian if and only if X is Hermitian. Now observe that

$$\det\left(X'\right) = \det\left(g\, X\, g^{\dagger}\right) = |\det g|^{2}\, \det X. \tag{11.165}$$

Therefore if $|\det g| = 1$, the transformation $X \mapsto g\, X\, g^{\dagger}$ preserves both Hermiticity and determinant of X. This means that if $|\det(g)| = 1$, the transformation $X \mapsto g\, X\, g^{\dagger}$ is in $\mathrm{O}(1,3)$. We shall now prove that this transformation is actually in $\mathrm{SO}^{\uparrow}(1,3)$.

Given $g \in M(2, \mathbb{C})$, we can define the mapping $X \mapsto g\, X\, g^{\dagger}$, which is defined on $M(2, \mathbb{C})$. Let us use the notation T_g for this map. Obviously $T_g(X)$ is a linear function of X. It should be noted that if ω is a complex number with unit modulus ($|\omega| = 1$) and we define $g' = \omega\, g$, then $T_{g'}$ and T_g are the same linear map, because

$$T_{g'}(X) = (\omega\, g)\, X\, (\omega\, g)^{\dagger} = \omega \bar{\omega}\, g\, X\, g^{\dagger} = T_g(X). \tag{11.166}$$

Now observe the following fact:

$$T_g \circ T_h(X) = T_g(T_h(X)) = T_g(h\, X\, h^{\dagger}) = g\left(h\, X\, h^{\dagger}\right) g^{\dagger} \tag{11.167}$$

$$= g\, h\, X\, h^{\dagger}\, g^{\dagger} = (g\, h)\; X\, (g\, h)^{\dagger} \tag{11.168}$$

$$= T_{gh}(X), \tag{11.169}$$

which means

$$T_g \circ T_h = T_{gh}. \tag{11.170}$$

A basis for $M(2, \mathbb{C})$ is given by

$$E_1 = \begin{bmatrix} 1 & 0 \\ 0 & 0 \end{bmatrix}, \quad E_2 = \begin{bmatrix} 0 & 1 \\ 0 & 0 \end{bmatrix}, \quad E_3 = \begin{bmatrix} 0 & 0 \\ 1 & 0 \end{bmatrix}, \quad E_4 = \begin{bmatrix} 0 & 0 \\ 0 & 1 \end{bmatrix}. \tag{11.171}$$

It is easy to calculate

$$T_g(E_1) = g\, E_1\, g^{\dagger} = \begin{bmatrix} a & b \\ c & d \end{bmatrix} \begin{bmatrix} 1 & 0 \\ 0 & 0 \end{bmatrix} \begin{bmatrix} \bar{a} & \bar{c} \\ \bar{b} & \bar{d} \end{bmatrix} \tag{11.172}$$

$$= \begin{bmatrix} a & 0 \\ c & 0 \end{bmatrix} \begin{bmatrix} \bar{a} & \bar{c} \\ \bar{b} & \bar{d} \end{bmatrix} = \begin{bmatrix} a\, \bar{a} & a\, \bar{c} \\ c\, \bar{a} & c\, \bar{c} \end{bmatrix} \tag{11.173}$$

$$= a\, \bar{a}\, E_1 + a\, \bar{c}\, E_2 + c\, \bar{a}\, E_3 + c\, \bar{c}\, E_4. \tag{11.174}$$

Similarly we get

$$T_g(E_2) = a\, \bar{b}\, E_1 + a\, \bar{d}\, E_2 + c\, \bar{b}\, E_3 + c\, \bar{d}\, E_4, \tag{11.175}$$

$$T_g(E_3) = b\, \bar{a}\, E_1 + b\, \bar{c}\, E_2 + d\, \bar{a}\, E_3 + d\, \bar{c}\, E_4, \tag{11.176}$$

$$T_g(E_4) = b\, \bar{b}\, E_1 + b\, \bar{d}\, E_2 + d\, \bar{b}\, E_3 + d\, \bar{d}\, E_4. \tag{11.177}$$

Writing $X = \sum_{i=1}^{4} X^i\, E_i$, we get $T_g(X) = \sum_{i=1}^{4} X^i\, T_g(E_i)$, which can be written in the matrix form $X' = T_g X$, where

$$T_g = \begin{bmatrix} a\bar{a} & a\bar{b} & b\bar{a} & b\bar{b} \\ a\bar{c} & a\bar{d} & b\bar{c} & b\bar{d} \\ c\bar{a} & c\bar{b} & d\bar{a} & d\bar{b} \\ c\bar{c} & c\bar{d} & d\bar{c} & d\bar{d} \end{bmatrix}. \tag{11.178}$$

By a straightforward matrix multiplication one can see that

$$T_g = \begin{bmatrix} a\bar{a} & a\bar{b} & b\bar{a} & b\bar{b} \\ a\bar{c} & a\bar{d} & b\bar{c} & b\bar{d} \\ c\bar{a} & c\bar{b} & d\bar{a} & d\bar{b} \\ c\bar{c} & c\bar{d} & d\bar{c} & d\bar{d} \end{bmatrix} = \begin{bmatrix} a & 0 & b & 0 \\ 0 & a & 0 & b \\ c & 0 & d & 0 \\ 0 & c & 0 & d \end{bmatrix} \begin{bmatrix} \bar{a} & \bar{b} & 0 & 0 \\ \bar{c} & \bar{d} & 0 & 0 \\ 0 & 0 & \bar{a} & \bar{b} \\ 0 & 0 & \bar{c} & \bar{d} \end{bmatrix}. \tag{11.179}$$

Now we can calculate the determinant of T_g. First we note that

$$\begin{vmatrix} a & 0 & b & 0 \\ 0 & a & 0 & b \\ c & 0 & d & 0 \\ 0 & c & 0 & d \end{vmatrix} = (a\,d - b\,c)^2 = \det^2(g), \tag{11.180}$$

$$\begin{vmatrix} \bar{a} & \bar{b} & 0 & 0 \\ \bar{c} & \bar{d} & 0 & 0 \\ 0 & 0 & \bar{a} & \bar{b} \\ 0 & 0 & \bar{c} & \bar{d} \end{vmatrix} = \left(\bar{a}\bar{d} - \bar{b}\bar{c}\right)^2 = \det^2 \bar{g}. \tag{11.181}$$

From these two relations we get

$$\det T_g = |\det g|^4 . \tag{11.182}$$

Therefore $\det(T_g) = 1$ if $|\det(g)| = 1$. This completes the proof that $T_g \in \mathrm{SO}(1,3)$.

Remark. For those who already know tensor algebra, the above relations read as follows:

$$T_g = g \otimes \bar{g} = (g \otimes \mathbb{I})(\mathbb{I} \otimes \bar{g}), \tag{11.183}$$

$$\det(T_g) = \det(g \otimes \mathbb{I}) \cdot \det(\mathbb{I} \otimes \bar{g}), \tag{11.184}$$

$$\det(g \otimes \mathbb{I}) = \det^2(g), \tag{11.185}$$

$$\det(\mathbb{I} \otimes \bar{g}) = \det^2(\bar{g}). \tag{11.186}$$

Now we must prove that $\tau(T_g) = 1$. To do this, we must find the sign of the coefficient of t in the formula for t'. To get it, we first note that $2\,c\,t = \mathrm{tr}(X)$, and we calculate

$$2ct' = \mathrm{tr}(X') = \mathrm{tr}(g\,X\,g^\dagger) = \mathrm{tr}(g^\dagger\,g\,X) \tag{11.187}$$

$$= \mathrm{tr}\left(\begin{bmatrix} \bar{a} & \bar{c} \\ \bar{b} & \bar{d} \end{bmatrix}\begin{bmatrix} a & b \\ c & d \end{bmatrix}\begin{bmatrix} ct+z & x-iy \\ x+iy & ct-z \end{bmatrix}\right) \tag{11.188}$$

$$= \mathrm{tr}\left(\begin{bmatrix} |a|^2+|c|^2 & \bar{a}b+\bar{c}d \\ \bar{b}a+\bar{d}c & |b|^2+|d|^2 \end{bmatrix}\begin{bmatrix} ct+z & x-iy \\ x+iy & ct-z \end{bmatrix}\right), \tag{11.189}$$

$$2ct' = 2ct\left(|a|^2+|b|^2+|c|^2+|d|^2\right) + \ldots. \tag{11.190}$$

This shows that the coefficient of t in the formula of t' is positive, *i.e.*, $\tau(T_g) = 1$. We have therefore established the following theorem.

Theorem. *The groups* PSL(2, $\mathbb{C}$) *and* $\mathrm{SO}^\uparrow(1,3)$ *are isomorphic.*

11.8 The Poincaré group

The isometry group of the Minkowski space-time is called the Poincaré[5] group. It is larger than the Lorentz group. Not only the Lorentz transformations leave the space-time line element $\mathrm{d}s^2$ invariant; any translation in time or space also leaves the line element invariant. Let us look at this in more detail. First we should clarify what we mean by isometry. In the following x means the point with coordinates x^μ, and X similarly means the point with coordinates X^μ.

Isometry. A transformation $f : \mathbb{M} \to \mathbb{M}$ sending $X \mapsto x$ is said to be an isometry of the space-time if for any two infinitesimally close events X_1 and X_2, which are transformed to x_1 and x_2, the space-time line element $\mathrm{d}s^2$ is the same. In other words, if X_1 is given by coordinates X^μ and X_2 is given by coordinates $X^\mu + \mathrm{d}X^\mu$, and similarly x_1 is given by x^μ and x_2 is given by $x^\mu + \mathrm{d}x^\mu$, then the following identity must be fulfilled:

$$\eta_{\mu\nu}\,\mathrm{d}X^\mu\,\mathrm{d}X^\nu = \eta_{\alpha\beta}\,\mathrm{d}x^\alpha \mathrm{d}x^\beta. \tag{11.191}$$

In general X^μs are functions of x^μs, and we have

$$\mathrm{d}X^\mu = \frac{\partial X^\mu}{\partial x^\alpha}\,\mathrm{d}x^\alpha. \tag{11.192}$$

The transformation is acceptable if it is invertible, which means that the determinant of the matrix $\partial X^\mu/\partial x^\alpha$ does not vanish. Now we write

$$\mathrm{d}s^2 = \eta_{\mu\nu}\,\mathrm{d}X^\mu\,\mathrm{d}X^\nu = \eta_{\mu\nu}\,\frac{\partial X^\mu}{\partial x^\alpha}\,\frac{\partial X^\nu}{\partial x^\beta}\,\mathrm{d}x^\alpha\,\mathrm{d}x^\beta. \tag{11.193}$$

[5] After Jules Henri Poincaré (1854–1912).

This is equal to $\eta_{\alpha\beta}\,dx^\alpha\,dx^\beta$ for arbitrary dx^α if and only if

$$\eta_{\alpha\beta} = \eta_{\mu\nu}\frac{\partial X^\mu}{\partial x^\alpha}\frac{\partial X^\nu}{\partial x^\beta}. \tag{11.194}$$

Differentiating this relation once more we get

$$\eta_{\mu\nu}\frac{\partial^2 X^\mu}{\partial x^\gamma\,\partial x^\alpha}\frac{\partial X^\nu}{\partial x^\beta} + \eta_{\mu\nu}\frac{\partial X^\mu}{\partial x^\alpha}\frac{\partial^2 X^\nu}{\partial x^\gamma\,\partial x^\beta} = 0. \tag{11.195}$$

To simplify notation, let us introduce

$$A_{\mu\beta} = \eta_{\mu\nu}\frac{\partial X^\nu}{\partial x^\beta}, \tag{11.196}$$

$$B^\mu_{\gamma\alpha} = \frac{\partial^2 X^\mu}{\partial x^\gamma\,\partial x^\alpha}. \tag{11.197}$$

The conditions (11.195) now read

$$A_{\mu\beta}\,B^\mu_{\gamma\alpha} + A_{\nu\alpha}\,B^\nu_{\gamma\beta} = 0. \tag{11.198}$$

Change the summation index ν to μ in the second term, then write this equation renaming the indices $\alpha \to \beta \to \gamma \to \alpha$,

$$A_{\mu\gamma}\,B^\mu_{\alpha\beta} + A_{\mu\beta}\,B^\mu_{\alpha\gamma} = 0, \tag{11.199}$$

and once more, again reshuffling the indices, and this time multiply the equation by -1. Then we obtain

$$-A_{\mu\alpha}\,B^\mu_{\beta\gamma} - A_{\mu\gamma}\,B^\mu_{\beta\alpha} = 0. \tag{11.200}$$

Now add these three equations and note that $B^\mu_{\gamma\alpha} = B^\mu_{\alpha\gamma}$ to get

$$A_{\mu\beta}\,B^\mu_{\gamma\alpha} = 0. \tag{11.201}$$

For any choice of γ and α (fix); this is a set of four equations ($\beta = 0, 1, 2, 3$) for the four unknowns $B^\mu_{\gamma\alpha}$, $\mu = 0, 1, 2, 3$. Since the determinant of the coefficients, $\det A_{\mu\beta}$, does not vanish, all the $B^\mu_{\gamma\alpha} = \frac{\partial^2 X^\mu}{\partial x^\gamma\,\partial x^\alpha}$ are zero. This means that the functions $X^\mu(x^0, x^1, x^2, x^3)$ are of the form

$$X^\mu = L^\mu{}_\nu\,x^\nu + a^\mu, \tag{11.202}$$

where $L^\mu{}_\nu = \partial X^\mu/\partial x^\nu$ is a constant matrix and a^μ is a constant vector. Now condition (11.194) says that transformation (11.202) is an isometry if and only if

$$L^\mu{}_\alpha\,L^\nu{}_\beta\,\eta_{\mu\nu} = \eta_{\alpha\beta}, \tag{11.203}$$

that is, if $L \in O(1,3)$ is a Lorentz transformation. It is easy to write the combination rule for two successive transformations:

$$X \to X' = MX + A, \tag{11.204}$$

$$X' \to X'' = NX' + B, \tag{11.205}$$

$$X \to X'' = NMX + (NA + B), \tag{11.206}$$

which shows that the Poincaré group is the semidirect product of the Lorentz group and the Abelian group of translations:

$$IO(1,3) = O(1,3) \ltimes \mathbb{R}^4. \tag{11.207}$$

One can find a matrix representation of the Poincaré group in exactly the same way as we did for the Euclidean group (§10.15). We define a 5D extended space-time with coordinates $(x^0, x^1, x^2, x^3, x^4)$. Let us denote the components of this vector by x^A, where an upper case Latin index takes value in the set $\{0, 1, 2, 3, 4\}$. Now consider the 5×5 matrices satisfying

$$L^\mu{}_\nu \in O(1,3), \quad L^\mu{}_4 = \alpha^\mu \in \mathbb{R}, \quad L^4{}_\mu = 0, \quad L^4{}_4 = 1, \tag{11.208}$$

or in matrix notation,

$$\left[L^A{}_B\right] = \begin{bmatrix} L^0{}_0 & L^0{}_1 & L^0{}_2 & L^0{}_3 & \alpha^0 \\ L^1{}_0 & L^1{}_1 & L^1{}_2 & L^1{}_3 & \alpha^1 \\ L^2{}_0 & L^2{}_1 & L^2{}_2 & L^2{}_3 & \alpha^2 \\ L^3{}_0 & L^3{}_1 & L^3{}_2 & L^3{}_3 & \alpha^0 \\ 0 & 0 & 0 & 0 & 1 \end{bmatrix}. \tag{11.209}$$

These matrices map the $x^4 = b$ hyperplane into itself, for

$$L^4{}_B x^B = L^4{}_\nu x^\nu + L^4{}_4 x^4 = 0 + b. \tag{11.210}$$

On this hyperplane

$$\begin{aligned} L^\mu{}_B x^B &= L^\mu{}_\nu x^\nu + L^\mu{}_4 x^4 \\ &= L^\mu{}_\nu x^\nu + \alpha^\mu b. \end{aligned} \tag{11.211}$$

Whatever the dimension of $x^4 = b$ be, $\alpha^1 b$, $\alpha^2 b$, and $\alpha^3 b$ must have the dimension of length, and $\alpha^0 b$ must have the dimension of time. We write everything in terms of

$$a^\mu = \alpha^\mu b. \tag{11.212}$$

The generators of these transformations are as follows.

Generators of Lorentz boosts:

$$\mathcal{T}_1 = \begin{bmatrix} 0 & c^{-2} & 0 & 0 & 0 \\ 1 & 0 & 0 & 0 & 0 \\ 0 & 0 & 0 & 0 & 0 \\ 0 & 0 & 0 & 0 & 0 \\ 0 & 0 & 0 & 0 & 0 \end{bmatrix}, \quad \mathcal{T}_2 = \begin{bmatrix} 0 & 0 & c^{-2} & 0 & 0 \\ 0 & 0 & 0 & 0 & 0 \\ 1 & 0 & 0 & 0 & 0 \\ 0 & 0 & 0 & 0 & 0 \\ 0 & 0 & 0 & 0 & 0 \end{bmatrix}, \quad \mathcal{T}_3 = \begin{bmatrix} 0 & 0 & 0 & c^{-2} & 0 \\ 0 & 0 & 0 & 0 & 0 \\ 0 & 0 & 0 & 0 & 0 \\ 1 & 0 & 0 & 0 & 0 \\ 0 & 0 & 0 & 0 & 0 \end{bmatrix}; \tag{11.213}$$

generators of spatial rotations:

$$\mathcal{S}_1 = \begin{bmatrix} 0 & 0 & 0 & 0 & 0 \\ 0 & 0 & 0 & 0 & 0 \\ 0 & 0 & 0 & -1 & 0 \\ 0 & 0 & 1 & 0 & 0 \\ 0 & 0 & 0 & 0 & 0 \end{bmatrix}, \quad \mathcal{S}_2 = \begin{bmatrix} 0 & 0 & 0 & 0 & 0 \\ 0 & 0 & 0 & 1 & 0 \\ 0 & 0 & 0 & 0 & 0 \\ 0 & -1 & 0 & 0 & 0 \\ 0 & 0 & 0 & 0 & 0 \end{bmatrix}, \quad \mathcal{S}_3 = \begin{bmatrix} 0 & 0 & 0 & 0 & 0 \\ 0 & 0 & -1 & 0 & 0 \\ 0 & 1 & 0 & 0 & 0 \\ 0 & 0 & 0 & 0 & 0 \\ 0 & 0 & 0 & 0 & 0 \end{bmatrix}; \tag{11.214}$$

generators of spatial translations:

$$\mathcal{P}_1 = \begin{bmatrix} 0 & 0 & 0 & 0 & 0 \\ 0 & 0 & 0 & 0 & 1 \\ 0 & 0 & 0 & 0 & 0 \\ 0 & 0 & 0 & 0 & 0 \\ 0 & 0 & 0 & 0 & 0 \end{bmatrix}, \quad \mathcal{P}_2 = \begin{bmatrix} 0 & 0 & 0 & 0 & 0 \\ 0 & 0 & 0 & 0 & 0 \\ 0 & 0 & 0 & 0 & 1 \\ 0 & 0 & 0 & 0 & 0 \\ 0 & 0 & 0 & 0 & 0 \end{bmatrix}, \quad \mathcal{P}_3 = \begin{bmatrix} 0 & 0 & 0 & 0 & 0 \\ 0 & 0 & 0 & 0 & 0 \\ 0 & 0 & 0 & 0 & 0 \\ 0 & 0 & 0 & 0 & 1 \\ 0 & 0 & 0 & 0 & 0 \end{bmatrix}; \tag{11.215}$$

and the generator of translation in time:

$$\mathcal{H} = \begin{bmatrix} 0 & 0 & 0 & 0 & 1 \\ 0 & 0 & 0 & 0 & 0 \\ 0 & 0 & 0 & 0 & 0 \\ 0 & 0 & 0 & 0 & 0 \\ 0 & 0 & 0 & 0 & 0 \end{bmatrix}. \tag{11.216}$$

By simple matrix multiplications we find the following commutation relations:

$$[\mathcal{S}_n, \mathcal{S}_m] = \epsilon_{nmk}\, \mathcal{S}_k, \tag{11.217}$$

$$[\mathcal{S}_n, \mathcal{T}_k] = \epsilon_{nmk}\, \mathcal{T}_k, \tag{11.218}$$

$$[\mathcal{T}_n, \mathcal{T}_m] = -\frac{1}{c^2}\epsilon_{nmk}\, \mathcal{S}_k, \tag{11.219}$$

$$[\mathcal{S}_n, \mathcal{P}_m] = \epsilon_{nmk}\, \mathcal{P}_k, \tag{11.220}$$

$$[\mathcal{T}_n, \mathcal{P}_m] = \frac{1}{c^2}\delta_{nm}\, H, \tag{11.221}$$

$$[\mathcal{S}_n, \mathcal{H}] = [\mathcal{P}_n, \mathcal{H}] = 0, \tag{11.222}$$

$$[\mathcal{T}_n, \mathcal{H}] = \mathcal{P}_n. \tag{11.223}$$

In physics it is customary to use the following complex generators and notation:

$$J_n = i\,\mathcal{S}_n, \quad K_n = i\,\mathcal{T}_n, \quad P_n = i\,\mathcal{P}_n, \quad H = i\,\mathcal{H}. \tag{11.224}$$

In terms of these "complex" generators the Lie algebra of the Poincaré group is given by the following commutation relations:

$$[J_n, J_m] = i\,\epsilon_{nmk}\, J_k, \tag{11.225}$$

$$[J_n, K_m] = i\,\epsilon_{nmk}\, K_k, \tag{11.226}$$

$$[K_n, K_m] = -i\,\frac{1}{c^2}\epsilon_{nmk}\, J_k, \tag{11.227}$$

$$[J_n, P_m] = i\,\epsilon_{nmk}\,P_k, \tag{11.228}$$

$$[K_n, P_m] = \frac{1}{c^2}\,i\,\delta_{nm}\,H, \tag{11.229}$$

$$[S_n, H] = [P_n, H] = 0, \tag{11.230}$$

$$[K_n, H] = i\,P_n. \tag{11.231}$$

Covariant notation. The algebra of the Poincaré group could be written in terms of the following tensors and vectors[6]:

$$J^{00} = J^{11} = J^{22} = J^{33} = 0, \tag{11.232}$$

$$J^{0n} - J^{n0} = -\varsigma\,K_n, \tag{11.233}$$

$$J^{nm} = -J^{mn} = \varsigma\,\epsilon_{mnk}\,J_k, \tag{11.234}$$

$$P^0 = \frac{H}{c^2}, \tag{11.235}$$

$$P^n = P_n. \tag{11.236}$$

Note that because of our convention for $x^0 = t$, not $c\,t$, P^0 has dimensions of mass! It is straightforward to check that the commutation relations read

$$i\left[J^{\mu\nu}, J^{\alpha\beta}\right] = \eta^{\nu\alpha}\,J^{\mu\beta} - \eta^{\mu\alpha}\,J^{\nu\beta} + \eta^{\mu\beta}\,J^{\nu\alpha} - \eta^{\nu\beta}\,J^{\mu\alpha}, \tag{11.237}$$

$$i\left[J^{\mu\nu}, P^{\alpha}\right] = \eta^{\nu\alpha}\,P^{\mu} - \eta^{\mu\alpha}\,P^{\nu}. \tag{11.238}$$

Here we present the checks for some special cases:

$$\begin{aligned}[S_1, T_2] &= \left[\varsigma\,J^{23}, -\varsigma\,J^{02}\right] = -\varsigma^2\left[J^{23}, J^{02}\right] \\ &= -i\left(-\eta^{22}\,J^{30}\right) = i\,\varsigma\,(\varsigma\,T_3) = i\,T_3,\end{aligned} \tag{11.239}$$

$$\begin{aligned}[T_1, T_2] &= \left[-\varsigma\,J^{01}, -\varsigma\,J^{02}\right] = \varsigma^2\left[J^{01}, J^{02}\right] \\ &= -i\left(-\eta^{00}\,J^{12}\right) = i\left(-\frac{\varsigma}{c^2}\right)(\varsigma\,S_3) = -\frac{i}{c^2}S_3,\end{aligned} \tag{11.240}$$

$$\begin{aligned}[T_1, H] &= \left[-\varsigma\,J^{01}, c^2\,P^0\right] = -\varsigma\,c^2\left[J^{01}, P^0\right] \\ &= -\varsigma\,c^2(-i)\left(-\eta^{00}\,P^1\right) = i\,P^1 = i\,P_1,\end{aligned} \tag{11.241}$$

$$\begin{aligned}[T_1, P_1] &= \left[-\varsigma\,J^{01}, P^1\right] = -\varsigma(-i)\left(\eta^{11}\,P^0\right) = P^0 \\ &= \frac{H}{c^2}.\end{aligned} \tag{11.242}$$

[6] Remember that the symbol ς is $+1$ for the space-like convention of the metric and -1 for the time-like convention.

Unitary representations. The Poincaré group is the isometry (symmetry) group of the Minkowski space-time. According to quantum mechanics, the state space of any system is a complex vector space. If one changes the system by a symmetry operator of the space-time, the state vector must change accordingly, so that the state space of a quantum system must be a unitary representation of the Poincaré group.[7] It is because of this that the unitary representations of the Poincaré group are very important in physics. For the representation to be unitary, the complex generators must be Hermitian.

The defining representation given by (11.213)–(11.216) is 5D. When multiplied by i, we get three Hermitian matrices $i\,\mathcal{S}_n$ and seven non-Hermitian matrices $i\,\mathcal{T}_n$, $-i\,\mathcal{P}_n$, and $-i\mathcal{H}$. Therefore this 5D representation is not unitary. In fact, the Poincaré group has no finite-dimensional *unitary* representation—all unitary representations of the Poincaré (and the Lorentz) group are infinite-dimensional. This is due to the fact that the Lorentz group is not "compact." The definition of "compact" is beyond the scope of this book.

Contraction to the Galilean group. The $c \to \infty$ limit of the commutation relations (11.225)–(11.231) of the Poincaré group gets the Lie algebra of the full Galilean group. The velocity of light appears in (11.227) and (11.229), and as $c \to \infty$ we get

$$[\,K_n,\, K_m\,] = 0, \tag{11.243}$$

$$[\,K_n,\, P_m\,] = 0. \tag{11.244}$$

These have simple interpretations: In the Galilean limit boosts commute with each other, and boosts commute with the spatial translations. Let us see this explicitly for the latter case. Consider the following two transformations:

$$x' = f(x) = x + vt, \tag{11.245}$$

$$x' = g(x) = x + a. \tag{11.246}$$

We now have

$$f \circ g(x) = f(g(x)) = g(x) + vt = (x + a) + vt = x + a + vt, \tag{11.247}$$

$$g \circ f(x) = g(f(x)) = f(x) + a = (x + vt) + a = x + a + vt. \tag{11.248}$$

However, there is a point which must be considered when we try to apply the contraction $c \to \infty$ to the generators of the unitary representation used in quantum mechanics to represent Poincaré transformation on the state space of massive particles.

Suppose $\left\{\tilde{\boldsymbol{J}}, \tilde{\boldsymbol{K}}, \tilde{\boldsymbol{P}}, \tilde{H}\right\}$ are the generators of the unitary representation on the state space of a massive particle of mass M. Then

$$\boldsymbol{J} = \hbar\tilde{\boldsymbol{J}}, \quad \boldsymbol{K} = \hbar\tilde{\boldsymbol{K}}, \quad \boldsymbol{P} = \hbar\tilde{\boldsymbol{P}}, \quad H = \hbar\tilde{H}, \tag{11.249}$$

[7] The unitarity is dictated by the fact that the norm (or absolute value) of the state vector representing the system must not change.

where $\hbar$ is the reduced Planck constant, satisfy the same algebra with the structure constants multiplied by $\hbar$. Now H has the interpretation of the energy and $\boldsymbol{P}$ has the interpretation of the momentum operators.[8] If the mass of the particle is M we have

$$H^2 = M^2 c^4 \mathbb{I} + P_1^2 + P_2^2 + P_3^2. \tag{11.250}$$

The $\mathbb{I}$ is there, because H and P_ns are operators—$\mathbb{I}$ denotes the identity operator in the representation space.

The $v \ll c$ limit of this relation is

$$H = M c^2 \mathbb{I} + \frac{1}{2M}\left(P_1^2 + P_2^2 + P_3^2\right). \tag{11.251}$$

Therefore in the limit $v \ll c$ the relation (11.229) becomes

$$\begin{aligned}
[K_n, P_m] &= \frac{\hbar}{c^2} i\, \delta_{nm} H \\
&\rightarrow \frac{\hbar}{c^2} i\, \delta_{nm} \left[M c^2 \mathbb{I} + \frac{1}{2M}\left(P_1^2 + P_2^2 + P_3^2\right)\right] \\
&= i\, \hbar M \delta_{nm} \mathbb{I} + \frac{\hbar}{2 M c^2}\left(P_1^2 + P_2^2 + P_3^2\right) \\
&\rightarrow i\, \hbar M \delta_{nm} \mathbb{I} \quad \text{if } \frac{1}{2} M v^2 \ll M c^2.
\end{aligned} \tag{11.252}$$

We conclude that in nonrelativistic quantum mechanics, for massive particles, the observables $\boldsymbol{J}$, $\boldsymbol{K}$, $\boldsymbol{P}$, and H satisfy the following commutation relations:

$$[J_n, J_m] = i\, \hbar \epsilon_{nmk} J_k, \tag{11.253}$$
$$[J_n, K_m] = i\, \hbar \epsilon_{nmk} K_k, \tag{11.254}$$
$$[K_n, K_m] = 0, \tag{11.255}$$
$$[J_n, P_m] = i\, \hbar \epsilon_{nmk} P_k, \tag{11.256}$$
$$[K_n, P_m] = i\, M \hbar \delta_{nm} \mathbb{I}, \tag{11.257}$$
$$[J_n, H] = [P_n, H] = 0, \tag{11.258}$$
$$[K_n, H] = i\, \hbar P_n. \tag{11.259}$$

Commutation relation (11.257) says that on the state space of the particle of mass M, the operators $\exp\left(\frac{i}{\hbar} \boldsymbol{v} \cdot \boldsymbol{K}\right)$, representing a boost, and $\exp\left(-\frac{i}{\hbar} \vec{a} \cdot \boldsymbol{P}\right)$, representing a translation, do not commute, although the corresponding operators acting on the space-time do commute.

[8] Remember that in quantum mechanics observables are Hermitian operators.

Jargon. The algebra we just obtained is called *a central extension* of the Galilean algebra. The $M\mathbb{I}$, which has different values for different particles, is the central element. It is said that "the symmetry group of the nonrelativistic quantum mechanics is the central extension of the Galilean group."

The Galilean group has no finite-dimensional unitary representation, but it does have finite-dimensional nonunitary representation: The defining representation is 5D, like the defining representation of the Poincaré group. The central extension of the Galilean group, however, has no finite-dimensional representation, not even a nonunitary one. The reason is that if K_n and P_m are finite-dimensional matrices, then $\operatorname{tr}[K_n, P_m] = 0$, but $\operatorname{tr} M\mathbb{I}$ is equal to M multiplied by the dimension of the representation.

Guide to Literature. The original article on the unitary representations of the Poincaré group is Wigner (1939). The contraction of groups was first studied by Inonu and Wigner (1953). For the representation theory of the Lorentz group, we refer the reader to Sexl and Urbantke (1992, pp. 229–260). For the representation theory of the Poincaré group we refer the reader to Sexl and Urbantke (1992, pp. 261–316) or Weinberg (1995, pp. 58–91).

References

Einstein, A., Lorentz, H.A., Weyl, H., Minkowski, H., 1952. The Principle of Relativity. Dover, New York.

Inonu, E., Wigner, E.P., 1953. On the contraction of groups and their representations. Proceedings of the National Academy of Sciences of the United States of America 39 (6), 510–524.

Sexl, R.U., Urbantke, H.K., 1992. Relativity, Groups, Particles. Special Relativity and Relativistic Symmetry in Field and Particle Physics. Springer-Verlag, Wien.

Weinberg, S., 1995. The Quantum Theory of Fields, vol. I, Foundations. Cambridge University Press, Cambridge.

Wigner, E., 1939. On the unitary representations of the inhomogeneous Lorentz group. Annals of Mathematics 40 (1), 149–204.

APPENDIX

Elementary analytic geometry

A.1 Linear change of coordinates

Let (x, y) be an orthonormal Cartesian coordinate system on the Euclidean plane. By orthonormal we mean two properties:

1. The axes x and y are orthogonal.
2. The unit of length used to assign x and y are the same.

Now let (X, Y) be defined by a pair of linear equations. For example, consider the following linear relations:

$$X = 2x - y, \tag{A.1}$$

$$Y = x + y. \tag{A.2}$$

It is easy to solve these two equations for x and y:

$$x = \frac{1}{3}X + \frac{1}{3}Y, \tag{A.3}$$

$$y = -\frac{1}{3}X + \frac{2}{3}Y. \tag{A.4}$$

By the x-axis we mean the set $y = 0$. Similarly, by the X-axis we mean the set $Y = 0$ (see Fig. A.1). To find the unit of length on the X-axis, we have to find the point $(X = 1, Y = 0)$. Similarly, the point $(X = 0, Y = 1)$ determines the unit of length on the Y-axis.

The set of points at distance 1 from the origin is, by definition, the unit circle, which in the orthonormal (x, y)-coordinates has the expression $x^2 + y^2 = 1$. This *same circle*, in the oblique (X, Y)-coordinates, has the expression

$$1 = x^2 + y^2 = \left(\frac{1}{3}X + \frac{1}{3}Y\right)^2 + \left(-\frac{1}{3}X + \frac{2}{3}Y\right)^2 \tag{A.5}$$

$$= \frac{2}{9}X^2 - \frac{2}{9}XY + \frac{5}{9}Y^2. \tag{A.6}$$

Though this expression is commonly called an ellipse, the precise statement is that this equation is an ellipse *if* the coordinates (X, Y) are orthonormal. That is, if we

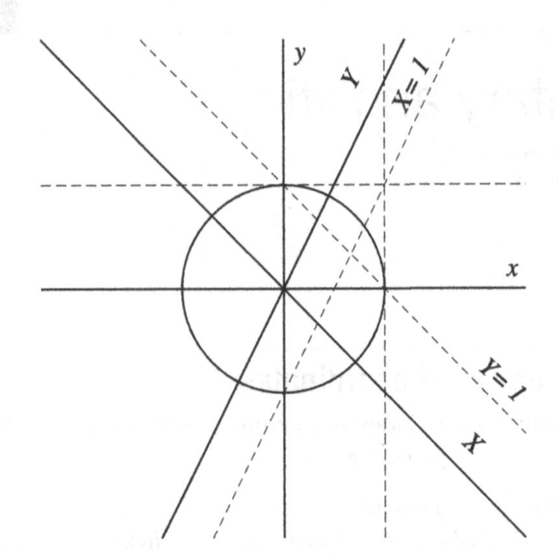

FIGURE A.1

(x, y) are orthonormal Cartesian. $(X, Y) = (2x - y, x + y)$ are Cartesian, but are not orthonormal. The X-axis is the set $Y = 0$. The Y-axis is the set $X = 0$. The lines $Y = 1$ and $X = 1$ are parallel, respectively, to the X-axis and the Y-axis. These lines help us to find both the directions and the unit lengths on the X- and Y-axes. The circle shown is the circle $x^2 + y^2 = 1$, which is the set of points at distance 1 from the origin. This same circle, in terms of the oblique coordinates (X, Y), has the algebraic expression $2X^2 - 2XY + 5Y^2 = 9$.

draw two perpendicular axes X and Y and we use the same unit of length on them, then this expression is an ellipse. On our nonorthonormal coordinates (X, Y) given above, this equation describes *the unit circle*!

Problem 46. Suppose (X, Y) is an orthonormal coordinate system. Write the equation of a circle with radius a and apply the transformation $(X, Y) \to (x, y)$ given above.

A.2 Rotations

Consider the following pair of equations:

$$X = x \cos\theta + y \sin\theta, \tag{A.7}$$

$$Y = -x \sin\theta + y \cos\theta, \tag{A.8}$$

where θ is a fixed angle (a real number). We suppose (x, y) are orthonormal Cartesian, as usual. The X-axis is, by definition, the set $Y = 0$, which from the second of the equations above is the line $y = x \tan\theta$. The Y-axis is the set $X = 0$, which is the line $y = -x \cot\theta$. Since $-\cot\theta \cdot \tan\theta = -1$, the X- and Y-axes are orthogonal. It can be easily seen that we have

$$x^2 + y^2 = X^2 + Y^2. \tag{A.9}$$

This means that the circle with center at the origin and radius R has the same equation in both (x, y)- and (X, Y)-coordinate systems. Note that the intersections of any of these four axes (x, y, X, or Y) with the unit circle show the unit of length on that axis.

A.3 Hyperbolic rotations

Let (x, y) be Cartesian coordinates on the Euclidean plane and let θ be a real number. Define another pair of coordinates (X, Y) as follows:

$$X = \ \ x \cosh\theta - y \sinh\theta, \tag{A.10}$$

$$Y = -x \sinh\theta + y \cosh\theta. \tag{A.11}$$

This change of coordinates is called a *hyperbolic rotation*. It is easy to see that

$$\begin{aligned} X^2 - Y^2 &= (x \cosh\theta - y \sinh\theta)^2 - (-x \sinh\theta + y \cosh\theta)^2 \\ &= x^2 - y^2, \end{aligned} \tag{A.12}$$

from which it follows that for this change of coordinates, the equation of the hyperbolas $x^2 - y^2 = \pm a^2$ has the same form in both (x, y)- and (X, Y)-coordinates.

Now let us define $v = \tanh\theta$. It is easy to find $\cosh\theta$ and $\sinh\theta$ in terms of v. We have

$$1 - v^2 = 1 - \tanh^2\theta = 1 - \frac{\sinh^2\theta}{\cosh^2\theta} = \frac{1}{\cosh^2\theta}, \tag{A.13}$$

$$\gamma(v) := \frac{1}{\sqrt{1 - v^2}} = \cosh\theta, \qquad \cosh\theta \geq 1 > 0, \tag{A.14}$$

$$\sinh\theta = \tanh\theta \cosh\theta = \frac{v}{\sqrt{1 - v^2}} = \gamma(v)\, v. \tag{A.15}$$

So we have

$$x = X \cosh\theta + Y \sinh\theta = \gamma(v)\,(X + v\,Y), \tag{A.16}$$

$$y = X \sinh\theta + Y \cosh\theta = \gamma(v)\,(v\,X + Y), \tag{A.17}$$

or

$$X = +x \cosh\theta - y \sinh\theta = \gamma(v)\,(x - v\,y), \tag{A.18}$$

$$Y = -x \sinh\theta + y \cosh\theta = \gamma(v)\,(y - v\,x). \tag{A.19}$$

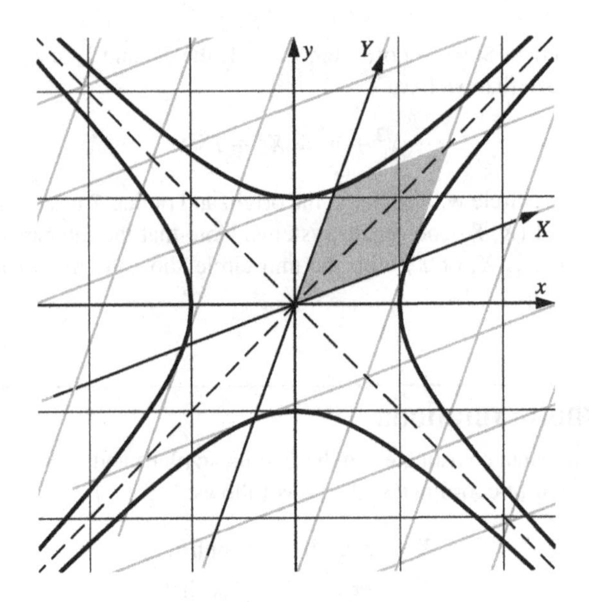

FIGURE A.2

The hyperbolic rotation of the coordinates of the plane. The x- and y-axes are as usual. The X-axis is the line $y = x \tanh\theta = v\,x$, and the Y-axis is the line $y = x \coth\theta = v^{-1}\,x$, for $\theta = 0.381$, or $v = \tanh(0.381) = 0.364$. The Euclidean angle of the X-axis with the x-axis is $20°$, and we have $\tan 20° = \tanh 0.381 = 0.364$. The hyperbolas $x^2 - y^2 = X^2 - Y^2 = \pm 1$ are drawn. The intersections of the coordinate axes with these hyperbolas determine the units of length on the axes. In the figure two meshes are drawn. The black (orthogonal) mesh is used to find the (x, y)-coordinates; the gray (oblique) mesh is used to find the (X, Y)-coordinates. A unit cell in the (X, Y)-mesh is shaded. The areas of the unit cells (both orthogonal and oblique) are equal. The shaded cell (the rhombus) is the result of the hyperbolic rotation of the unit rectangle. Note that the lines $y = \pm x$ are the same as the lines $Y = \pm X$. Together, these lines are called the light-cone (this terminology will be explained later).

To investigate the geometric interpretation of this linear change of coordinates, we consider the x- and y-axes being perpendicular to each other (see Fig. A.2). After that, we must specify the X- and Y-axes. The X-axis is the set $Y = 0$. From the above equations, it is the set $y = v\,x$. Since $v < 1$ (we consider positive v), this is a line with a slope less than 1. The Y-axis is the set $X = 0$. From the above equations, it is the set $x = v\,y$ or $y = v^{-1}\,x$, which is a line with a slope greater than 1.

Next we must find the unit of lengths on the X- and Y-axes. This is done using the invariant quadratic $X^2 - Y^2 = x^2 - y^2$. In the (x, y)-plane the set $x^2 - y^2 = \pm 1$ consists of two hyperbolas. These hyperbolas intersect the x-axis at the points $(\pm 1, 0)$ and the y-axis at the points $(0, \pm 1)$. Since these hyperbolas have the same equation in terms of X and Y, the same argument shows that the intersection of the X-axis with these hyperbolas consists of two points: $(X = \pm 1, Y = 0)$. The intersection of the Y-axis with the same hyperbolas consists of two points ($X = 0, Y = \pm 1$). The procedure is explained in Fig. A.2.

Problem 47. Consider the hyperbolic rotation.

1. What happens if θ is negative?
2. What happens if the x- and y-axes are not orthogonal?
3. Write a version of the hyperbolic rotation whose invariant hyperbolas are $x\,y =$ constant.
4. Write a version of the hyperbolic rotation whose invariant hyperbolas are $(x/a)^2 - (y/b)^2 =$ constant.

Guide to Literature. Hyperbolic functions and their similarity with trigonometric functions are taught in every textbook on calculus. See, for example, Thomas et al. (2014, p. 432).

A.4 Hyperbolic quadratic forms

The equations

$$\frac{x^2}{a^2} - \frac{y^2}{b^2} = \left(\frac{x}{a} - \frac{y}{b}\right)\left(\frac{x}{a} + \frac{y}{b}\right) = \pm\alpha^2, \tag{A.20}$$

in which α is a constant, describe a pair of hyperbolas (see Fig. A.3). These hyperbolas have two asymptotes—two intersecting lines

$$\frac{x}{a} - \frac{y}{b} = 0, \qquad \frac{x}{a} + \frac{y}{b} = 0. \tag{A.21}$$

The set of these two intersecting lines is given by the equation

$$\frac{x^2}{a^2} - \frac{y^2}{b^2} = 0, \tag{A.22}$$

that is, by setting $\alpha = 0$ in the equation defining the hyperbolas.

The equations

$$\frac{x^2 + y^2}{a^2} - \frac{z^2}{b^2} = \pm\alpha^2 \tag{A.23}$$

for constant α describe a set of hyperbolas of revolution. This could be easily seen once we change to cylindrical coordinates:

$$\rho = \sqrt{x^2 + y^2}, \quad \phi = \tan^{-1}\frac{y}{x}. \tag{A.24}$$

In this coordinate system we have

$$\frac{\rho^2}{a^2} - \frac{z^2}{b^2} = \pm\alpha^2. \tag{A.25}$$

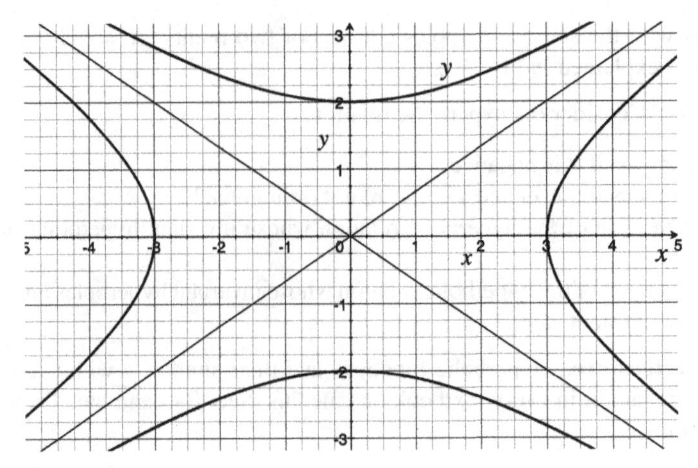

FIGURE A.3

Hyperbolas $(x/a)^2 - (y/b)^2 = \pm 1$ for $a = 3$ and $b = 2$.

Rotating this around the z-axis, we get two hyperboloids of revolution:

$$\frac{x^2 + y^2}{a^2} - \frac{z^2}{b^2} = +\alpha^2, \qquad \text{two sheets,} \tag{A.26}$$

$$\frac{x^2 + y^2}{a^2} - \frac{z^2}{b^2} = -\alpha^2, \qquad \text{one sheet.} \tag{A.27}$$

Setting $\alpha \to 0$, we get the equation of a cone,

$$\frac{x^2 + y^2}{a^2} - \frac{z^2}{b^2} = 0, \tag{A.28}$$

which is simply the rotation of the lines

$$\frac{\rho^2}{a^2} - \frac{z^2}{b^2} = 0 \tag{A.29}$$

around the z-axis, which is a cone with vertex at the origin.

Generalizing, the equations

$$\frac{x^2 + y^2 + z^2}{a^2} - \frac{w^2}{b^2} = \pm\alpha^2 \tag{A.30}$$

describe 3D hyperboloid spaces. To see this, we use spherical polar coordinates for the 3D space (x, y, z) and rewrite these equations as

$$\frac{r^2}{a^2} - \frac{w^2}{b^2} = \pm\alpha^2. \tag{A.31}$$

In the 2D space of (r, w) this is exactly like the 3D version with the obvious replacement $(\rho, z) \to (r, w)$. The cone $(r/a) = \pm(w/b)$, which is the limit of $\alpha \to 0$, is the asymptote of these hyperboloids.

Reference

Thomas, G.B., Weir, M.D., Hass, J., Heil, C., 2014. Thomas' Calculus, 13th edition. Pearson, Chicago.

APPENDIX

B Active vs. passive rotations

When an apparatus is rotated, we get a new apparatus, which may or may not behave as the previous apparatus before rotation. When we rotate the coordinate system, it has absolutely no physical effect on the behavior of the apparatus; only *our mathematical formulas* describing the behavior of the apparatus have changed. If the space is isotropic, rotating the whole system will not change the behavior of the system. Using the rotation of the coordinate system, this could be stated another way, as follows: If the space is isotropic, the equations governing the system retain their form if we rotate the coordinates.

The terms "active" and "passive" transformations are used in various texts, mostly written by physicists and engineers. However, there is a reasonable tendency to avoid this terminology. Various authors, mostly mathematicians and theoretical physicists, prefer to use "transformation" instead of "active transformation" and "change of coordinate" instead of "passive transformation."

B.1 The rotation matrix

Let $B = \{\hat{e}_1, \hat{e}_2, \hat{e}_3\}$ be three orthonormal unit vectors such that $\hat{e}_1 \times \hat{e}_2 = \hat{e}_3$. Such a set is called a right-handed orthonormal basis. Let $B' = \{\hat{e}'_1, \hat{e}'_2, \hat{e}'_3\}$ be another right-handed orthonormal basis. All these vectors and any other vectors we are considering share the same base point, which we denote by O. The elements of B' could be written in terms of the basis B as follows:

$$\hat{e}'_j = \sum_{i=1}^{3} R_{ij}\,\hat{e}_i. \tag{B.1}$$

Note that the summation is on the *first*[1] index of R_{ij}, which means

$$\hat{e}'_1 = R_{11}\,\hat{e}_1 + R_{21}\,\hat{e}_2 + R_{31}\,\hat{e}_3, \tag{B.2}$$
$$\hat{e}'_2 = R_{12}\,\hat{e}_1 + R_{22}\,\hat{e}_2 + R_{32}\,\hat{e}_3, \tag{B.3}$$
$$\hat{e}'_3 = R_{13}\,\hat{e}_1 + R_{23}\,\hat{e}_2 + R_{33}\,\hat{e}_3. \tag{B.4}$$

[1] This is the standard convention being used in mathematics and physics.

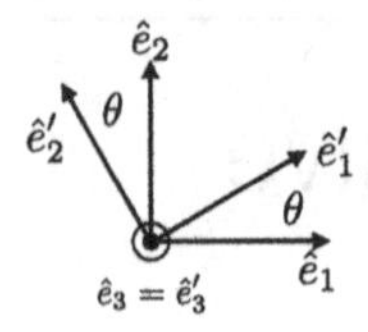

FIGURE B.1 Two orthonormal bases.

Two orthonormal bases $B = \{\hat{e}_1, \hat{e}_2, \hat{e}_3\}$ and $B' = \{\hat{e}'_1, \hat{e}'_2, \hat{e}'_3\}$. B' is the result of the rotation of B by θ around $\hat{e}_3$.

An example is shown in Fig. B.1, from which we see that for this rotation, which is around $\hat{e}_3$ by angle θ,

$$\hat{e}'_1 = +\cos\theta\, \hat{e}_1 + \sin\theta\, \hat{e}_2, \tag{B.5}$$
$$\hat{e}'_2 = -\sin\theta\, \hat{e}_1 + \cos\theta\, \hat{e}_2, \tag{B.6}$$
$$\hat{e}'_3 = \hat{e}_3. \tag{B.7}$$

Thus *the matrix* of this rotation is

$$R = \begin{pmatrix} R_{11} & R_{12} & R_{13} \\ R_{21} & R_{22} & R_{23} \\ R_{31} & R_{32} & R_{33} \end{pmatrix} = \begin{pmatrix} \cos\theta & -\sin\theta & 0 \\ \sin\theta & \cos\theta & 0 \\ 0 & 0 & 1 \end{pmatrix}. \tag{B.8}$$

B.2 Rotation of vectors

Now think of a mapping, a function, that sends $\hat{e}_1 \to \hat{e}'_1$, $\hat{e}_2 \to \hat{e}'_2$, and $\hat{e}_3 \to \hat{e}'_3$. Such a mapping is called a rotation of the basis B. A very good interpretation of this mapping is as follows: Consider a rigid body (a box for example), attach three vectors of B to the rigid body, and rotate the rigid body while fixing the origin O. It is easy to see that as a result of this mapping, any vector is being rotated. For example, the vector

$$\mathbf{A} = \sum_{j=1}^{3} a_j \hat{e}_j \tag{B.9}$$

in Fig. B.2 is mapped to

$$\mathbf{A}' = \sum_{j=1}^{3} a_j \hat{e}'_j. \tag{B.10}$$

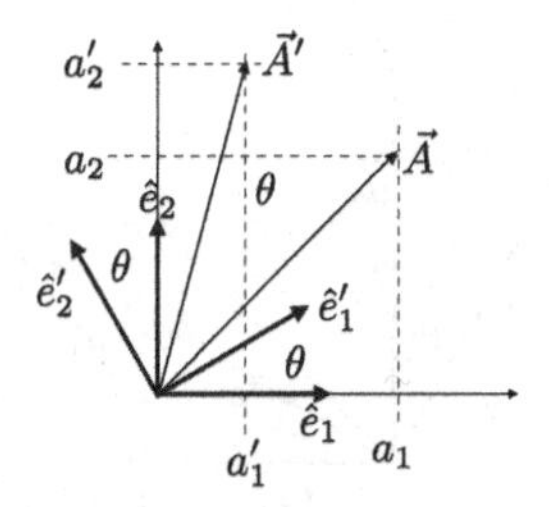

FIGURE B.2 Rotating vectors around the x_3-axis.

Rotation by θ around $\hat{e}_3 = \hat{e}_1 \times \hat{e}_2$ maps any vector to another vector. The vector $\vec{A}$ is mapped to $\vec{A}'$.

Using (B.1) we have

$$\vec{A}' = \sum_{j=1}^{3} a_j \sum_{i=1}^{3} R_{ij}\, \hat{e}_i \tag{B.11}$$

$$= \sum_{i=1}^{3} \left(\sum_{j=1}^{3} R_{ij}\, a_j \right) \hat{e}_i, \tag{B.12}$$

showing that

$$a'_i = \sum_{j=1}^{3} R_{ij}\, a_j, \tag{B.13}$$

which means

$$\begin{pmatrix} a'_1 \\ a'_2 \\ a'_3 \end{pmatrix} = \begin{pmatrix} R_{11} & R_{12} & R_{13} \\ R_{21} & R_{22} & R_{23} \\ R_{31} & R_{32} & R_{33} \end{pmatrix} \begin{pmatrix} a_1 \\ a_2 \\ a_3 \end{pmatrix}. \tag{B.14}$$

We have

$$\vec{A}' = \sum_{i=1}^{3} a'_i\, \hat{e}_i. \tag{B.15}$$

Equation (B.14) is the form of the rotation of vectors. Acting on the column vector

$$A = \begin{pmatrix} a_1 \\ a_2 \\ a_3 \end{pmatrix}, \tag{B.16}$$

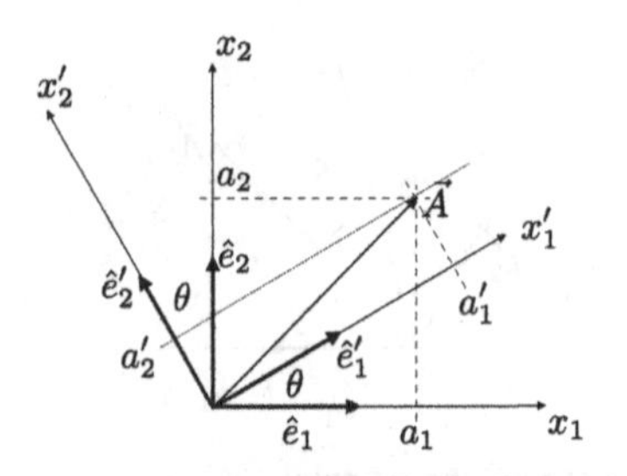

FIGURE B.3 Rotating the coordinate system around the x_3-axis.

Rotation of the coordinate system around $\hat{e}_3 = \hat{e}_1 \times \hat{e}_2$ by θ. The vector $\vec{A}$ could be described by its components in either of the coordinate systems.

which represents the vector $\boldsymbol{A}$ in the basis B, we get the column matrix

$$A' = \begin{pmatrix} a_1' \\ a_2' \\ a_3' \end{pmatrix}, \tag{B.17}$$

which represents the vector $\boldsymbol{A}'$ in *the same basis* B! This relation is written in the compact form

$$A' = R\,A, \qquad \text{rotation of the vector.} \tag{B.18}$$

B.3 Rotation of the coordinate system

Consider the same relation between B and B', and construct two coordinate systems on them (see Fig. B.3). A vector $\boldsymbol{A}$ could be described by its components in either the (x_1, x_2, x_3)-coordinates or the (x_1', x_2', x_3')-coordinates:

$$\boldsymbol{A} = \sum_{i=1}^{3} a_i\,\hat{e}_i = \sum_{j=1}^{3} a_j'\,\hat{e}_j' \tag{B.19}$$

$$= \sum_{j=1}^{3} a_j' \sum_{i=1}^{3} R_{ij}\,\hat{e}_i \tag{B.20}$$

$$= \sum_{i=1}^{3} \left(\sum_{j=1}^{3} R_{ij}\,a_j' \right) \hat{e}_i, \tag{B.21}$$

which means that

$$a_i = \sum_{j=1}^{3} R_{ij} a_j'. \tag{B.22}$$

We see that the *same* vector $\boldsymbol{A}$ is described by two different column matrices A and A', which specify its components in the two coordinate systems,

$$A = \begin{pmatrix} a_1 \\ a_2 \\ a_3 \end{pmatrix}, \qquad A' = \begin{pmatrix} a'_1 \\ a'_2 \\ a'_3 \end{pmatrix}, \tag{B.23}$$

and we have

$$\begin{pmatrix} a_1 \\ a_2 \\ a_3 \end{pmatrix} = \begin{pmatrix} R_{11} & R_{12} & R_{13} \\ R_{21} & R_{22} & R_{23} \\ R_{31} & R_{32} & R_{33} \end{pmatrix} \begin{pmatrix} a'_1 \\ a'_2 \\ a'_3 \end{pmatrix}, \tag{B.24}$$

or

$$A = R\,A', \qquad \text{rotation of the coordinate system.} \tag{B.25}$$

Since the matrix R is orthogonal, we have $R^{-1} = R^T$, and this equation could be written as

$$A' = R^T A. \tag{B.26}$$

Remark. When reading and using a textbook, one should pay attention to the conventions. For example, Goldstein (1980, p. 133, eq. 4-12′) writes the rotation of coordinate by $x'_j = a_{ij}\,x_j$, which in our notation means $A' = R^{-1} A$. So Goldstein's "rotation matrix a" is actually the inverse of the rotation matrix defined by (B.1). Our convention is the same as that of Sakurai (1985, p. 153).

References

Goldstein, H., 1980. Classical Mechanics, 2nd edition. Addison-Wesley, Reading, MA.
Sakurai, J.J., 1985. Modern Quantum Mechanics. Addison-Wesley, Redwood City, CA.

APPENDIX

Projective (Möbius) action

The rotation group SO(2) has a natural action on the (X, Y)-plane, which is given by

$$R(\theta): \begin{pmatrix} X \\ Y \end{pmatrix} \mapsto \begin{pmatrix} X' \\ Y' \end{pmatrix} = \begin{pmatrix} \cos\theta & -\sin\theta \\ \sin\theta & \cos\theta \end{pmatrix} \begin{pmatrix} X \\ Y \end{pmatrix}. \tag{C.1}$$

If C is a circle with center at the origin $(0, 0)$, any such transformation maps C to itself. We say that the orbits of this action of SO(2) are circles with center at O. There is another action of SO(2) on $\mathbb{R}^2$, which is called the projective or Möbius[1] action. Under this action, a specific line in the $\mathbb{R}^2$-plane is left invariant, that is, this line is being mapped into itself. In other words, the orbits of this action are more complicated, and one of them is a line. We now give a brief introduction to this action and then see its generalization to the Lorentz group.

Let us begin with a simple but general enough example. Consider the mapping $f : (x, y) \mapsto (X, Y)$ as defined by

$$X = \frac{x}{1+x}, \quad Y = \frac{y}{1+x}. \tag{C.2}$$

Obviously this function is not defined on the line $x = -1$, which we denote by δ. It is easy to see that the inverse mapping $f^{-1} : (X, Y) \mapsto (x, y)$ is given by

$$x = \frac{X}{1-X}, \quad y = \frac{Y}{1-X}. \tag{C.3}$$

The function f^{-1} is not defined on the line $X = 1$, which we denote by Δ (see Fig. C.1).

Let R be a rotation of the (X, Y)-plane, defined by (C.1), and define

$$S(\theta) = f^{-1} \circ R(\theta) \circ f, \tag{C.4}$$

where $\circ$ means the combination of mappings. This is a transformation on the (x, y)-plane: f maps (x, y) to (X, Y), then $R(\theta)$ rotates it to (X', Y'), and finally f^{-1} maps this (X', Y') to (x', y') in the original (x, y)-plane. Note also the combination rule:

$$S_2 \circ S_1 = \left(f^{-1} \circ R_2 \circ f\right) \circ \left(f^{-1} \circ R_1 \circ f\right)$$

---

[1] After August Ferdinand Möbius (1790–1868).

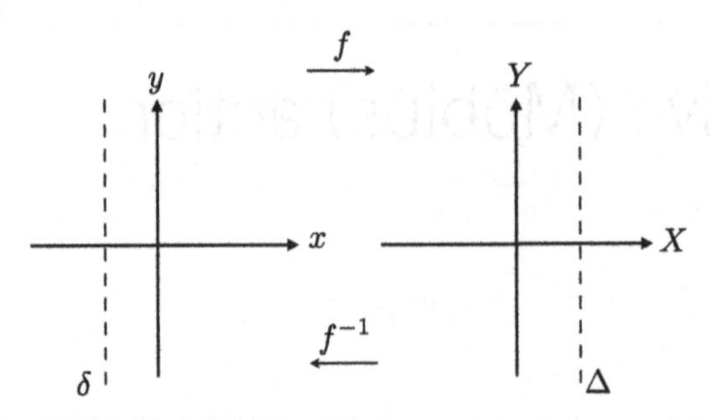

FIGURE C.1 A projective similarity transformation.

The functions f and f^{-1} are defined by (C.2) and (C.3).

$$\begin{aligned} &= f^{-1} \circ \left(R_2 \circ f \circ f^{-1} \circ R_2\right) \circ f \\ &= f^{-1} \circ R_2 \circ R_1 \circ f \\ &= f^{-1} \circ (R_2 R_1) \circ f, \end{aligned} \tag{C.5}$$

where $R_2 R_1$ is the usual matrix multiplication. This rule means that $S(\theta)$ is a representation of the group SO(2). We say it is a projective or Möbius representation, because $S(\theta)$ acts on the real projective plane. Note that this action is not linear!

Let us return to our example. It is easy to find the mathematical form of this transformation explicitly:

$$x' = \frac{x \cos\theta - y \sin\theta}{1 + x(1 - \cos\theta) + y \sin\theta}, \tag{C.6}$$

$$y' = \frac{x \sin\theta + y \cos\theta}{1 + x(1 - \cos\theta) + y \sin\theta}. \tag{C.7}$$

We note that the set $x = -1$, that is, the line δ, is being mapped to itself:

$$x' = \frac{-1 \cos\theta - y \sin\theta}{1 - (1 - \cos\theta) + y \sin\theta} = -1, \tag{C.8}$$

$$y' = \frac{-\sin\theta + y \cos\theta}{\cos\theta + y \sin\theta} = \frac{y - \tan\theta}{1 + y \tan\theta}. \tag{C.9}$$

Since $\tan\theta = \tan(\theta + \pi)$, the set of all rotations maps δ to δ twice! But in some sense we can say that the set δ is invariant under $S(\theta)$; by this we do not mean that every point of δ is left invariant, but we mean that $S(\theta)$ sends the set δ to itself.

In the (X, Y)-plane the circle $X^2 + Y^2 = \mu$ is mapped to itself under $R(\theta)$ (see Fig. C.2). We say that the orbits of the R-action of the SO(2) on the (X, Y)-plane are concentric circles. In the (x, y)-plane this set is given by the equation

$$\frac{x^2 + y^2}{(1 + x)^2} = \mu, \tag{C.10}$$

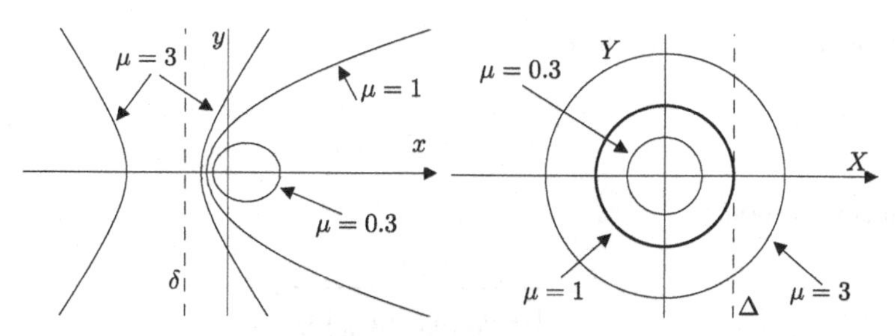

FIGURE C.2 Orbits of the action of the rotation group.

On the (X, Y)-space (right), the action is the usual rotation given by (C.1); the orbits are circles. Note that the circle $X^2 + Y^2 = 1$ is tangent to the line Δ. On the (x, y)-space (left), the action is the projective action defined by (C.4); the orbits are conics. The parabola $\mu = 1$ corresponds to the touching circle $X^2 + Y^2 = 1$ of the (X, Y)-space. The "invariant" line δ is the limit of the $\mu \to \infty$ hyperbola. (Reprinted with permission from N. Jafari and A. Shariati, Phys. Rev. D. 84, 065038 (2011).)

which is an ellipse (if $\mu < 1$), a parabola (if $\mu = 1$), or a hyperbola (if $\mu > 1$). For $\mu \to \infty$ we get the line δ. These are the orbits of the S-action of the SO(2) on the (x, y)-plane. Note that $\mu \to \infty$ is a limiting hyperbola of two branches. It is because of this that $S(\theta)$ for all $\theta \in [0, 2\pi]$ covers δ twice.

***Generalizing to* SO(1, 3).** If a is a 4-vector (a column matrix) and H is the Minkowski metric, $H = [\eta_{\mu\nu}]$, define

$$f : x \mapsto X = \frac{x}{1 + a^T H x}, \tag{C.11}$$

with the inverse

$$f^{-1} : X \mapsto x = \frac{X}{1 - a^T H X}, \tag{C.12}$$

where x and X are column matrices with entries x^μ and X^μ. The set δ in this case is given by the equation

$$1 - a^T H X = 0, \tag{C.13}$$

which in index notation is

$$\eta_{\mu\nu} a^\mu x^\nu = 1. \tag{C.14}$$

This is a hyperplane of dimension 3 in $\mathbb{M}^4$ (or $\mathbb{R}^4$), normal, in the Minkowski sense, to the vector a^μ, so that if a^μ is time-like, the set δ is a space-like hyperplane; if a^μ is space-like, the set δ is a time-like hyperplane; and if a^μ is light-like (null), the set δ is light-like.

Now if $L \in \mathrm{SO}(1, 3)$ is a Lorentz transformation, define the projective action of L on the x-space as follows:

$$S = f^{-1} \circ L \circ f. \tag{C.15}$$

It is straightforward to show that

$$x' = \frac{L\,x}{1 + a^T H\,(\mathbb{I} - L)\,x}, \tag{C.16}$$

or in index notation,

$$x'^{\mu} = \frac{L^{\mu}_{\ \nu}\,x^{\nu}}{1 + a_{\alpha}\left(\delta^{\alpha}_{\beta} - L^{\alpha}_{\ \beta}\right)x^{\beta}}. \tag{C.17}$$

By construction, this transformation maps δ to δ.

Another property of this projective Lorentz transformation is this: Defining

$$D = 1 + a^T H\,(\mathbb{I} - L)\,x, \tag{C.18}$$

we have

$$\begin{aligned} x'^T H x' &= \frac{x^T L^T}{D} H \frac{Lx}{D} \\ &= \frac{x^T\,(L^T H L)\,x}{D^2} \qquad\qquad L^T H L = H \\ &= \frac{x^T\,H\,x}{D^2}, \end{aligned} \tag{C.19}$$

which shows that $x^T H x = 0$ if and only if $x'^T H x' = 0$, or in index notation,

$$\eta_{\mu\nu} x^{\mu} x^{\nu} = 0 \quad \Leftrightarrow \quad \eta_{\mu\nu} x'^{\mu} x'^{\nu} = 0. \tag{C.20}$$

Transformation (C.17) is known in the literature as the Fock–Lorentz transformation. Fock arrived at this transformation while investigating the most general transformation, which preserves the fronts of the wave equation, that is, condition (C.20), relaxing the linearity condition.

Doubly special relativity. One can apply the same procedure, not to the space-time, but to the momentum space with coordinates $p^{\mu} = (E, \boldsymbol{p})$. This leads to transformations known in the literature as the Magueijo–Smolin transformations, first introduced by Magueijo and Smolin to construct theories with both Lorentz invariance and an invariant energy scale. An example of such a transformation to be considered as a boost along the x-axis is

$$E' = \frac{\gamma\,(E - v\,p_x)}{1 + \alpha\,(\gamma - 1)\,E - \alpha\,\gamma\,v\,p_x}, \tag{C.21}$$

$$p_x' = \frac{\gamma\,(p_x - c^{-2}\,v\,E)}{1 + \alpha\,(\gamma - 1)\,E - \alpha\,\gamma\,v\,p_x}, \tag{C.22}$$

$$p_y' = \frac{p_y}{1 + \alpha\,(\gamma - 1)\,E - \alpha\,\gamma\,v\,p_x}, \tag{C.23}$$

$$p'_z = \frac{p_z}{1 + \alpha(\gamma - 1)E - \alpha\gamma v p_x}, \tag{C.24}$$

$$\gamma = \frac{1}{\sqrt{1-\beta^2}}, \quad \beta = \frac{v}{c}, \tag{C.25}$$

where α is a parameter of dimensions $1/\text{energy}$.[2]

The motivation is that if $E \ll \alpha^{-1}$, then the denominator in the above transformation is approximately 1, and we get the usual Lorentz transformation. Maybe at very high scales of energies the true transformations are these projective transformations.

These theories are called doubly special relativity because they have two invariants—the speed of light, c, and this energy scale.

Guide to Literature. For the original derivation by Fock see Fock (1964, pp. 403–410), but note that his notation is somewhat difficult to follow. Magueijo–Smolin transformation was introduced in Magueijo and Smolin (2002). This article is recommended to anyone who wants to understand the fundamentals of doubly special relativity theories. For the motivation of modifying special relativity in this direction see Amelino-Camelia (2002). For a review of doubly special relativity see Kowalski-Glikman (2005). For a geometric interpretation of the projective action of the Lorentz group and the structure of the invariant sets, see Jafari and Shariati (2011).

References

Amelino-Camelia, G., 2002. Special treatment. Nature 418 (6893), 34–35.

Fock, V., 1964. The Theory of Space, Time and Gravitation, 2nd revised edition. Pergamon Press, Oxford.

Jafari, N., Shariati, S., 2011. A projective interpretation of some doubly special relativity theories. Physical Review D 84, 065038.

Kowalski-Glikman, J., 2005. Introduction to doubly special relativity. In: Kowalski-Glikman, J., Amelino-Camelia, G. (Eds.), Planck Scale Effects in Astrophysics and Cosmology. In: Lecture Notes in Physics, vol. 669. Springer-Verlag, Berlin.

Magueijo, J., Smolin, L., 2002. Lorentz invariance with an invariant energy scale. Physical Review Letters 88, 190403.

[2] The combination $a_\mu L^\mu_{\ \nu} p^\nu$ must be dimensionless, and for a time-like a^μ, the corresponding invariant set δ is a surface of constant energy.

APPENDIX

Metric, curvature, and geodesics of a surface

D.1 Induced metric of a surface

We know that the line element of the Euclidean space $\mathbb{R}^3$ in an orthonormal Cartesian coordinate system $(x^1, x^2, x^3) = (x, y, z)$ is

$$ds^2 = dx^2 + dy^2 + dz^2 \tag{D.1}$$

$$= \sum_{i=1}^{3} \left(dx^i\right)^2. \tag{D.2}$$

This is nothing but the Pythagorean theorem: The distance between two nearby points (x, y, z) and $(x + dx, y + dy, z + dz)$ is ds.

From the definition of the spherical polar coordinates

$$x = r \sin\theta \cos\phi, \tag{D.3}$$

$$y = r \sin\theta \sin\phi, \tag{D.4}$$

$$z = r \cos\theta, \tag{D.5}$$

it follows by simple differentiation that

$$ds^2 = dr^2 + r^2 \left(d\theta^2 + \sin^2\theta \, d\phi^2\right). \tag{D.6}$$

Now, consider the sphere $r = a$, which we denote by Σ. Consider two nearby points (a, θ, ϕ) and $(a, \theta + d\theta, \phi + d\phi)$ on Σ. These two points are in $\mathbb{R}^3$, so, noting that $dr = 0$, the distance between them is given by

$$ds^2 = a^2 \left(d\theta^2 + \sin^2\theta \, d\phi^2\right). \tag{D.7}$$

This is the length of a straight line segment, the chord, in $\mathbb{R}^3$ joining the points. If $d\theta$ and $d\phi$ are infinitesimal, this chord (straight line segment) could be considered as laying on Σ. We say that Σ is embedded in $\mathbb{R}^3$, and (D.7) is the induced line element or induced metric of Σ.

The generalization is that for any surface Σ embedded in $\mathbb{R}^3$, one can find coordinates $q = (q^1, q^2)$ that uniquely specify a point

$$x = x(q), \quad y = y(q), \quad z = z(q) \tag{D.8}$$

on Σ, and the induced line element on Σ is

$$\mathrm{d}s^2 = \sum_{i,j=1}^{2} g_{ij}\,\mathrm{d}q^i\,\mathrm{d}q^j, \tag{D.9}$$

where

$$g_{ij} = \begin{pmatrix} g_{11}(q) & g_{12}(q) \\ g_{21}(q) & g_{22}(q) \end{pmatrix}, \qquad g_{12} = g_{21}, \tag{D.10}$$

is a symmetric 2×2 matrix called the (induced) metric of Σ. Here g_{ij} is computed as follows:

$$\mathrm{d}s^2 = \sum_{i=1}^{3} \left(\mathrm{d}x^i\right)^2 \tag{D.11}$$

$$= \sum_{i=1}^{3} \left(\sum_{j=1}^{2} \frac{\partial x^i}{\partial q^j}\,\mathrm{d}q^j \right) \left(\sum_{k=1}^{2} \frac{\partial x^i}{\partial q^k}\,\mathrm{d}q^k \right) \tag{D.12}$$

$$= \sum_{j,k=1}^{2} \left(\sum_{i=1}^{3} \frac{\partial x^i}{\partial q^j} \frac{\partial x^i}{\partial q^k} \right) \mathrm{d}q^j\,\mathrm{d}q^k, \tag{D.13}$$

which means

$$g_{ij}(q) = \sum_{i=1}^{3} \frac{\partial x^i}{\partial q^j} \frac{\partial x^i}{\partial q^k}. \tag{D.14}$$

D.2 Gaussian curvature

By metric (or intrinsic) properties of a surface we mean those properties which are derived from the lengths of curves and angles between curves (at the intersections). The matrix g_{ij} is called *the metric* of the surface, because all metric properties of the surface are derived from it.[1] One such property is the Gaussian curvature.[2] Here we present an operational definition of the Gaussian curvature which is very useful to understand the meaning of the sign of the Gaussian curvature.

Let $q = (q^1, q^2)$ be a point on Σ. By a circle of radius r with center at q, we mean all points on Σ which are a distance r from q. This is determined by the metric. Now we can calculate the circumference of C, which is also determined by the metric.

[1] A surface has also *extrinsic* properties which are not derivable from the metric. One example is the *mean curvature*.

[2] After Johann Carl Friedrich Gauss (1777–1855).

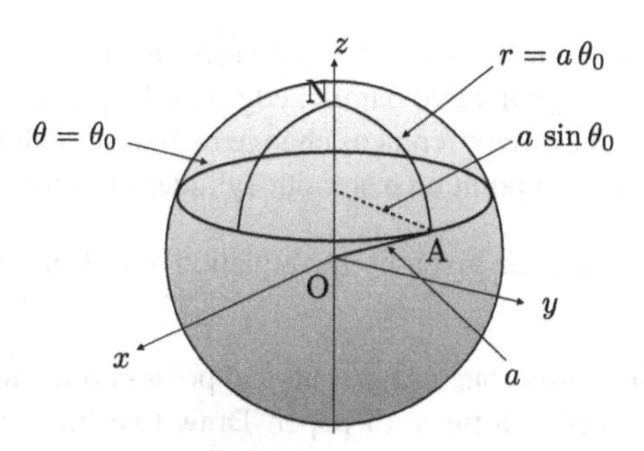

FIGURE D.1 Gaussian curvature of a sphere.

O is the center of the sphere, N is the north pole with coordinate $\theta = 0$. The parallel $\theta = \theta_0$ is a planar circle (in the plane $z = a\cos\theta_0$) of radius $a\sin\theta_0$. The arc from N to A is on a meridian with arc length $r = a\,\theta_0$. The parallel $\theta = \theta_0$ is a *circle of radius r on the sphere*—any point on this parallel is a distance $a\,\theta_0$ from the north pole.

Let us denote the circumference by $L(r)$. The Gaussian curvature is the following number with dimensions of $1/\mathrm{L}^2$:

$$\kappa = \lim_{r\to 0} \frac{6}{r^2}\left(\frac{2\pi r - L(r)}{2\pi r}\right). \tag{D.15}$$

Clearly, for the Euclidean plane $L(r) = 2\pi r$, and $\kappa = 0$ identically, at any point. Now let us calculate κ for a sphere of radius a. Let us use spherical coordinates (θ, ϕ) on the sphere, and let us calculate the circumference of a circle of radius r, centered at the point $\theta = 0$. It is easy to see that this is the parallel $\theta = \theta_0$ (see Fig. D.1). The distance of a point on the parallel θ_0 from the north pole is $r = a\,\theta_0$. The perimeter of this parallel is clearly $2\pi a \sin\theta_0$. Also note that the parallel $\theta = \theta_0$ is a circle of radius $a \sin\theta_0$ in the $z = a\cos\theta_0$ plane. Therefore we have

$$L(r) = 2\pi a \sin\theta_0 = 2\pi a \sin\frac{r}{a} \qquad \theta_0 = \frac{r}{a} \tag{D.16}$$

$$= 2\pi a\left(\frac{r}{a} - \frac{1}{3!}\frac{r^3}{a^3} + \frac{1}{a^5}(\ldots)\right). \tag{D.17}$$

From this it is easy to see that for a sphere of radius a the Gaussian curvature κ at point $\theta = 0$ is $1/a^2$. By symmetry, the Gaussian curvature of the sphere is the same at any point. So by this definition, a sphere is a surface of constant *positive* curvature $1/a^2$.

The curvature of a surface Σ at a point q could be a negative number. If so, the circumference of the "circle" C_r would be more than $2\pi r$. The curvature on the surface of a saddle is negative.

The following two experimental problems show that a surface could be somehow curved, yet its Gaussian curvature could be zero.

Problem 48. Draw on a paper a set of small concentric circles. Now make a smooth cylinder using this paper, gluing two parallel edges of the paper. Note that neither the radii of the circles nor the circumferences change by this smooth rolling of the paper. Using this, calculate κ for a point on a smooth cylinder of radius R.

Problem 49. Consider a cone given by the equation $\theta = \theta_0$ in spherical polar coordinates.

(1) Write the metric of this cone using spherical polar coordinates (r, θ, ϕ).
(2) Let O be the center of a piece of paper. Draw two lines, passing through O, making an angle of $\psi = 254$ degrees. Draw also a circle with center at O. Now cut the paper along the two lines, such that the edges of the paper make the angle ψ with each other. Now glue these edges so that a cone is formed, with O as its vertex. What is the angle θ_0 for this cone?
(3) Prove that the curvature of a cone is 0 at any point other than the vertex.
(4) What could be said about the curvature of a cone at its vertex?

D.3 Motion of a free particle

Now let us consider a particle moving on Σ. At time t, the particle is at point $q^i(t)$ on Σ, which means at point x^i in $\mathbb{R}^3$. Since x^i is a function of q and q is a function of t, by the chain rule, the components of the velocity vector in $\mathbb{R}^3$ are

$$v^i = \frac{dx^i}{dt} = \sum_{j=1}^{2} \frac{\partial x^i}{\partial q^j} \dot{q}^j, \qquad \dot{q}^j = \frac{dq^j}{dt}. \tag{D.18}$$

Let us calculate the square of the length of v^i:

$$|v|^2 = \sum_{i=1}^{3} \left(v^i\right)^2 = \sum_{i=1}^{3} \left(\sum_{j=1}^{2} \frac{\partial x^i}{\partial q^j} \dot{q}^j\right)\left(\sum_{k=1}^{2} \frac{\partial x^i}{\partial q^k} \dot{q}^k\right) \tag{D.19}$$

$$= \sum_{j,k=1}^{2} \left(\sum_{i=1}^{3} \frac{\partial x^i}{\partial q^j} \frac{\partial x^i}{\partial q^k}\right) \dot{q}^j \dot{q}^k \tag{D.20}$$

$$= \sum_{j,k=1}^{2} g_{ij}(q)\, \dot{q}^j \dot{q}^k. \tag{D.21}$$

If M is the mass of the particle, the kinetic energy of the particle moving on Σ is

$$T = \frac{1}{2} M\, |v|^2 = \frac{M}{2} \sum_{j,k=1}^{2} g_{jk}(q)\, \dot{q}^i \dot{q}^j. \tag{D.22}$$

For a sphere in (θ, ϕ) coordinates, we have

$$T = \frac{M a^2}{2} \left(\dot{\theta}^2 + \sin^2 \theta \, \dot{\phi}^2 \right). \tag{D.23}$$

D.4 Geodesic motion

Notation. In the remaining parts of this appendix we use Einstein's summation convention. All repeated indices are summed on the set $\{1, 2\}$.

Consider a particle of mass M that is constrained to be on the smooth surface Σ, but otherwise free. This means that the surface exerts a constraint force on M, constraining the particle to be on Σ, but there is no other force. In particular, there is no friction, which means no tangential force, so that the constraint force is always normal to Σ. The constraint force is not known a priori. Only having found the equations of motion could we find this force. Now the problem is this: Suppose at time $t = t_0$ we know that the particle is at point q_0 on Σ, and we also know its velocity at $t = t_0$ is $\dot{q}_0$. What is the position (and velocity) of this particle at later (or earlier) times? To find the equations of motion, we can use either of the following methods:

1. Find the acceleration $\boldsymbol{a}$, decompose it to vectors normal to Σ and tangent to Σ, and set the tangential components equal to zero (since there is no tangential force and, according to Newton's second law, acceleration is proportional to force).
2. Write the Lagrangian of the particle and derive the Euler–Lagrange equations.

These methods are equivalent, and each has some benefits in understanding the problem. Here we use the Lagrange method.

Guide to Literature. Euler–Lagrange equations are taught in texts on classical mechanics. See, for example, Goldstein (1980, pp. 35–45).

Since there is no potential energy, the Lagrangian is simply the kinetic energy $L = T$. Before writing the Euler–Lagrange equations, we note the following very important fact, which is very simple to prove, but since it is very important, we state it as a theorem.

Theorem. *The equations of motion of a particle constrained on a surface Σ are independent of the mass of the particle.*

Therefore dropping M (or assuming it is the unit mass) we write

$$L = \frac{1}{2} g_{ij}(q) \dot{q}^i \dot{q}^j. \tag{D.24}$$

We also note that since L as a function of velocities is homogeneous of degree 2 and independent of time t, the total energy, which is equal to L, is a constant of motion.

That is, we know that

$$\frac{\mathrm{d}}{\mathrm{d}t}\left(\frac{1}{2}g_{ij}(q)\dot{q}^i\dot{q}^j\right)=0. \tag{D.25}$$

This has a very simple mechanical interpretation: Since the force acting on the particle is normal to the surface, while the motion is tangent to the surface, no work is done on the particle, and the kinetic energy (or the velocity) of the particle is constant. This does not mean that $\dot{q}$ is constant, because the coefficients $g_{ij}(q)$ depend on q.

We now compute the Euler–Lagrange equations:

$$\frac{\partial L}{\partial q^i}=\frac{1}{2}\frac{\partial g_{jk}}{\partial q^i}\dot{q}^j\dot{q}^k=\frac{1}{2}g_{jk,i}\dot{q}^j\dot{q}^k, \tag{D.26}$$

where we have used the notation

$$f_{,i}=\frac{\partial f}{\partial q^i}. \tag{D.27}$$

Moreover,

$$\frac{\partial L}{\partial \dot{q}^i}=\frac{\partial}{\partial \dot{q}^i}\frac{1}{2}g_{jk}(q)\dot{q}^j\dot{q}^k \tag{D.28}$$

$$=\frac{1}{2}\left(g_{jk}\frac{\partial \dot{q}^j}{\partial \dot{q}^i}\dot{q}^k+g_{jk}\dot{q}^j\frac{\partial \dot{q}^k}{\partial \dot{q}^i}\right) \tag{D.29}$$

$$=\frac{1}{2}\left(g_{jk}\delta_i^j\dot{q}^j+g_{jk}\dot{q}^j\delta_i^k\right) \tag{D.30}$$

$$=\frac{1}{2}\left(g_{ik}\dot{q}^k+g_{ji}\dot{q}^j\right)=\frac{1}{2}2g_{ij}\dot{q}^j \tag{D.31}$$

$$=g_{ij}(q)\dot{q}^j, \tag{D.32}$$

$$\frac{\mathrm{d}}{\mathrm{d}t}\frac{\partial L}{\partial \dot{q}^i}=g_{ij}(q)\ddot{q}^j+\left(\frac{\mathrm{d}}{\mathrm{d}t}g_{ij}(q)\right)\dot{q}^j, \tag{D.33}$$

$$\frac{\mathrm{d}}{\mathrm{d}t}g_{ij}(q)=\frac{\partial g_{ij}}{\partial q^k}\dot{q}^k=g_{ij,k}\dot{q}^k. \tag{D.34}$$

Putting everything together, it is easy to see that the Euler–Lagrange equations

$$\frac{\mathrm{d}}{\mathrm{d}t}\frac{\partial L}{\partial \dot{q}^i}-\frac{\partial L}{\partial q^i}=0 \tag{D.35}$$

are

$$g_{ij}\ddot{q}^j+g_{ij,k}\dot{q}^j\dot{q}^k-\frac{1}{2}g_{jk,i}\dot{q}^j\dot{q}^k=0, \tag{D.36}$$

or

$$\ddot{g}_{ij}q^j+\frac{1}{2}\left(2g_{ij,k}-g_{jk,i}\right)\dot{q}^j\dot{q}^k, \tag{D.37}$$

which noting that

$$g_{ij,k}\,\dot q^j\,\dot q^k = g_{ik,j}\,\dot q^k\,\dot q^j, \tag{D.38}$$

we prefer to write in the following form:

$$g_{ij}\,\ddot q^j + \frac{1}{2}\left(g_{ij,k} + g_{ik,j} - g_{jk,i}\right)\dot q^j\,\dot q^k = 0. \tag{D.39}$$

The metric g_{ij} is invertible, and we denote the entries of the inverse matrix by g^{ij}, that is,

$$g^{ik}\,g_{kj} = \delta^i_j. \tag{D.40}$$

It is now easy to see that the Euler–Lagrange equations for a particle moving freely on Σ are

$$\ddot q^k + \Gamma^k_{ij}\,\dot q^i\,\dot q^j = 0, \tag{D.41}$$

where

$$\Gamma^k_{ij} = \frac{1}{2}g^{kn}\left(-g_{ij,n} + g_{ni,j} + g_{jn,i}\right). \tag{D.42}$$

Here, Γ^k_{ij} are called the components of the *connection* or the Christoffel symbols. The set of differential equations (D.41) is called the equations of the geodesic motion.

D.5 Geodesics—shortest paths

If the surface Σ is the plane $z = 0$ with coordinates $q^1 = x^1$ and $q^2 = x^2$, then the metric $g_{ij} = \delta_{ij}$, with all partial derivatives equal to zero, and we have $\Gamma^k_{ij} = 0$ and the equations of geodesic motion are $\ddot x^i = 0$, which simply mean motion with zero acceleration along a straight line. The straight lines of the plane are called geodesics of the plane. The generalization is simple: The path of a particle freely moving on a surface Σ are the geodesics of Σ. These geodesics are the shortest paths, which we now prove.

Consider a particle moving along the curve C. The motion is given by functions $q^i(t)$, defined on the time interval $[t_0, t_1]$. At time t_0 the particle is at q_0, and at time t_1 it is at q_1. The length of C is

$$\ell(C) = \int_{t_0}^{t_1} v(t)\,\mathrm{d}t = \int_{t_0}^{t_1} \sqrt{g_{ij}\,\dot q^i\,\dot q^j}\,\mathrm{d}t \tag{D.43}$$

$$= \int_{q_0}^{q_1} \mathrm{d}s, \tag{D.44}$$

where

$$\mathrm{d}s^2 = g_{ij}(q)\,\mathrm{d}q^i\,\mathrm{d}q^j. \tag{D.45}$$

Therefore to find the shortest path between q_0 and q_1, all we have to do is to write the Euler–Lagrange equations for the "Lagrangian"

$$L = \sqrt{g_{ij}\,\dot{q}^i\,\dot{q}^j}. \tag{D.46}$$

To find the differential equations we define $f = g_{nm}\,\dot{q}^n\,\dot{q}^m$. We then have $L = \sqrt{f}$, and we can write the Euler–Lagrange equations:

$$\frac{\partial L}{\partial q^n} = \frac{1}{2}\frac{1}{\sqrt{f}}\,g_{km,n}\,\dot{q}^k\,\dot{q}^m, \tag{D.47}$$

$$\frac{\partial L}{\partial \dot{q}^n} = \frac{1}{2}\frac{1}{\sqrt{f}}\,2\,g_{kn}\,\dot{q}^k, \tag{D.48}$$

$$\begin{aligned}\frac{\mathrm{d}}{\mathrm{d}t}\frac{\partial L}{\partial \dot{q}^n} - \frac{\partial L}{\partial q^n} &= \frac{\mathrm{d}}{\mathrm{d}t}\left(\frac{1}{\sqrt{f}}\,g_{kn}\,\dot{q}^k\right) - \frac{1}{2}\frac{1}{\sqrt{f}}\,g_{km,n}\,\dot{q}^k\,\dot{q}^m \\ &= \frac{1}{\sqrt{f}}\left\{\frac{\mathrm{d}}{\mathrm{d}t}\left(g_{kn}\,\dot{q}^k\right) - \frac{1}{2}\,g_{km,n}\,\dot{q}^k\,\dot{q}^m\right\} \\ &\quad + g_{kn}\,\dot{q}^k\,\frac{\mathrm{d}}{\mathrm{d}t}\frac{1}{\sqrt{f}} \\ &= 0.\end{aligned} \tag{D.49}$$

Let us simplify the term in the curly brackets:

$$\begin{aligned}\frac{\mathrm{d}}{\mathrm{d}t}\left(g_{kn}\,\dot{q}^k\right) - \frac{1}{2}\,g_{km,n}\,\dot{q}^k\,\dot{q}^m &= g_{kn}\,\ddot{u}^k + g_{kn,m}\,\dot{q}^m\,\dot{q}^k - \frac{1}{2}g_{km,n}\,\dot{q}^k\,\dot{q}^m \\ &= g_{nk}\ddot{q}^k + \frac{1}{2}\left(-g_{km,n} + g_{nk,m} + g_{mn,k}\right)\dot{q}^n\,\dot{q}^m.\end{aligned} \tag{D.50}$$

This term is zero *if* the geodesic motion equations $\ddot{q}^k + \Gamma^k_{nm}\dot{q}^n\,\dot{q}^m = 0$ are satisfied. We know that if this pair of equations are satisfied, then $f = g_{nm}\,\dot{q}^n\,\dot{q}^m$ is constant, so that $\frac{\mathrm{d}}{\mathrm{d}t}\frac{1}{\sqrt{f}} = 0$, and therefore the above Euler–Lagrange equations are satisfied. This means that the trajectory of a geodesic motion is a curve of extremum length.

The converse of the preceding statement is not true—a curve could be an extremum-length curve, but not a geodesic motion. For example, on the plane, the geodesic motion is on a straight line with constant velocity. An accelerated motion of a particle on a straight line is not a geodesic motion—it does not satisfy $\ddot{q}^i = 0$—but the path, a straight line, is still a curve of minimum length.

D.6 Generalization

What we have seen in the previous sections has two main generalizations. First, one can consider higher-dimensional spaces by letting the indices be in the larger set $1 \le i \le D$, where D is the dimension of the space. It should be noted that all that we need is a $D \times D$, symmetric positive definite metric g_{ij}. Now the geodesic equations are a set of D equations.

The second generalization is to consider space-time. For this, the metric $g_{\mu\nu}$ is not positive definite: Diagonalizing $\varsigma\, g_{\mu\nu}$ at any event, we must have one negative and three positive eigenvalues. This guarantees that $\mathrm{d}s^2 = g_{\mu\nu}\, \mathrm{d}x^\mu\, \mathrm{d}x^\nu$ is the line element of a space-time.

We now give some comments about the geodesic motion and geodesics of the space-time.

1. In the geodesic equation for a curve on space, t (time) is a real variable parameterizing the geodesic. When we consider geodesics of the space-time, we should distinguish the temporal variable x^0 from the variable parameterizing the geodesic. If we denote this parameter by σ, the geodesic "motion" equation is

$$\frac{\mathrm{d}^2 x^\mu}{\mathrm{d}\sigma^2} + \Gamma^\beta_{\mu\alpha} \frac{\mathrm{d}x^\alpha}{\mathrm{d}\sigma} \frac{\mathrm{d}x^\beta}{\mathrm{d}\sigma} = 0. \tag{D.51}$$

2. A curve in space-time could be time-like, space-like, or light-like, according to the character of its tangent, which could be different at different points. Geodesics are time-like, space-like, or light-like.
3. A time-like geodesic is a curve of maximum proper time.
4. Considering time-like geodesics, we have two Lagrangians:

$$L = -M\,c^2 \sqrt{-\frac{\varsigma}{c^2} g_{\mu\nu} \frac{\mathrm{d}x^\mu}{\mathrm{d}\sigma} \frac{\mathrm{d}x^\nu}{\mathrm{d}\sigma}}, \tag{D.52}$$

$$\tilde{L} = \frac{\varsigma}{2} M\, g_{\mu\nu} \frac{\mathrm{d}x^\mu}{\mathrm{d}\sigma} \frac{\mathrm{d}x^\nu}{\mathrm{d}\sigma}. \tag{D.53}$$

Let us see what these Lagrangians are for the Minkowski space-time with Cartesian coordinates when $g_{\mu\nu} = \eta_{\mu\nu}$. It is easy to see that

$$L = -M\,c^2 \sqrt{\left(\frac{\mathrm{d}t}{\mathrm{d}\sigma}\right)^2 - \frac{1}{c^2}\left(\frac{\mathrm{d}x}{\mathrm{d}\sigma}\right)^2 + \left(\frac{\mathrm{d}y}{\mathrm{d}\sigma}\right)^2 + \left(\frac{\mathrm{d}z}{\mathrm{d}\sigma}\right)^2}, \tag{D.54}$$

$$\tilde{L} = \frac{M}{2}\left[\left(\frac{\mathrm{d}x}{\mathrm{d}\sigma}\right)^2 + \left(\frac{\mathrm{d}y}{\mathrm{d}\sigma}\right)^2 + \left(\frac{\mathrm{d}z}{\mathrm{d}\sigma}\right)^2 - c^2\left(\frac{\mathrm{d}t}{\mathrm{d}\sigma}\right)^2\right]. \tag{D.55}$$

Noting that

$$\begin{aligned} \int L\, \mathrm{d}\sigma &= -M\,c^2 \int \sqrt{\mathrm{d}t^2 - \frac{1}{c^2}\left[\mathrm{d}x^2 + \mathrm{d}y^2 + \mathrm{d}z^2\right]} \\ &= -M\,c^2 \int \mathrm{d}\tau, \end{aligned} \tag{D.56}$$

we see that L is simply the Lagrangian we saw in §5.2.
5. To obtain equations of motion, it is usually easier to use $\tilde{L}$.

Reference

Goldstein, H., 1980. Classical Mechanics, 2nd edition. Addison-Wesley, Reading, MA.

APPENDIX

Answers to selected problems

Solution to Problem 4 (page 15).

For the velocity, we have

$$c' = c\sqrt{1-2\beta\,\cos\theta+\beta^2} \simeq c(1-\beta\,\cos\theta) \tag{E.1}$$

$$= c - v\,\cos\theta. \tag{E.2}$$

For the angle θ we have three formulas:

$$\cos\theta' = \frac{\cos\theta-\beta}{\sqrt{1-2\beta\,\cos\theta+\beta^2}}, \tag{E.3}$$

$$\sin\theta' = \frac{\sin\theta}{\sqrt{1-2\beta\,\cos\theta+\beta^2}}, \tag{E.4}$$

$$\tan\theta' = \frac{\sin\theta}{\cos\theta-\beta}. \tag{E.5}$$

Expanding in powers of β we get

$$\cos\theta' \simeq \cos\theta - \beta\,(1-\cos\theta), \tag{E.6}$$

$$\sin\theta' \simeq \sin\theta + \beta\,\sin\theta\,\cos\theta, \tag{E.7}$$

$$\tan\theta' \simeq \frac{\sin\theta}{\cos\theta-\beta} = \tan\theta - \frac{\tan\theta}{\cos\theta}\beta. \tag{E.8}$$

These formulas give the same value only for $\beta \ll 1$, or better to say, to order β^2. If β is not $\ll 1$, then these formulas do not get the same values. For example, if $\beta = 0.1$, then $\beta^2 = 0.01\,\text{rad} = 0.5°$. Now if $\theta = 20°$ the exact formulas give $\theta' = 22.16°$, while from the above approximate formulas we get, respectively, 22.99°, 21.97°, and 22.16°. We see that the third formula, $\tan\theta' = \sin\theta/(\cos\theta-\beta)$, gives the best approximation.

Solution to Problem 8 (page 42).

The geometry is shown in Fig. E.1. At time t_0, the object S emits a light, which is detected by the astronomer located at O at time

$$t_0' = t_0 + \frac{D}{c}. \tag{E.9}$$

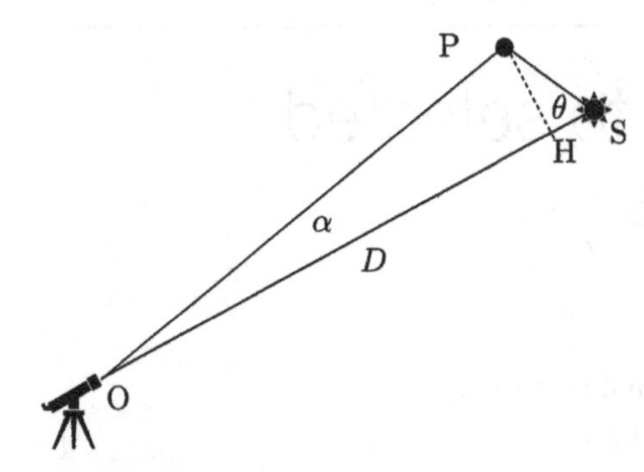

FIGURE E.1 The geometry of Problem 8.

Object S ejects object P. An observer at O sees two objects separating. α is the angular distance of the objects S and P. D is the distance of the system from the observer.

The object P, ejected from S at time t_0, is at P at time $t_0 + \delta t$, where $|SP| = u\,\delta t$. The light emitted at this time from P reaches the observer at time

$$t_1' = t_0 + \delta t + \frac{D - u\,\delta t\,\cos\theta}{c} = t_0' + \Delta t. \tag{E.10}$$

The increment of time as measured by the astronomer at O is then

$$\Delta t = t_1' - t_0' = \delta t\left(1 - \frac{u}{c}\cos\theta\right). \tag{E.11}$$

The length

$$|HP| = D\,\alpha = u\,\delta t\,\sin\theta \tag{E.12}$$

is observable by the astronomer at O (if the distance D is known). Therefore the "velocity" the astronomer assigns to this motion is

$$v = \frac{D\alpha}{\Delta t} = \frac{u\,\sin\theta}{1 - \frac{u}{c}\cos\theta}. \tag{E.13}$$

For $u \to c^-$ (which means u approaches c from below) we get

$$v \simeq c\,\frac{\sin\theta}{1 - \cos\theta} = c\,\cot\frac{\theta}{2}. \tag{E.14}$$

For $\theta = 12°$ we get $v \simeq 9.5\,c$.

Solution to Problem 10 (page 50).

The world-lines of the Earth and the Sun in space-time are

$$x = y = z = 0, \quad t \in \mathbb{R}, \quad \text{Sun}, \tag{E.15}$$

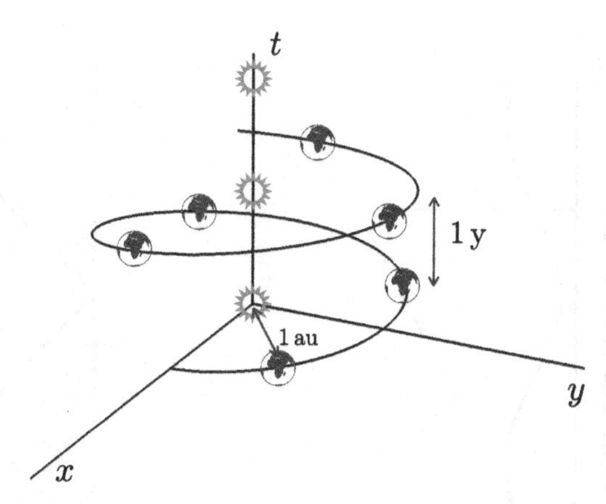

FIGURE E.2 World-line of the Sun–Earth system of Problem 10.

The vertical axis is the time t. The distance from the Sun to the Earth is $1\,\text{au} = 1.5 \times 10^{11}\,\text{m}$, while the period of revolution is $T = 1\,\text{y} = 3.16 \times 10^{7}\,\text{m}\,\text{s}^{-1}$. In these units the velocity of light is $6.3 \times 10^{4}\,\text{au}\,\text{y}^{-1}$. This means that if we want to draw the world-line of a light signal emitted from the Sun, its slope is $\lesssim 10^{-3}$, and it is represented by a line almost parallel to the $t = 0$ plane in the figure.

$$x = a\,\cos\omega t, \quad y = a\,\sin\omega t, \quad z = 0, \quad t \in \mathbb{R}, \quad \text{Earth}, \tag{E.16}$$

where

$$a = 1\,\text{au} = 1.5 \times 10^{11}\,\text{m}, \qquad \omega = \frac{2\pi}{T}, \qquad T = 1\,\text{y}. \tag{E.17}$$

These world-lines are drawn in Fig. E.2. Note that the velocity of light is

$$c = 3.0 \times 10^{8}\frac{\text{m}}{\text{s}} \cdot \frac{1\,\text{au}}{1.5 \times 10^{11}\,\text{m}} \cdot \frac{3.16 \times 10^{7}\,\text{s}}{1\,\text{y}} = 6.3 \times 10^{4}\,\frac{\text{au}}{\text{y}}. \tag{E.18}$$

So the slope of all light rays in this figure is $\sim 10^{-4}$, which means very close to the $t = 0$ plane.

The world-line of a harmonic oscillator with equation $x = x(0) + a\,\sin(\omega t)$ for $x(0) = 1.5$, $a = 1$, and $\omega = 0.5$ is drawn in Fig. E.3a. The dashed line with slope $+1$ is the world-line of a light ray. The slope of the line tangent to the world-line at time $t = 0$ is $\frac{1}{v} > 1$. For our numerical values, $v = 0.5$ and the slope is $+2$.

The world-line of a light ray reflecting between two parallel mirrors is drawn in Fig. E.3b.

Solution to Problem 11 (page 56).

The solution is drawn in Fig. E.4. We use the system of units such that $c = 1$, and we take $v = 0.6$ and $L = 4.5$.

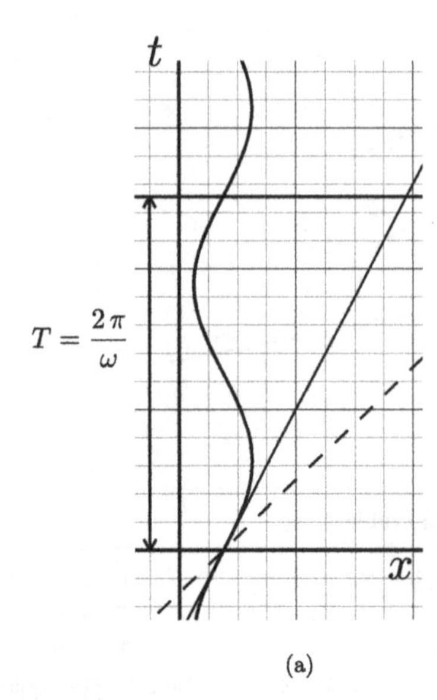

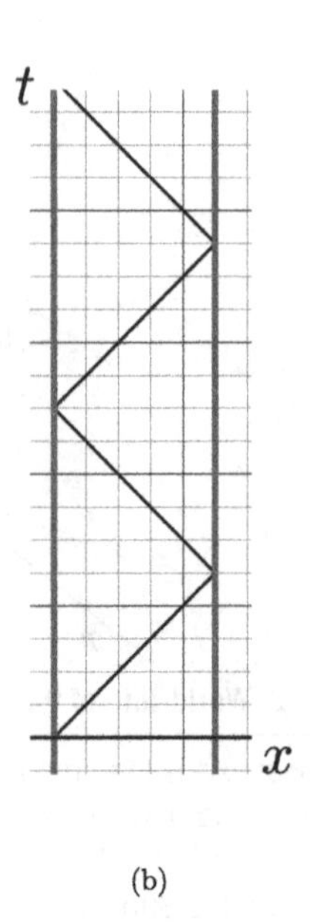

FIGURE E.3 World-lines of Problem 10.

(a) The world-line of a simple harmonic oscillator. The dashed line shows a light ray emitted at this event. The solid line is tangent to the world-line and its slope is $1/v$. (b) The world-line of a light ray reflected between two parallel mirrors (if the angle of incidence is 0).

The (t, x)-coordinates are as usual. When clock B is going to the right, it is stationary in the system K' for which the t' axis is given by $t = x/v$. The intersection of the invariant hyperbola $t^2 - x^2 = 1$ with this line gives the unit of time in the K' frame. In the figure two other hyperbolas, $t^2 - x^2 = 4$ and $t^2 - x^2 = 9$, are also drawn. Clock B shows the proper time τ. When clock B is returning toward A, it is stationary in a third frame, K'', which is moving with the same velocity v but in the $-x$-direction. If measured by clocks stationary relative to A, the unit of time in this frame, K'', is equal to the unit of time in K', because this depends only on the absolute value of the velocity.[1] As shown in the figure, clock A sends light signals at instants $t = 1, 2, 3, 4, 5, 6$, and 7, and at instant $t = 7.5$ both clocks are at the same position. Clock B sends light signals at instants $\tau = 1, 2, 3, 4$, and 5, and at $\tau = 6$ the two clocks are at the same position.

Solution to Problem 15 (page 62).

The Galilean and Einsteinian formulas are, respectively,

[1] Note that we are not saying that the unit of time in K'' is the same as in K' if measured by clocks in K' or K''!

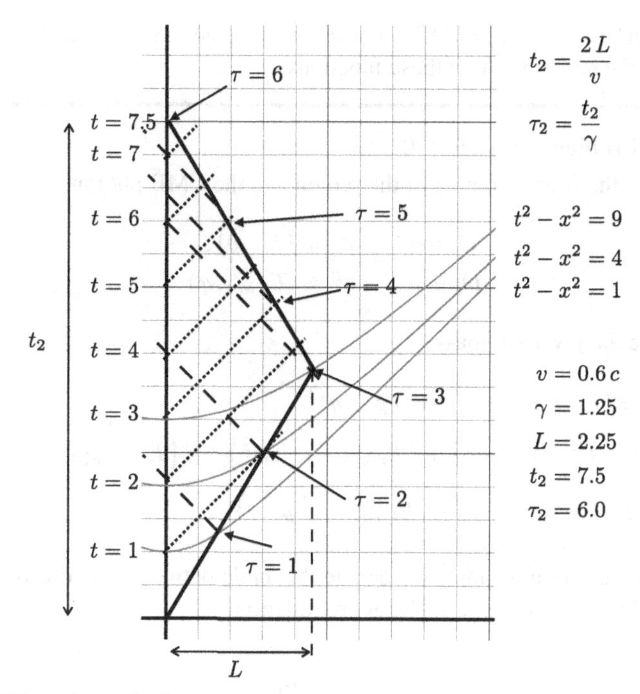

FIGURE E.4 World-lines of twin clocks of Problem 11.

The intersection of the invariant hyperbolas with the world-line of clock B, when it is moving to the right, shows instants of proper time τ. When B is returning, because its velocity is the same, the unit of time on its world-line is the same distance apart as when it was moving to the right. This round trip takes $t_2 = 7.5$ and $\tau_2 = 6$.

$$\theta'_{\mathrm{G}} = \tan^{-1}\left(\frac{\sin\theta}{\cos\theta - \beta}\right), \tag{E.19}$$

$$\theta'_{\mathrm{E}} = 2\tan^{-1}\left(\sqrt{\frac{1+\beta}{1-\beta}}\tan\frac{\theta}{2}\right). \tag{E.20}$$

Using a calculator, it is easy to find the values ($\beta = 0.1$).

$$\begin{array}{ccccc} \theta & 5.00^\circ & 45.00^\circ & 85.00^\circ & 90.00^\circ \\ \theta'_{\mathrm{G}} & 5.55^\circ & 49.35^\circ & -89.26^\circ & -84.29^\circ \\ \theta'_{\mathrm{E}} & 5.53^\circ & 49.21^\circ & 90.74^\circ & 95.74^\circ \end{array} \tag{E.21}$$

For small angles, the two formulas give almost the same values. But for angles near 90°, the Galilean prediction is quite different. The point is that for $\theta_0 = \cos^{-1}\beta$ the denominator in the Galilean formula vanishes. For angles greater than this critical value, the Galilean formula predicts a negative angle. The interesting feature of these functions is that for $|\theta| < \cos^{-1}\beta$ they are very close to each other. For example, for

$\beta = 0.1$ we have $\cos^{-1} 0.1 = 84.3°$. For $\theta = 83°$ we get $\theta_G = 88°$ and $\theta_E = 89°$. We recommend the reader to draw these functions.

Solution to Problem 27 (page 110).

We first write the four momenta of the proton and the CMB photon,

$$\text{proton} \qquad p^\mu = \left(E_p, \boldsymbol{p}\right), \tag{E.22}$$

$$\text{photon} \qquad q^\mu = (E_{CMB}, \boldsymbol{q}), \tag{E.23}$$

and calculate the invariant mass,

$$\begin{aligned} M^2 c^4 &= \left(E_p + E_{CMB}\right)^2 - c^2 (\boldsymbol{p} + \boldsymbol{q})^2 \\ &= \left(E_p^2 - c^2 |\boldsymbol{p}|^2\right) + \left(E_{CMB}^2 - c^2 |\boldsymbol{q}|^2\right) + 2\left(E_p E_{CMB} - c^2 \boldsymbol{p} \cdot \boldsymbol{q}\right) \\ &\simeq m_p^2 c^4 + 0 + 2\left(E_p E_{CMB} - c^2 \boldsymbol{p} \cdot \boldsymbol{q}\right), \end{aligned} \tag{E.24}$$

where m_p is the proton's mass. We denote the angle between $\boldsymbol{p}$ and $\boldsymbol{q}$ by α. We use the identity $|\boldsymbol{q}| = c^{-1} E_{CMB}$ and the approximation

$$|\boldsymbol{p}| \simeq \frac{1}{c} E_p, \tag{E.25}$$

which is valid for ultrarelativistic velocities. Then the above relation becomes

$$M^2 c^4 = m_p^2 c^4 + 2 E_p E_{CMB} (1 - \cos\alpha), \tag{E.26}$$

or

$$E_p = \frac{\left(M^2 - m_p^2\right) c^4}{2 E_{CMB}(1 - \cos\alpha)}. \tag{E.27}$$

Obviously, the minimum energy of the proton corresponds to the situation when a pair (p, π^0) is being created, at rest in the COM frame and $\alpha = 180°$. In these conditions we have

$$M = m_p + m_{\pi^0} = 938 + 135 = 1073\,\text{MeV}\,c^{-2}. \tag{E.28}$$

Thus we get

$$E_p^{min} = \frac{\left(M^2 - m_p^2\right) c^2}{4 E_{CMB}} = \frac{(2 m_p + m_{\pi^0})\, m_{\pi^0}\, c^2}{4 E_{CMB}} \tag{E.29}$$

$$\simeq 1 \times 10^{20}\,\text{eV}. \tag{E.30}$$

Remark. The probability of the interaction $p + \gamma \to p + \pi^0$ at the invariant mass $1073\,\text{MeV}/c^2$ is small, while at $M = 1232\,\text{MeV}/c^2$ it is higher. Therefore to get the

GZK limit, one should use $m = 1232\,\mathrm{MeV}/c^2$, which leads to $\sim 3 \times 10^{20}\,\mathrm{eV}$. See Weinberg (2008, pp. 107–108).

Solution to Problem 30 (page 134).

Let us define

$$x^0 = p, \quad x^1 = q, \quad x^2 = u, \quad x^3 = v \tag{E.31}$$

and consider the following change of coordinates:

$$cT = \alpha\,(p + q + u + v)\,, \tag{E.32}$$
$$X = p + q + u - v, \tag{E.33}$$
$$Y = p + q - u + v, \tag{E.34}$$
$$Z = p - q + u + v. \tag{E.35}$$

It is easy to see that

$$\varsigma\,\mathrm{d}s^2 = -c^2\,\mathrm{d}T^2 + \mathrm{d}X^2 + \mathrm{d}Y^2 + \mathrm{d}Z^2 \tag{E.36}$$

$$\begin{aligned} &= \left(3 - \alpha^2\right)\left(\mathrm{d}p^2 + \mathrm{d}q^2 + \mathrm{d}u^2 + \mathrm{d}v^2\right) \\ &\quad + 2(1-\alpha)\,(\mathrm{d}p\,\mathrm{d}q + \mathrm{d}p\,\mathrm{d}u + \mathrm{d}p\,\mathrm{d}v) \\ &\quad - 2(1+\alpha)\,(\mathrm{d}q\,\mathrm{d}u + \mathrm{d}q\,\mathrm{d}v + \mathrm{d}u\,\mathrm{d}v)\,, \end{aligned} \tag{E.37}$$

meaning that

$$\mathrm{sgn}\left(\varsigma\, g_{\mu\mu}\right) = \begin{cases} -1 & \alpha > \sqrt{3}, \\ 0 & \alpha = \sqrt{3}, \\ 1 & \alpha < \sqrt{3}. \end{cases} \tag{E.38}$$

Therefore all of the coordinates $x^\mu = (p, q, u, v)$ are time-like if $\alpha > \sqrt{3}$; all are null if $\alpha = \sqrt{3}$; and all are space-like if $\alpha < \sqrt{3}$.

Remark. The linear-algebraic explanation of this result is very simple: In $(1{+}D)$-dimensional Minkowski space-time one can always find $1 + D$ linearly independent vectors, all of which are time-like, all of which are null, or all of which are space-like. The $(1+1)$-dimensional case is shown in Fig. E.5. The 2-vectors $\{a_1, a_2\}$ are linearly independent and both are space-like. The 2-vectors $\{b_1, b_2\}$ are linearly independent and both are time-like. The 2-vectors $\{c_1, c_2\}$ are linearly independent and both are null.

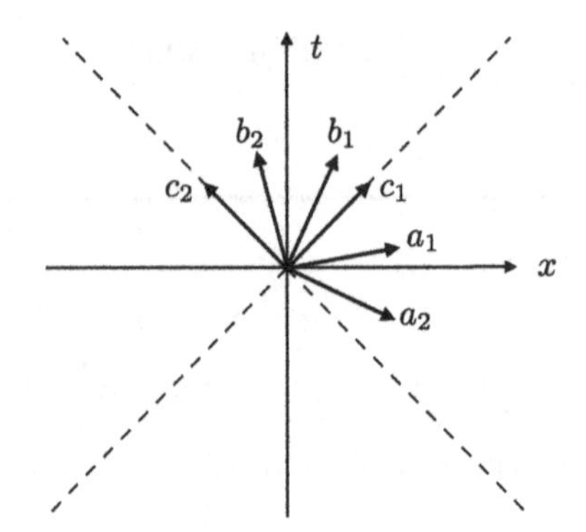

FIGURE E.5 Time-like, space-like, and light-like basis for the $(1+1)$-dimensional Minkowski space-time.

Vectors $\boldsymbol{a}_1$ and $\boldsymbol{a}_2$ are linearly independent and space-like. Vectors $\boldsymbol{b}_1$ and $\boldsymbol{b}_2$ are linearly independent and time-like. Vectors $\boldsymbol{c}_1$ and $\boldsymbol{c}_2$ are linearly independent and null (light-like).

Solution to Problem 35 (page 154).

We set $c = 1$ and use $x_0 = 1$ and $v = 0.6$. From $x = \xi \cosh\theta$ and $t = \xi \sinh\theta$, we have

$$x = 1, \qquad \xi = \frac{1}{\cosh\theta}, \tag{E.39}$$

$$x = 1 + 0.6t, \qquad \xi = \frac{1}{\cosh\theta - 0.6\sinh\theta}, \tag{E.40}$$

$$x = 1 + t, \qquad \xi = e^{\theta}, \tag{E.41}$$

$$x = 1 - t, \qquad \xi = e^{-\theta}. \tag{E.42}$$

These world-lines are drawn in Fig. E.6. They are interpreted as follows:

1. $x = 1$ is the world-line of a particle stationary in the inertial frame K. In the Rindler frame this particle is accelerating—falling toward the horizon $\xi = 0$.
2. $x = 1 + 0.6t$ is the world-line of a particle moving with velocity 0.6 in the inertial frame K, being at $x = 1$ for $t = 0$. In the Rindler frame this particle first goes to the right, but the acceleration causes it to turn back and finally fall to the horizon $\xi = 0$.
3. $x = 1 + t$ is the world-line of a light moving to the right. In the Rindler frame this world-line emerges from the horizon $\xi = 0$ for the Rindler time $\theta \to -\infty$.
4. $x = 1 - t$ is the world-line of a light moving to the left. In the Rindler frame this world-line goes to the horizon $\xi = 0$ for the Rindler time $\theta \to \infty$.

One could say that in the Rindler frame, there is a gravitational field attracting free particles to $\xi = 0$.

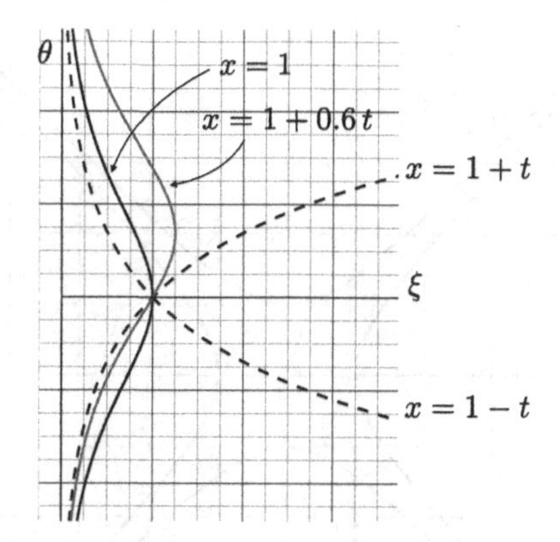

FIGURE E.6 Geodesics of the Rindler frame.

Dashed curves are world-lines of light (the light-cone). Solid curves are time-like geodesics. In the Cartesian (t, x)-coordinates these are just time-like lines. In the Rindler (θ, ξ)-coordinates these world-lines are being attracted toward the horizon $\xi = 0$.

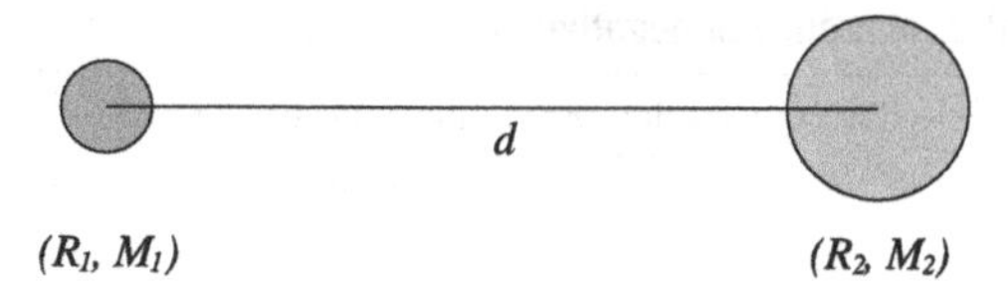

FIGURE E.7 Two spherical objects of Problem 41.

Two spherical objects (stars or planets) are a distance d apart. Since R_1, $R_2 \ll d$, the Newtonian potential at the surface of each sphere is $-G\,M_i/R_i$.

Solution to Problem 41 (page 175).

The objects are shown in Fig. E.7. Since $d \gg R_1$ and R_2, when we are calculating the Newtonian potential at the surface of one object we can neglect the contribution due to the other objects' gravity. The Newtonian potential of a spherical object is $\phi(r) = -G\,M/r$. Therefore

$$\frac{|\phi_1|}{|\phi_2|} = \frac{M_1}{M_2} \cdot \frac{R_2}{R_1} \simeq 6 \times 10^3. \tag{E.43}$$

From this we conclude that $\phi_1 - \phi_2 \simeq \phi_1$, and thus

$$\frac{\nu_2 - \nu_1}{\nu_1} \simeq \frac{\phi_1}{c^2} = -\frac{G\,M_1}{R_1\,c^2} \simeq 3 \times 10^{-6}. \tag{E.44}$$

In a real situation the objects are not stationary, but moving. If $D \ll v\,t$, where v is the typical velocity and t is the time of flight of the photons, the change in D caused by the motions of the objects is not important. However, the Doppler shift due to the motion of the source and the observer, which is of order $\beta = v/c$, must be considered.

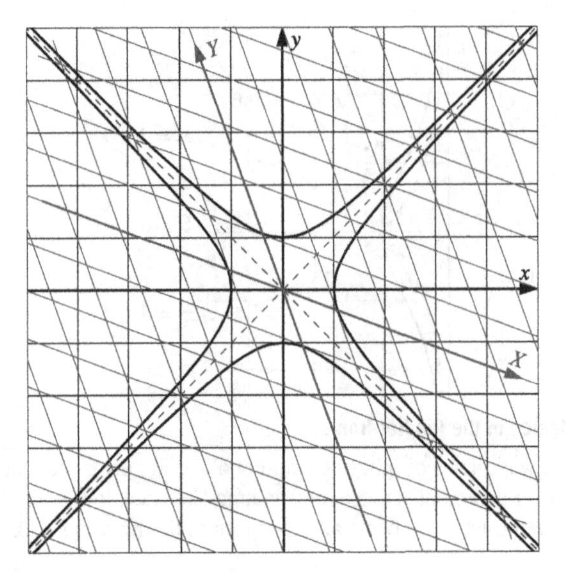

FIGURE E.8 Hyperbolic rotation with negative angle.

Two meshes to calculate the coordinates of points in the two coordinate systems (x, y) and (X, Y) are drawn.

Solution to Problem 47 (page 287).

Part 1. The situation is shown in Fig. E.8. In this figure the Euclidean angle between the x- and X-axes is 20°, and since $\tan 20° = 0.364 = \tanh(0.381)$, the hyperbolic angle of this rotation is -0.381 (dimensionless).

Part 2. The situation is shown in Fig. E.9. The dashed lines are the lines $x = \pm y$, which are the same as the lines $X = \pm Y$. The invariant hyperbolas are as usual given by

$$x^2 - y^2 = \alpha = X^2 - Y^2. \tag{E.45}$$

Therefore the intersections of the hyperbolas $x^2 - y^2 = \pm 1$ with the X- and Y-axes show the unit length on these axes. In Fig. E.9 the Euclidean angle between the dot-dash line and the x-axis is 10°. Since $\tan 10° = 0.176 = \tanh(0.178)$, the x-axis is the result of *hyperbolic* rotation of the dot-dash line by $\theta_1 = 0.178$. Also, the Euclidean angle between the dot-dash line and the X-axis is 20°, and since $\tan 20° = \tanh(0.381)$, the X-axis is the result of hyperbolic rotation of the dot-dash line by $\theta_2 = 0.381$. Therefore the X-axis is the result of the hyperbolic rotation corresponding to the hyperbolic angle $\theta_2 - \theta_1 = 0.203$. This can also be calculated if a line parallel to the y-axis is drawn, resulting in a hyperbolic-right angle triangle $\triangle$OAB (see Fig. E.9), in which we have

$$\tanh\theta = \frac{|\mathrm{AB}|}{|\mathrm{OB}|} = 0.2 \quad \Leftrightarrow \quad \theta = 0.203. \tag{E.46}$$

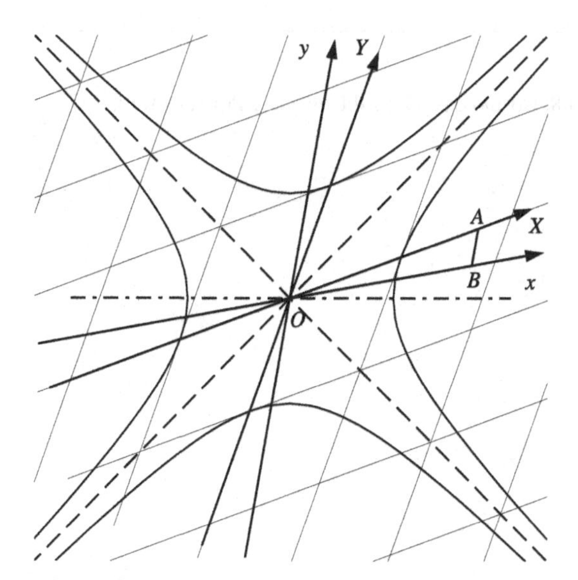

FIGURE E.9 Hyperbolic rotation when the axes are not perpendicular.

The axes x and y are not perpendicular. Still the hyperbolic rotation is defined as usual. The dashed lines are given by $x = \pm y$, or equivalently by $X = \pm Y$. The triangle $\triangle$OBA is used to find the hyperbolic angle between the x- and X-axes. Note that AB is parallel to the y-axis, while OB is parallel to the x-axis, and $\tanh\theta = |\text{AB}|\,/\,|\text{OB}|$.

Part 3. We note that $X^2 - Y^2 = (X - Y)(X + Y)$, so if in the formulas of the hyperbolic rotation we use $x = X + Y$ and $y = X - Y$ as the independent variables, we get the required transformation, which reads

$$x = X + Y \to e^{\theta}(X + Y) = e^{\theta}\, x, \quad y = X - Y \to e^{-\theta}(X - Y) = e^{-\theta} y. \tag{E.47}$$

The required transformation is

$$(x, y) \to (e^{\theta}\, x, e^{-\theta}\, y). \tag{E.48}$$

Part 4. We can write this transformation if we rename $x \to x/a$, $y \to y/b$, $X \to X/a$, and $Y \to Y/b$ in the usual form of the transformation. We then get

$$X = x\, \cosh\theta + y\, \frac{a}{b}\, \sinh\theta, \tag{E.49}$$

$$Y = x\, \frac{b}{a}\, \sinh\theta + y\, \cosh\theta. \tag{E.50}$$

Reference

Weinberg, S., 2008. Cosmology. Oxford University Press, Oxford.

Index

Printed in the United States
by Baker & Taylor Publisher Services